# Basic Electronics and Computer Applications

# Basic Electronics and Computer Applications

**(Contains Assignments for Lab Sessions)**

**Written as per AICTE Syllabus for First Semester of B. Pharma**

**Rajiv Khanna**
**Technical Director**
**Bell Infotech Systems (P) Ltd.**

PUBLISHING FOR ONE WORLD

**NEW AGE INTERNATIONAL (P) LIMITED, PUBLISHERS**

*(formerly Wiley Eastern Limited)*

New Delhi • Bangalore • Chennai • Cochin • Guwahati • Hyderabad
Jalandhar • Kolkata • Lucknow • Mumbai • Ranchi

Visit us at www.newagepublishers.com

Published by New Age International (P) Ltd., Publishers
First Edition: 2007



**ISBN : 978-81-224-2144-6**

**Rs. 175.00**

**C-07-05-1565**

Printed in India at Mohan Lal Printers, Delhi.
Typeset by author

**PUBLISHING FOR ONE WORLD**

**NEW AGE INTERNATIONAL (P) LIMITED, PUBLISHERS**
***(formerly Wiley Eastern Limited)***
4835/24, Ansari Road, Daryaganj, New Delhi - 110002
Visit us at **www.newagepublishers.com**

## Book Highlights
## (An Edge Over Others)

Although many books are available in the market for B. Pharma students, but this book has an edge over others. Following are the highlights, which have silently made it the best book of the market:

(i) This book is written exactly as per syllabus suggested by AICTE.

(ii) Syllabus specifies three main streams for learning. First stream suggests topics related to basic electronics. Second stream suggests computer topics and third stream suggests topics for computer lab sessions. This book is a single-window-solution for all the three streams. It covers all the three aspects. Students need not purchase separate book for each stream. This book will be sufficient for them.

(iii) Language and the style of writing are the prime highlights of this book. Language is easy, no jargons are used and all the technical words are explained well, before they are introduced. Thus no confusion of any kind prevails in the book. Readers get best returns for their time invested in this book.

(iv) Computer language portion of the book is not only content rich but is also program rich. After explaining a topic or command, many programs are introduced, so as to explain the practical aspects of the topic.

(v) All the topics are explained in very fine manner. Even the students of average intelligence can follow the explanation and learn the subject in easy manner.

(vi) While dealing with programming languages, most of the books directly write program, without explaining the problem, for which the program is written. But in this book, each problem is explained in clear words, before the program is written. This approach helps in understanding the logic of the following program, clearly.

(vii) All the programs of this book are working programs. Readers can directly type these programs on their computer and work with them. This will make the learning fast.

(viii) Wherever necessary, results of the program are also included after the program. This approach helps in understanding the dynamics of the program.

(ix) Syllabus doesn't suggest theory sessions for softwares like MS-Word, MS-Excel, MS-PowerPoint etc. but it suggests practical sessions for these subjects. To avoid all sorts of inconvenience to the students, we have explained all important theoretical aspects of these subjects, before suggesting practical assignments for them. This makes the book complete in itself

(x) Lab assignments suggested in the last six chapters of the book are quite logical and well defined. All the assignments put together take care of all the practical aspects of the subject. Their completion ensures good knowledge and sharp skills.

In short we would say that all the efforts have been made to create perfect learning experience through this book. We are sure that readers will definitely find it interesting and better than any other book, available in the market.

Author

# PREFACE

Modern advancements and developments in the field of computer have brought the world into new era. Computer has emerged as a powerful tool that is continuously transforming the shape of our working methodologies and day-to-day services. There is hardly any sphere of life left now, in which computers are not being used. Be it any area of operations, computer knowledge is must. This is the reason, why most of the education institutions have introduced computer as a subject in their course curriculum. AICTE suggests study of basic electronics and computer applications in first semester of B. Pharma studies.

This book is specially written for B. Pharma students of first semester. It is exactly as per syllabus suggested by AICTE and takes care of all the subjects, topics and practical assignments mentioned therein. However, at some places, sequences of the topics have been slightly changed with the objective of explaining the topics in gradual manner and making the learning easy. Readers will definitely appreciate this change, when they go through the book.

The language of the book is very easy and never becomes barrier in learning. We have followed American English pattern in this book. Some of the spellings like colour, centre etc. may appear as color, center etc. They are wrong as per British pattern but are correct in American pattern. We are sure that students and teachers will appreciate this change because software packages mentioned in the syllabus also follow the same language pattern.

This book titled, **"Basic Electronics and Computer Applications"** comprises of 22 chapters. First 16 chapters are for theory sessions, while last 6 chapters are for lab sessions. Each theory chapter explains the subject in details and interesting manner. To make the explanation more meaningful, lots of examples and figures have been included in each chapter. This aspect has emerged as main strength of the book.

At the end of each chapter, a question bank, titled "Exercises" is there. All the questions put together, cover all important aspects of the topic. Solving these questions correctly, on self-knowledge basis, will generate confidence and ensure good marks in the examination.

We are sure that this book will serve as perfect teaching guide for the teachers and good reference book for the students. It is expected that they will take full advantage of our knowledge and experience.

Although enough care has been taken to make the book error free but some mistakes may have gone unnoticed. Feedbacks and suggestions in this regard, on our address, will be highly appreciated.

Rajiv Khanna

# SYLLABUS

***SEMESTER-I***

## Basic Electronics and Computer Applications

## Theory

**1. Basic Electronics**:

Semiconductors (1), PN Junction diode (18), LED (22), Photodiode and its uses (23-24), Rectifiers [half wave, full wave/ with filters] (26-32), Transistors configuration (36), Transistor amplifiers (44), Introduction to integrated circuits (60), Photocells (26) and photo multiplier tubes (25).

**2. Computers:**

**2.1 Introduction to computers (71)**

*History of computer (71) development and respective generations (77) :* Abacus(72), Napier's Bones (72), Slide Rule (73), Pascal's Calculators(73), Need to use computers applications in pharmacy and in general (304).

*Computer Classification (90):*

Mainframe (91), Minicomputer (91) and Microcomputers (90), Comparison of Analog & Digital computers (80), Hardware & Software (92), Calculator and Computer (93).

**2.2 Operating Systems (96)**

Introduction to types of operating systems (98), UNIX (111), MS-DOS (104), etc. RAM (88), ROM (88), Virtual Memory (89) etc.

**2.3 Type of Languages (123):**

Conventional Languages (126), Their advantages (126), Limitations (127), C (127), Pascal (143), FORTRAN (151), Programming of these languages (127-157).

**2.4 Introduction to Computer Networks (160):**

Architecture of seven layers of communications (168).

**2.5 Introduction to Data Structure (214):**

Queues (217), List (220), Trees (226), Binary Trees (227), Algorithm (180), Flowchart (184), Structured Systems Analysis and Designed (193), Ingress-SQL (234), Gateway (173) etc.,

**Note:** *Software packages like Word Star, Lotus 123 etc. are obsolete now and have no relevance in present scenario. Rather their counter parts like MS-Word, MS-Excel etc. are popular these days. To make our readers familiar with the latest softwares, we have explained operational details of MS-Word, MS-Excel, MS-PowerPoint etc. in chapter 20, 21 and 22 of this book. Suggested practical assignments for the lab sessions are general in nature. It is suggested that they should be done, using latest versions of these softwares. However if one is too rigid for older softwares, he / she can always complete those assignments, using age-old softwares too.*

# Table Of Contents

# BASIC ELECTRONICS

## CHAPTER 1

# *Semiconductors*

In today's world, electronics has become a part of our life. From simple digital watches to complex equipments like calculators television, computers, etc. can be found in most of the houses. Electronic equipments have become so common that now life seems to be miserable without them. But, here the question is, what is electronics? The word electronic has been derived from two other word i.e. **electron** and **mechanics**. The word electronics clearly indicates that it is the science based on the mechanics of electron. In this science, we study the behavior of electron under different conditions that are externally created. Its behavior is then exploited to form wide variety of devices and equipments, which are of our use. All those devices, in which conduction takes place due to the movement of electron, are called electronic equipments.

## HISTORY OF ELECTRONICS

Electron was identified in 1909 by American physicist named A Millikan. After this another American inventor Thomas A. Edison noticed that current of electrons flowed from one electrode to another, if the second one was relatively positively charged. Later this discovery led to electron tubes.

Electron tubes were quite useful for manufacturing X-ray tubes, radio signal transmitter etc. Later improvements were made in electron tubes and many improved versions of vacuum tubes came into existence. Vacuum tubes were quite popular during first half of the 20$^{th}$ century. Many useful devices like, radio, television, first computer etc, were made using vacuum tubes.

Vacuum tubes were large in size, they generated lot of heat and their reliability was poor. Cost of production of vacuum tubes was quite high. As a result of all these shortcomings, an alternative of vacuum tube was very much required.

In 1947, at Bell laboratories, scientist, John Bardeen, Walter H. Brattain and William B. Shockley invented transistor. Transistor was smaller in size; it generated less heat and was more reliable than vacuum tubes. It was also cheaper in cost. Due to all these advantage, by 1960, transistors almost replaced vacuum tubes in most of the electronic devices. More complex and sophisticated electronic equipments were now developed, using transistors.

Independent research work of Jack Kilby of Texas Instruments, Jean Hoerni and Robert Noycel of Fairchild Semiconductor Corporation led to invention of Integrated Circuit (IC). Earlier ICs contained about 10 transistors in them. It was called Small Scale Integration (SSI).

In the next step of miniaturization, came the era of Medium Scale Integration (MSI). About 100 transistors were packed in a small chip. Gradually these numbers increased and by 1970 about 1000 transistors were packed in the same space. In 1971, first microprocessor came into existence. Invention of microprocessor brought revolution in electronics and computers.

After this, came the era of Large Scale Integration (LSI). In LSI, about 30,000 transistors were placed in single chip. Now it is the era of Very Large Scale Integration (VLSI), in which it has become possible to pack millions of transistors in a single chip.

With all these development taking place, electronic devices gradually became smaller and more sophisticated. Their scopes of operations also became large. Computers, medical equipment, aero systems, defense equipments, communication equipments etc. are quite advanced electronic devices. They all make use of modern electronic components.

## ELECTRONIC COMPONENTS

There are only 5 basic electronic components. All the electronic circuits, however simple or complex they may be, are made using these 5 components. These components are:

1. Resistors
2. Capacitors
3. Inductors
4. Tube devices
5. Semiconductor devices

These electronic components have been broadly classified into two categories i.e. Active electronic components and passive electronic components. Those electronic components that require power for their functioning fall into the category of active electronic components. For example, tube devices and transistor devices require power for their functioning. Thus they are categorized as active electronic components.

Those electronic components, which do not require power for their functioning fall into the category of passive electronic components. For example, resistors, capacitors and inductors do not require power for their functioning. Thus they are categorized as passive electronic components.

A block diagram, showing categorization of electronic components is shown here figure 1.1

**Electronic Components**

| **Passive Components** | **Active Components** |
|---|---|
| 1. Resistors | 1. Tube Devices |
| 2. Capacitors | 2. Semiconductors devices |
| 3. Inductors | |

**Figure 1.1**

## Resistor

A resistor is a two terminal electronic component that poses opposition, when current flows through it. When current flows through it, voltage drop occurs across it.

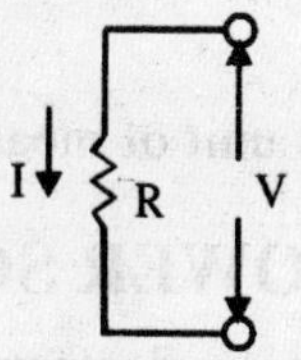


**Figure 1.2**

For example, if V is the voltage drop across the resistor and I is the current flowing through it then the resistance offered by the resistor will be $R = V/I$.

Its unit of measurement is Ohm. Resistors are commonly used at all those places in the circuit, where potential drop has to occur.

## Capacitors

Capacitor is an electronic component, which is used for storing charge in it. It comprises of closely placed conducting plates.

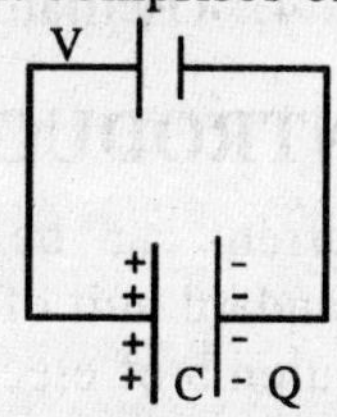


**Figure 1.3**

When voltage is applied to the capacitor, charge of equal amount but of opposite polarity gets built on each plate.

If potential V is applied to the capacitor and Q charge gets developed on the plate then capacitance of the capacitor will be $C = Q/V$.

Its unit of measurement is Farad. Capacitors pose high resistance (infinite) to DC current but low resistance to AC. That means AC can pass through them while DC can't. This is the reason why they are used in all those circuits, where DC component has to be blocked.

## Inductors

Inductors are basically coil shaped wires. They work on the principle that when current flows through coiled wire, magnetic field gets generated around it. This magnetic field opposes any change in current that tries to take place within the wire. This opposition is called inductance. In other words, it can be said that the electronic component, which produces inductance is called inductor. Inductors pose high resistance to AC, but low resistance to DC. Thus it tries to let the DC component of a signal, pass through it while it tries to block the AC component. If Φ is the flux that passes through the inductor and i is the current, which flows through the inductor then inductance L will be:

$$L = \Phi/i$$

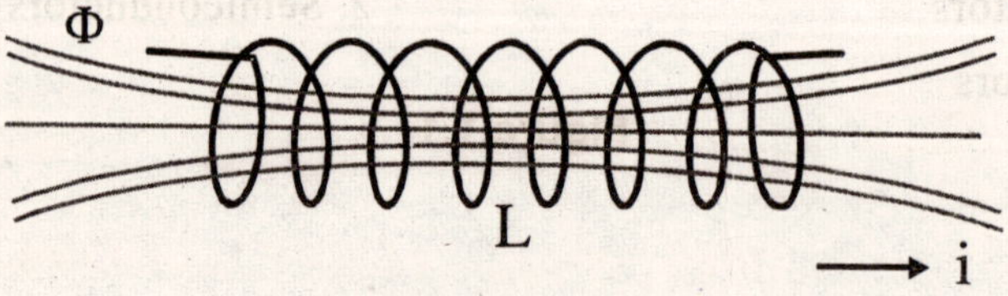


**Figure 1.4**

Its unit of measurement is Henry (H).

# POWER SOURCES

Every electronic circuit requires power for its functioning. Power can be supplied to the circuit, using some power equipment. Following are few power sources, which are commonly used for activating electronic circuits:

1. Battery (It provides low voltage DC).
2. Generator (It provides high voltage DC).
3. Alternator (It provides high voltage AC).
4. Oscillator (It supplies AC voltage).

# INTRODUCTION TO CURRENT

Current can be defined as flow of electrically charge carriers like electrons or holes. Standard unit of current is ampere, symbolized by A. One ampere of current represents one coulomb of electrical charge (i.e. $6.24 \times 10^8$ charge carriers) moving past a specific point in one second.

## Type of Current

Electric currents can be broadly classified into two categories:

1. DC current
2. AC current

DC is acronym for Direct Current. Electric current that flows only in one direction (i.e. opposite to the direction of flow of electrons) is called Direct Current or DC. For example, the type of current that we get from the batteries is DC current.

Nature of Direct Current, obtained from different sources is illustrated in the charts, shown in figure 1.5

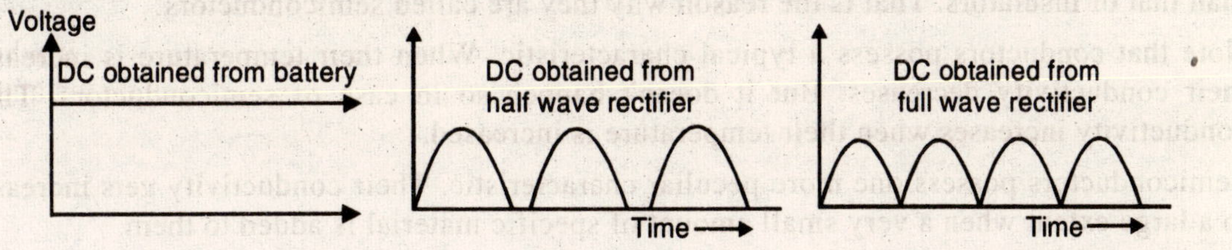


**Figure 1.5**

AC is acronym for Alternating Current. Electric current that reverses its direction, usually many a times per second, is called AC. For example, the type of current that we get in our houses is an example of Alternating Current.

Nature of AC current obtained from difference power sources is illustrated in charts that are shown in figure 1.6.

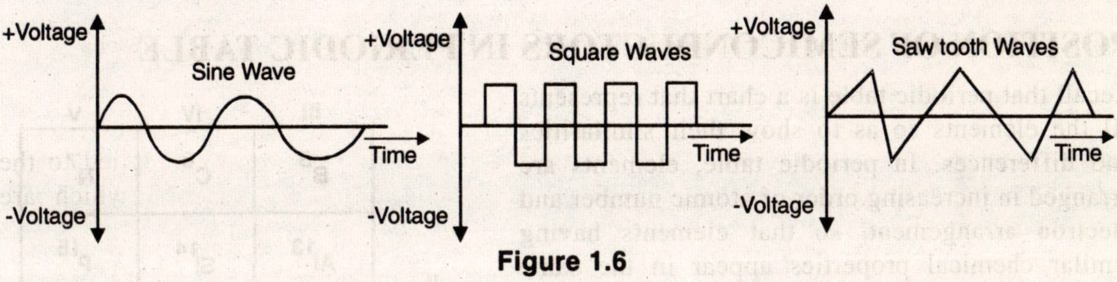


**Figure 1.6**

# ELECTRONICS AND SEMICONDUCTOR COMPONENTS

You know that semiconductor components are one of the 5 types of components that are used in electronic circuits. As the name indicates, semiconductor device are fabricated, using semiconductor material. In fact there are numerous semiconductor devices like P-N function diodes, P-N-P junction triodes, Integrated Circuits etc. that are commonly used in electronic equipments.

Now the question is, what semiconductors are? What are their characteristics and why they are used in fabricating commonly used electronic components? To answer these questions, we will first describe few fundamental aspects of semiconductor components.

# WHAT ARE SEMI CONDUCTORS

Semiconductors are materials that have conductivity between conductors and non-conductors. For example, metals like copper, iron, silver etc. are good conductors. Electric current can very easily flow through them. This is the reason why they are rated as conductors.

Non metallic elements and their compounds like ceramic, plastic, rubber etc. are bad conductors. Electricity cannot pass through them. That is why they are called insulators.

Elements like Germanium and Silicon are neither good conductors nor good insulators. At room temperature these materials have conductivity less than that of conductors and more than that of insulators. That is the reason why they are called semiconductors.

Note that conductors possess a typical characteristic. When their temperature is increased their conductivity decreases. But it doesn't happen so in case of semiconductors. Their conductivity increases when their temperature is increased.

Semiconductors possess one more peculiar characteristic. Their conductivity gets increased to a large extent when a very small amount of specific material is added to them.

## IMPURITY AND DOPING

Material, which is added to the semiconductors for increasing their conductivity, is often referred to as impurity and the process of adding the impurity is called doping.

To understand the characteristics of semiconductors and the functioning of semiconductor components, it is necessary to recall the position of semiconductor elements in periodic table and understand their atomic structure.

## POSITION OF SEMICONDUCTORS IN PERIODIC TABLE

Recall that periodic table is a chart that represents all the elements so as to show their similarities and differences. In periodic table, elements are arranged in increasing order of atomic number and electron arrangement, so that elements having similar chemical properties appear in the same column of the periodic table. Rows of the periodic tables are called groups or periods.

Few elements of group II, III, IV and V are illustrated in partial periodic table, shown in figure 1.7.

| II | III | IV | V |
|---|---|---|---|
| | $B^{5}$ | $C^{6}$ | $N^{7}$ |
| | $Al^{13}$ | $Si^{14}$ | $P^{15}$ |
| $Zn^{30}$ | $Ga^{31}$ | $Ge^{32}$ | $As^{33}$ |
| $Cd^{48}$ | $In^{49}$ | $Sn^{50}$ | $Sb^{51}$ |
| $Hg^{80}$ | | | |

**Figure 1.7**

Note that alphabets written in the blocks of the periodic table represent elements. For example, C represents carbon. Si represents silicon. Ge represents germanium. Numbers written in the blocks represent, total number of elements for the element. For example, 6 written in C block states that carbon atom consists of 6 electrons in it. Similarly 14 written in Si block represents that there are 14 electrons in silicon atom.

## STRUCTURE OF AN ATOM

Atom basically consists of three types of particles i.e. Protons, Neutrons and Electrons. Protons and Neutrons together remain confined in the central part of the atom, called Nucleus. Electrons revolve around nucleus in circular orbits.

Protons are positively charged. Electrons are negatively charged and neutrons carry no charge. Neutrons are neutral. A proton has the same amount of charge as an electron has. Since number of electrons in an atom are always equal to the number of protons in it hence atom is always neutral. The number of protons in atom is called its atomic number.

Not all the electrons of an atom revolve in the same orbit. As the number of electrons in the atom increase, electrons are placed in multiple orbits, called 1, 2, 3, 4 and so on.

Maximum number of electrons that can be placed in an orbit, often referred to as shell, are governed by the formula $2n^2$. Here n is the shell number. For example, first shell (i.e. n=1) will accommodate $2 \times 1^2 = 2$ electrons in it. Second orbit will accommodate $2 \times 2^2 = 8$ electrons in it. Similarly third orbit will accommodate $2 \times 3^2 = 18$ electrons in it and so on. But remember there are exceptions to this rule. The outermost orbit cannot accommodate more than 8 electrons in it.

Atomic structure of few elements is shown in figure 1.8

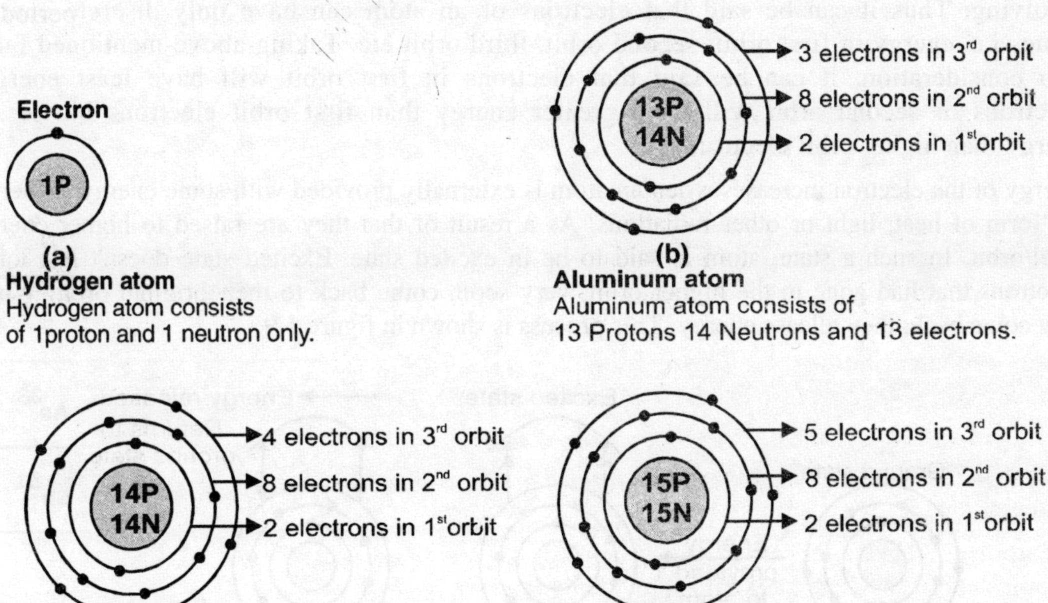


(a)
**Hydrogen atom**
Hydrogen atom consists of 1proton and 1 neutron only.

(b)
**Aluminum atom**
Aluminum atom consists of 13 Protons 14 Neutrons and 13 electrons.

(c)
**Silicon atom**
Silicon atom consists of 14 protons, 14 neutrons and 14 electrons.

(d)
**Phosphorous atom**
Phosphorous atom consists of 15 protons, 15 neutrons and 15 electrons.

**Figure1.8**

Electrons present in outermost orbit are called valance electrons. For example, hydrogen atom has one valance electron in it. Similarly aluminum, silicon and phosphorous have 3, 4 and 5 valance electrons in them.

## EXCHANGE OF ELECTRONS BETWEEN ATOMS

Inner orbit electrons do not leave the atom but outer electrons do that. Outer orbit electrons do not remain confined to the same atom. They leave the atom and go into the outer orbit of another atom. A vacancy thus created is filled by some other orbit electron, of some other atom. Thus other orbit electrons keep on shifting their place from atom to atom. Such electrons are called free electrons. It is due to the free electrons that an element conducts electricity.

## CONCEPT OF ELECTRON ENERGY

All the obits of the atom do not have the same amount of energy associated with them. An orbit closer to the nucleus has lesser energy than the one, which is farther. For example, third orbit has greater energy than second orbit but lesser than fourth.

Modern physics states that an electron has the same energy as the orbit, in which it is revolving. Thus it can be said that electrons of an atom can have only discrete energy values i.e. energy of first orbit, second orbit, third orbit etc. Taking above-mentioned facts into consideration, it can be said that electrons in first orbit will have least energy. Electrons of second orbit will have greater energy than first orbit electrons but lesser energy than higher orbit electrons.

Energy of the electron increases when an atom is externally provided with some energy, either in the form of heat, light or other radiations. As a result of that they are raised to higher energy level/orbit. In such a state, atom is said to be in excited state. Excited state doesn't last long. Electrons that had gone to the higher orbit, very soon, come back to their original orbit. When they come back, they release energy. This process is shown in figure 1.9.

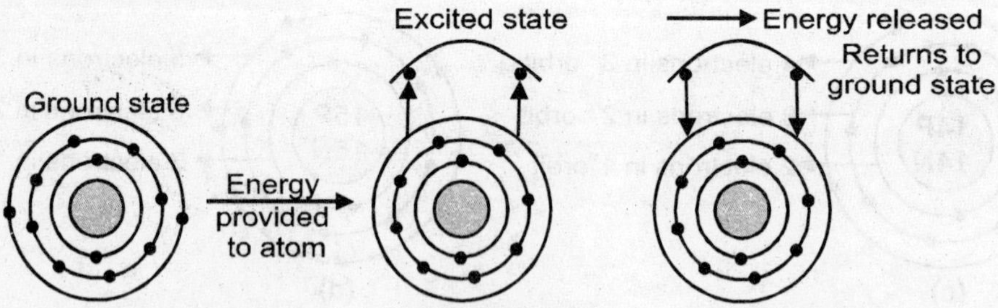


Figure 1.9

## CONCEPT OF ENERGY BAND

You are now familiar with electron energy or orbit energy levels. You know that orbits are discrete energy levels. That means electrons can either have this level of energy (that means certain fixed amount) or any other energy level (that means another fixed amount). It cannot have any thing in between. So far so good. But, now the question is what happens when numerous atoms come near each other to form a solid? In such a situation, orbit of an electron is not only influenced by its own nucleus but also by nuclei and electrons of all other atoms also. Thus energy levels no more remain so discrete. In fact they acquire the shape of a band. Energy band diagram for Aluminum is given in figure 1.10.

The highest energy band, in which electrons remain tied up to their atoms is called valence band. Electrons of valence band do not move from here to there when electric field is applied over them.

An energy band, higher than valence band, in which electrons become free to move within the metal when electric field is applied over them, is called conduction band.

Refer figure 1.10 and note that there is an energy gap between valence band and conduction band. Say the energy gap is z. When z amount of energy is provided to valence band electrons, they jump from valence band to conduction band. Thus conduction of electricity becomes possible in silicon (semiconductor), when its temperature is increased slightly.

Note that conduction of electricity is possible when electrons move in conduction band.

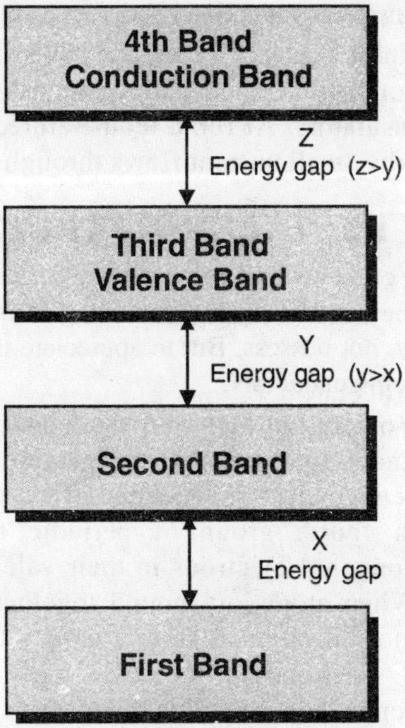


Energy Band Diagram For Silicon

**Figure 1.10**

## Energy Band Comparison

To be more familiar with energy bands, let's compare band diagrams of metals, insulators and semiconductors. This comparison is shown in figure 1.11

Refer figure 1.11(a) and note that in conductors (metals), valence band overlaps conduction band. Thus at room temperature itself, valence band electrons freely move between valence band and conduction band. When electric current is applied to conductors, these electrons move within conduction band and electric conduction takes place.

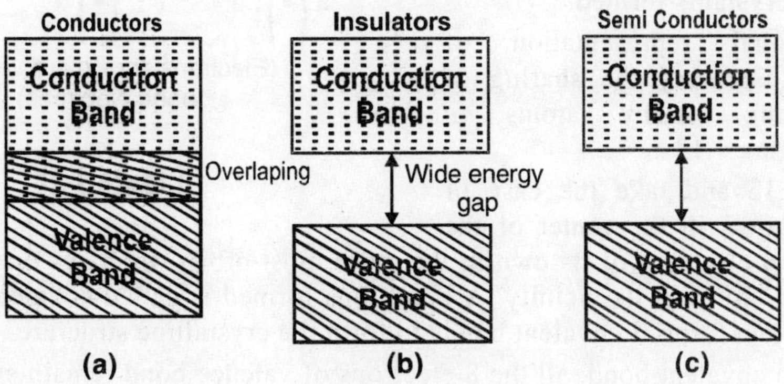


**Figure 1.11**

Refer figure 1.11(b) and note that in insulators, energy gap between valence band and conduction band is quite large. At room temperature valence band electrons remain quite distant from conduction band. When electric current is applied to insulators, due to absence of electrons in conduction band, electric current doesn't flow through them. On the other hand, if temperature of the insulator is raised to such an extent that some of the valence electrons pass the energy gap and jump to conduction band then a small amount of conduction become possible in them.

Refer figure 1.11(c) and note that even in semiconductors an energy gap exits between conduction band and valence band. But this gap is much lesser than the energy gap of insulators. At room temperature, few electrons can move to conduction. This is the reason why small current flows through them when electric current is applied to them.

## CRYSTALLINE STRUCTURE OF SEMICONDUCTORS

You know that semiconductors are not very good conductors. But by making some arrangement, they can be made good conductors and that too with few peculiar qualities that even conductors do not possess. But to appreciate this quality, we have to be familiar with crystalline structure of semiconductors.

You are familiar with semiconductors and know that silicon, germanium etc. are semiconductor elements. They are placed in fourth group of periodic table and contain 4 electrons in their valence band. When atoms are bound together to form molecules, each atom attempts to acquire 8 electrons in its outer most orbit. In semiconductors, this tendency is so strong that the atom tries to get 4 extra electrons in its outermost orbit by sharing 4 electrons with 4 nearby atoms. These sharing atoms align themselves in a uniform three-dimensional pattern and semiconductor crystal is formed.

Two dimensional representation of Silicon crystal, showing the sharing of electrons among nearby atoms is illustrated in figure 1.12

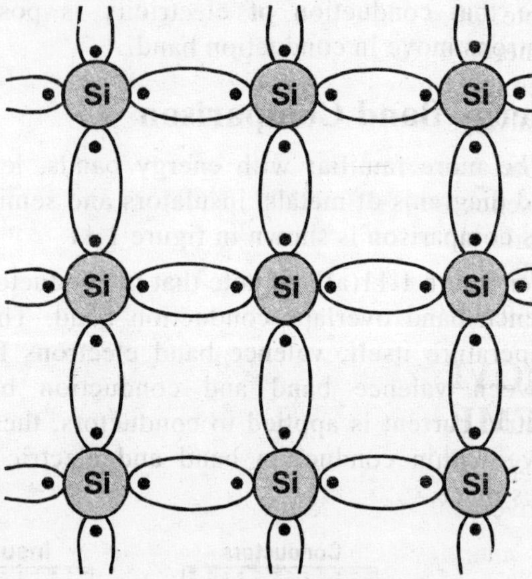


(Electrons shown are valence band electrons)

**Figure 1.12**

Refer figure 1.12 and take the case of Silicon atom shown in the center of the diagram. It has 4 electrons of its own in the outer most orbits. It shares one electron from each of the four atoms in its vicinity. A bond thus formed is called covalent bond. In this way, a silicon atom forms 4 covalent bonds to form the crystalline structure.

Note that due to covalent bond, all the 8 electrons of valence bond remain so tightly bound to the atom that no free electrons are available for conducting the electricity. Thus semiconductor behaves as perfect insulator, at absolute zero temperature.

## WHAT ARE INTRINSIC SEMICONDUCTOR

Semiconductor in its purest form (no impurity of any kind present in it) is called intrinsic semiconductor or undoped semiconductor. First step for making semiconductor is to acquire intrinsic semiconductor.

As mentioned above, semiconductors at absolute zero remain in the form shown in figure 1.12 and there are no free electrons for conducting the electricity. They behave as insulators at absolute zero temperature. At room temperature, some electrons get into excited state. They break the covalent band and move to valence band.

When an electron breaks the covalent bond and moves to conduction band, a vacancy is created in the broken covalent bond. This vacancy is called hole. Creation of free electrons and holes is shown in figure1.13.

Note that whenever a free electron is generated, a hole is also created. Holes and free electrons are always created in pair. Hole is always positively charged.

Note that electron is a particle while hole is absence of the particle. Electron can move while absence of electron at place, called hole, can shift from one place to another. Both the cases contribute to the conduction of electricity.

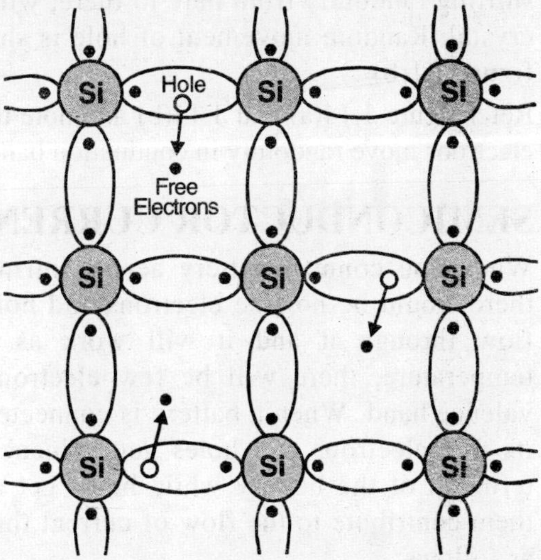

**Figure 1.13**

## MOVEMENT OF FREE ELECTRONS AND HOLES IN SEMICONDUCTORS

Free electrons and holes, both move within the crystal. The cause, nature and place of their movement are explained below.

As mentioned in previous section, at room temperature, some electrons participating in covalent bonding get excited and move to conduction band. They move within the crystal. They collide with nucleus or other free electrons. Their path then deviates. Thus they randomly move within the crystal. Path of a free electron is illustrated in figure 1.14(a).

As mentioned in the previous section, whenever a covalent bond breaks due to thermal excitement, a free electron exits the valance band and goes to the conduction band, leaving a positive charge, called hole, in valence band. The atom, from which electron exits becomes unstable because it has less than 8 electrons in its outermost orbit now. Due to the positive charge (hole) created in this process, it has great tendency to get the electron from somewhere and complete the orbit. An atom, present under similar conductions, may release free electron due to thermal agitation. This electron may not go to the conduction band. It may fill the vacancy that was created in previous atom. When it will do the hole will get filled. But at the same time, another hole will get created in the atom, which released the electron. If you view the process globally, hole basically gets shifted from first atom to second atom.

Now this hole will be filled by electron of some other atom and hole will shift to that atom. In other words, you can say that hole will keep on shifting randomly from here to there, within the crystal. Random movement of hole is shown in figure 1.1(b)

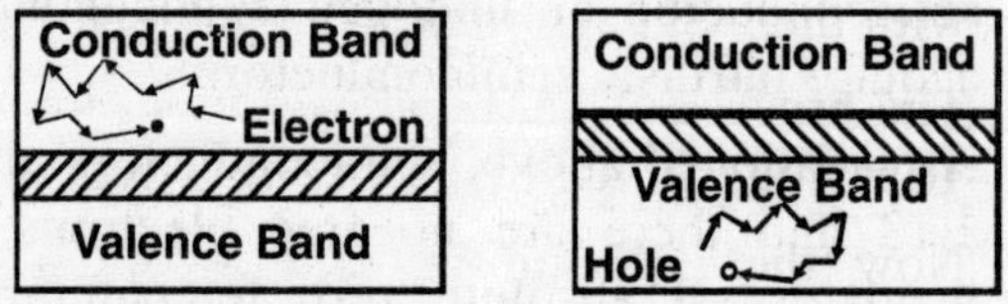


(a) (b)

Figure 1.14

Refer figure 1.14(a) and 1.14(b) and note that free electrons move randomly in conduction band, while holes move randomly in valence band.

# SEMICONDUCTOR CURRENT

When you connect battery across intrinsic semiconductor, at absolute zero temperature, there would be no free electrons and holes to move around. Thus electric current will not flow through it and it will work as perfect insulator. On the other hand, at room temperature, there will be few electrons in conduction band and few holes to drift in valence band. When a battery is connected to intrinsic semiconductor, at room temperature, its free electrons and holes get influenced by it. Electrons get attracted towards positive terminal of the battery while holes get attracted towards negative terminal. Thus both of them contribute to the flow of current through semiconductor. Total current can be defined as follows:

$$TotalCurrent, = I_n + I_p$$

Here $I_n$ is the current that is due to the movement of free electrons in conduction band. $I_P$ is the current that is due to the movement of holes in valence band.

Note that although electrons and protons move in opposite direction but currents due to them get added. This is because electron is negatively charged particle. Current due to its movement flows in the direction, opposite to its movement. Holes are positively charged. Current due to their movement flows in the same direction, in which they move. Thus both the current i.e. $I_n$ and $I_P$ flow in the same direction so they get added together.

Concept of semiconductor current is shown in figure 1.15.

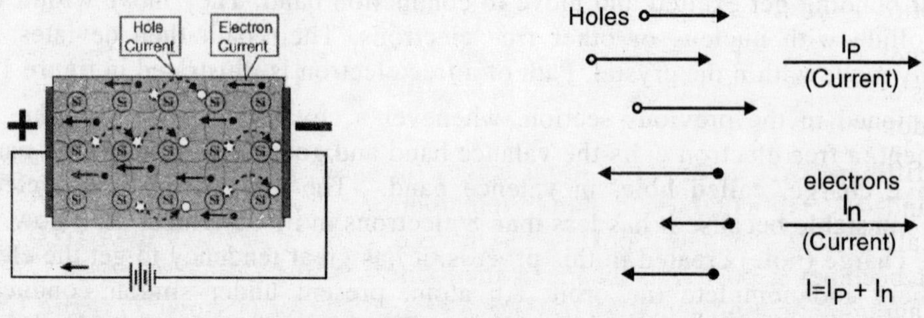


(a) (b)

Figure 1.15

# EXTRINSIC SEMICONDUCTORS AND DOPING

You know that intrinsic semiconductors are pure semiconductors. You are already familiar with their characteristics. Impure semiconductors are called extrinsic semiconductors.

The process of deliberately adding impurities to intrinsic semiconductors is called doping. Thus you can say that a doped semiconductor is called extrinsic semiconductor.

Now the question is, why doping is done? Well, doping improves semiconductor's conductivity and provides few peculiar characteristics to it. Different types of doped semiconductors are described below.

## N-Type Semiconductors

Those semiconductors, which are doped with the material that add excess electrons for the conduction of electricity, are called N-type semiconductors.

To understand the concept of N-type semiconductors, consider the case of silicon. Say a very small amount of phosphorus (about 1 part in one million) is added to it.

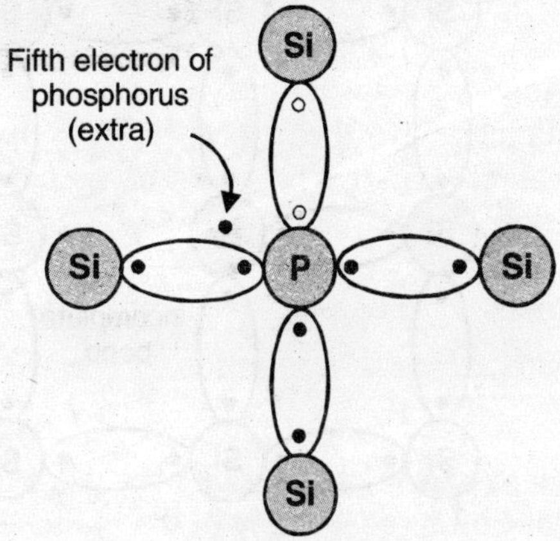


Figure 1.16

Refer figure 1.7 and note that phosphorus belongs to fifth group of periodic table and contains five electrons in its valence band. If the amount of Phosphorus added to silicon is very small then it can be assumed that in doped semiconductor, each phosphorus atom remains surrounded by 4 silicon atoms, as shown in figure 1.16.

Refer figure 1.16 and note that phosphorus atom has five valence electrons. Four of these electrons participate in forming covalent bonds with nearby atoms of silicon.

Since fifth electron doesn't participate in covalent bond formation hence it remains free. At room temperature, it gets excited and jumps to conduction band. Thus it contributes to the conduction of electricity.

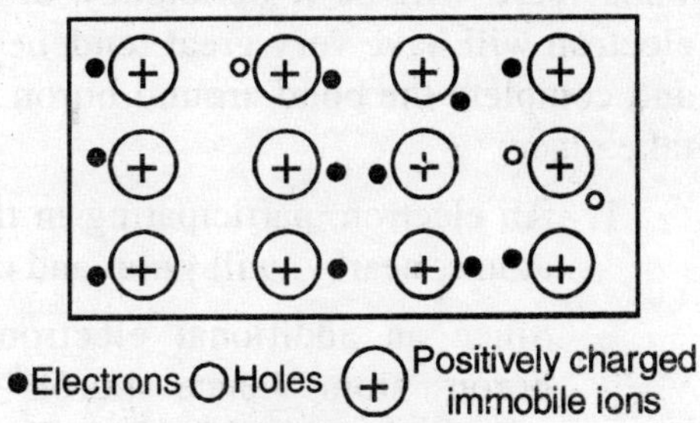


Figure 1.17

Since such type of impurity denotes one electron from its each atom in conduction band hence it is called donor type impurity.

When fifth electron of impurity atom leaves the atom and goes to conduction band, remaining part, called core, becomes positively charged. Thus it becomes a positively changed immobile ion. Diagrammatic representation of N-type semiconductor is shown in figure 1.17

Note block circles represent free electrons. White circles represent holes and bigger circles with positive (+) sign in them represented positive changed immobile ions, from which electron has already made exit to valence band.

## P-Type Semiconductors

Those semiconductors, which are doped with the material that creates extra holes for the conduction of electricity, are called P-types semiconductors.

To understand the concept of P-type semiconductors, consider the case of silicon. Say very small amount of Boron (one part in one million) is added to silicon. Refer figure 1.7 and note that boron belongs to the third group of periodic table and contains 3 electrons in its valence orbit. It the amount of boron added to silicon is very small then three electrons of boron will from covalent bond with 3 nearby silicon atoms. The fourth silicon atom near it will remain incomplete, as shown in figure 1.18 (a)

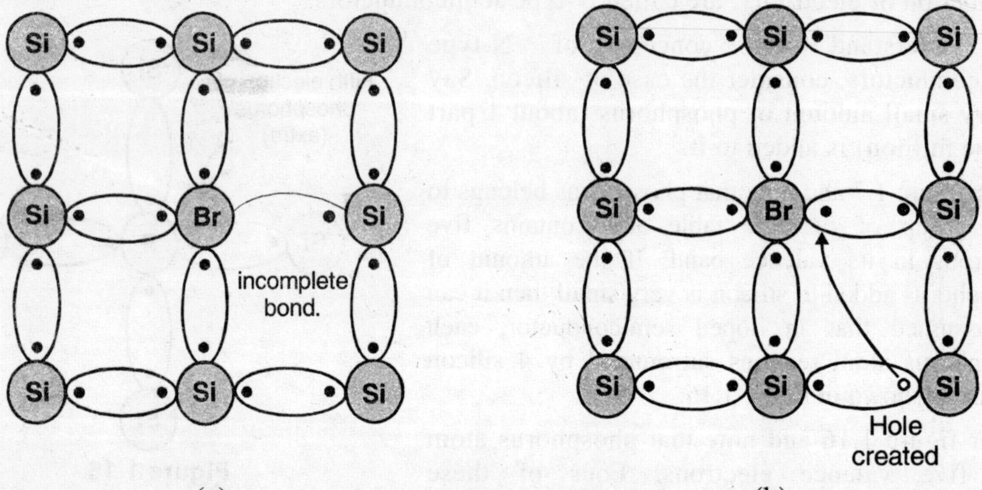


**Figure 1.18**

Thus there will be a deficiency of one electron around the boron atom. This unpaired electron will have very great tendency to get on electron from the neighboring silicon atom and complete the bond around boron. Due to this great tendency, following two things will take place:

1. An electron, participating in the formation of the covalent bond between two silicon atoms, nearby, will jump and complete the covalent bond, as shown in figure 1.18 (b)

   Since an additional electron will be placed in boron atom hence boron core will become negatively charged. It will work as negatively charged immovable ion.

2. A hole will be created in silicon atom, which gives its electron to complete the bond structure around boron.

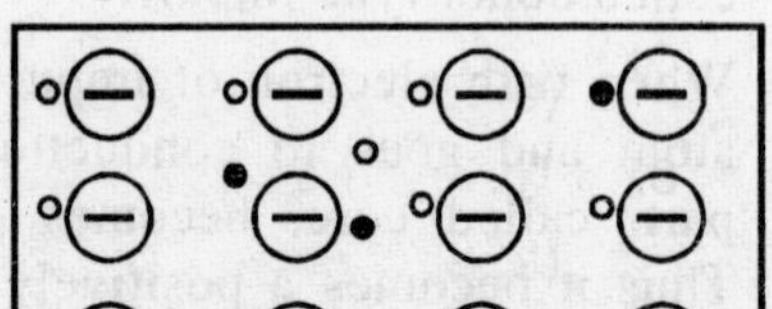

**Figure 1.19**

Since each atom of such type of impurity accepts one electron from nearby silicon atom hence it is called acceptor type impurity.

Diagrammatic representation of P-type semiconductor is shown in figure 1.19. Black circles represent electrons, while blank circles represent holes and bigger circles, with negative signs (-) in them represent negatively charged ions.

Semiconductors are mainly used for fabricating electronic components. Few commonly used electronic components that are fabricated, using semiconductor materials are described in next chapters.

# EXERCISES CHAPTER 1

## Short Type Questions

### A. Select most appropriate answers for following questions:

1. Who discovered electron?
   (a) Thomas A. Edison (b) A Millikan
   (c) John Bardeen (d) Walter h. Brattain
2. What do we mainly study in electronics?
   (a) Composition of atoms of different elements.
   (b) Electron's behavior under different external conditions.
   (c) The method of fabrication of electronic components.
   (d) Semiconductors and their devices.
3. Which of the following is not a power source?
   (a) Battery (b) Alternator (c) Amplifier (d) Oscillator
4. Which of the following will conduct electricity?
   (a) Semiconductor at absolute zero temperature.
   (b) Semiconductor at room temperature.
   (c) Insulator at absolute zero.
   (d) Insulator at room temperature.
5. Which of the following element is an example of semiconductor?
   (a) Phosphorus (b) Aluminum (c) Germanium (d) Boron
6. To which group of periodic table, semiconductors belong?
   (a) Second (b) Third (c) Fourth (d) Fifth
7. Which of the following possesses negative charge?
   (a) Electron (b) Proton (c) Neutron (d) Hole

8. Which of the following contributes to the flow of current in semiconductors?
   (a) Electrons in conduction band (b) Holes in conduction band
   (c) Both of the above (d) None of the above
9. Which of the following can be used as donor type impurity to dope intrinsic semiconductors?
   (a) Silicon (b) Boron (c) Carbon (d) Germanium
10. In which of the following situation, will a hole get created?
    (a) When an electron exits to conduction band.
    (b) When a covalent bond is formed.
    (c) When an electron returns from conduction band to valence band.
    (d) In none of the above condition.

## B. Fill in the blanks:

11. ..........and........are the two elements, which are commonly used as semiconductors.
12. Semiconductors possess lesser conductivity than............but...............conductivity than insulators.
13. Deliberate attempt of adding impurities in intrinsic semiconductors is called..........
14. To create N-type semiconductor, very small amount of...............................is added to..................
15. Pure semiconductors are also called...........................................semiconductors.
16. Adding small amount of boron to silicon creates.......................type semiconductor.
17. In diagrammatic representation of semiconductors, black circles represent .......................... while white circles represent ...............................
18. While conducting the current, free electrons move in ................band while holes drift in ..............band.
19. Donor type impurities generate.......................................for conduction of current.
20. Acceptor type impurities are used for making...................type of semiconductors.

## C. State, true or false for following statements:

21. Electronics is the study of electron's mechanics under different conductions that are externally created.
22. Oscillator is power source, which supplies DC voltage.
23. Active electronic components do not require power for their functioning.
24. Conductivity of semiconductors is decreased as the temperature is increased.
25. Conductivity of conductors is increased as the temperature is decreased.
26. Semiconductors are placed in fifth group of periodic table.
27. The formula 2n+1 governs the number of electrons that can be accommodated in $n^{th}$ orbit of an atom.
28. When external energy is provided to an atom, it jumps from conduction band to valence band.
29. When electron moves in some direction, current flows in opposite direction.

30. Adding very small amount of boron with silicon will form P-type semiconductor.

## D. Answer the following questions in short:

31. How many valences electrons will be there in germanium atom?
32. Out of valence band electron and conduction band electron, which will have greater energy?
33. When a battery is connected to extrinsic semiconductor, towards which terminal of the battery, will electrons and holes get attracted to?
34. Which type of charge is associated with holes?
35. Below which band lies the forbidden energy gap in semiconductors?
36. What do you call the process of adding impurities in intrinsic semiconductors?
37. In which ratio, impurities are added to intrinsic semiconductors to form extrinsic semiconductors?
38. Which type of charge will appear on core atom, if an electron exits for valence band?
39. What do you call the electrons of outermost orbit of an atom?
40. In P-type semiconductors, which contributes most for the conduction of current?
41. Name any two elements that are semiconductors.

### Detailed Answer Type Questions

## E. Answer the following questions in detail:

42. What is electronics? Write a brief history of electronics.
43. Differentiate between active and passive electronic components. Name the electronic components that fall into these categories.
44. What are semiconductors? Describe at least two characteristics of semiconductors.
45. What is doping? Why it is done? Differentiate between intrinsic and extrinsic semiconductors.
46. Define the following terms:
    (a) Conduction band (b) Forbidden energy gap (c) Valence band.
47. What are holes? When do they get created? What is their role in semiconductors?
48. Describe the crystalline structure of silicon.
49. What are N-type semiconductors? How they are created? How they are diagrammatically represented?
50. What are P-type semiconductors? How they are created? How they are diagrammatically represented?
51. Describe the main difference between N-type semiconductors and P-type semiconductors.
52. Describe the process that takes place when a battery is connected to intrinsic semiconductor at room temperature.
53. Differentiate between conductors, semiconductors and insulators. Draw energy band diagram to illustrate the difference.

****

# BASIC ELECTRONICS

# CHAPTER 2

## PN Junction Diodes

You are familiar with P-type and N-type semiconductors. Truly speaking, independently they are of very little use. But when they are combined together, combination behaves very differently than either type of the semiconductor. It is the combination of P-type and N-type semiconductors that is used in fabricating different types of electronic components. When P-type semiconductor is joined with N-type semiconductor in such a way that joint retains the crystalline structure then the combination is called a PN junction. In this chapter we will mainly describe the details and characteristics of PN junction.

### INTERNAL DYNAMICS OF PN JUNCTION

As mentioned above, PN junction is formed by smoothly joining P-type and N-type semiconductors. PN junction is illustrated in figure 2.1.

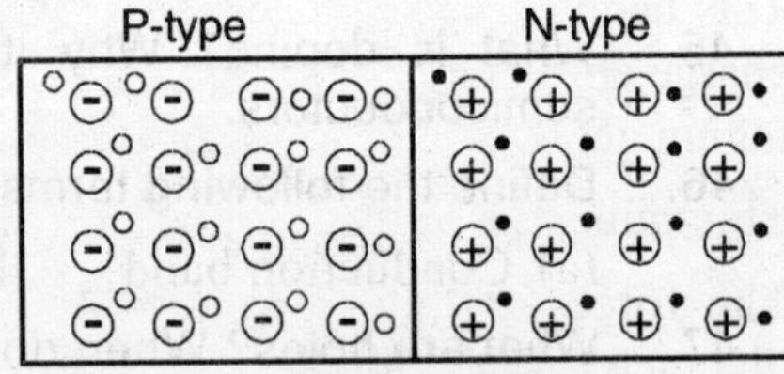

**Figure 2.1**

Refer this figure and note that P-type material has holes as its majority carriers. It also has negatively charged immobile impurity ions. It is quite obvious that holes can drift from one place to another within the crystal while charged ions cannot. On the other hand, N-type material has electrons as majority carriers. It also has positively charged immobile impurity ions. It is quite obvious that electrons can move within the crystal, while charged ions cannot.

*Note that to keep the discussion simple, at present we are not talking anything about minority carriers that remain present in both the regions (i.e. electrons in P region and holes in N type region).*

Initially when P-type and N-type semiconductors are joined together, following activities automatically take place:

1. Few free electrons from N region (that are near the junction) crossover to P region and meet the holes. When they meet, they cancel each other's charge and appear to be vanished. Thus negative charge (due to the presence of negatively charged immobile ions), starts building up near the junction in P region. This region is shown in figure 2.2

2. Similarly few holes from P region (that are near the junction) crossover to N region and meet the electrons. When they meet, they cancel each other's charge and appear to be vanished. Thus positive charge (due to the presence of positively charged immobile ions) starts building up near the junction in N region. This region is shown in figure 2.2

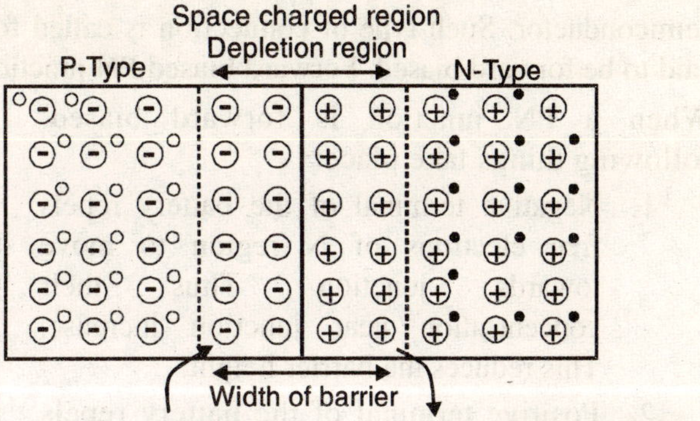


**Figure 2.2**

3. Since a column of positive charge gets developed in N region and a column of negative charge gets developed in P region hence further flow of holes and electrons stops.
4. Since near the junction, holes get eaten up by electrons (in P region) and electrons get eaten up by holes (in N region) hence a region depleted of moving charge gets created on both the sides of the junction. This region is called depletion region. Depletion region is illustrated in figure 2.2. Since this region contains immobile positively and negatively charged ions hence this region is also called space charged region.

This happens because holes, trying to drift from P region to N region get repelled by positively charge column that gets created in N region. Similarly electrons trying to drift from N region to P region get repelled by negatively charged column that gets created in P region.

This puts stop to any further self-movement of carriers (holes and electrons).

The force of repulsion that gets created near the junction is called barrier. The physical distance, over which the barrier extends, is called the width of the barrier.

The potential difference that gets created between the two ends of the barrier is called the height of the barrier. To give you a rough idea of the barrier height, for silicon PN junction barrier height is generally 0.7 volts, while for germanium PN junction, it is approximately 0.3 volts.

In short, it can be said that when PN junction is formed barrier voltage is automatically developed, across the junction, if no battery is connected.

## FORWARD BIASED PN JUNCTION

You are now familiar with the internal dynamics of PN junction and know that a potential barrier gets created near the junction, which stops further drifting of electrons and holes across the junction.

Now let's see, what happens when battery is connected to PN junction in such a way that positive terminal is connected to P type semiconductor and negative terminal is connected to N type semiconductor. Such type of connection is called forward biased connection and PN junction is said to be forward biased. Forward biased PN junction is illustrated in figure 2.3.

When a PN junction is forward biased, following things take place:

1. Negative terminal of the battery repels free electrons of N region to move towards junction. Thus their concentration near junction increase. This reduces the barrier height.
2. Positive terminal of the battery repels holes of P region to move towards the junction. Thus their concentration near the junction increases. This further reduces the barrier height.
3. Due to reduced barrier height more electrons and holes cross the junction.
4. Electrons that cross the barrier, drift towards the positive terminal of the battery.
5. Holes that cross the barrier, drift towards the negative terminal of the battery.
6. Due to electrons moving towards positive terminal and holes moving towards negative terminal, electric current flows.

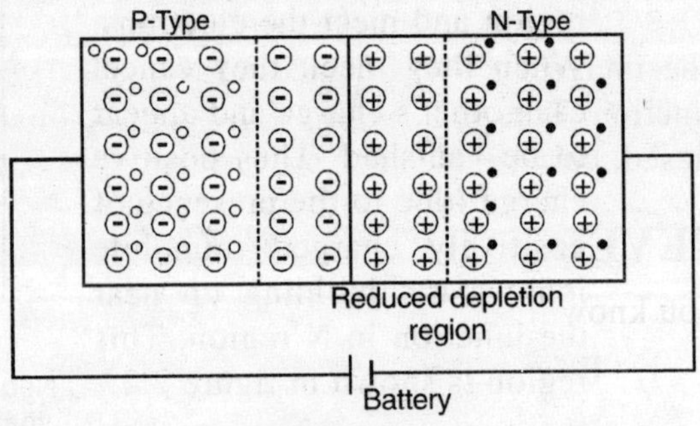

**Figure 2.3**

In short, it can be said that when PN junction is forward biased, current flows through it.

If the battery voltage is increased, barrier height is further reduced causing more electrons and holes to move across the junction. This will increase the magnitude of the flowing current.

## REVERSE BIASED PN JUNCTION

When a battery is connected to the PN junction in such a way that positive terminal of the battery is connected to the N-type and negative terminal is connected to P-type then such a connection is called reverse biased connection and PN junction is said to be reversed biased. Reverse biased PN junction is shown in figure 2.4

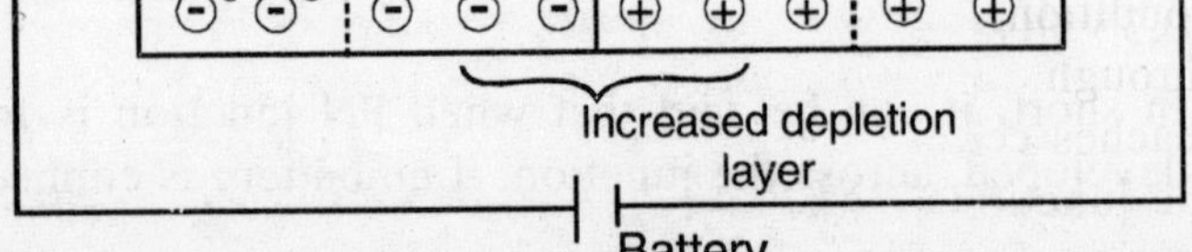

**Figure 2.4**

When PN junction is reverse biased, following events take place:

1. Electrons in N region get attracted towards positive terminal of the battery. Thus some of the atoms, near the junction, loose their electrons, leaving positively charged ions there. This widens the barrier width and height of the barrier increases.

2. Holes in P region get attracted towards negative terminal of the battery. Thus some of the atoms near the junction loose their holes, leaving negatively charged ions there. This further widens the width of the barrier and increases the height of the barrier.

The increased barrier makes it more difficult for electrons and holes to drift across the junction. Thus the current doesn't flow. In short it can be said that if PN junction is reverse biased, current doesn't flow through it.

## REVERSE BREAKDOWN OF PN JUNCTION

You know that when PN junction is formed, following things exist within PN junction:

1. Electrons as majority carriers within N region.
2. Few holes as minority carriers within N region.
3. Holes as majority carriers within P region.
4. Few electrons as minority carriers within P region.

Here the point to be noted is that when PN junction is reversed biased few electrons (minority carriers) from P side and few holes (minority carriers) from N side cross over the junction, so very small current flows through the circuit.

If the voltage is increased to some extent the current doesn't increase significantly. However, beyond certain voltage, crystal structure breaks and current increases abruptly. This situation is called reverse breakdown of PN junction. The voltage, at which breakdown occurs, is called breakdown voltage. V-I characteristic of PN junction is illustrated in figure 2.5 here. V represents voltage and I represents current.

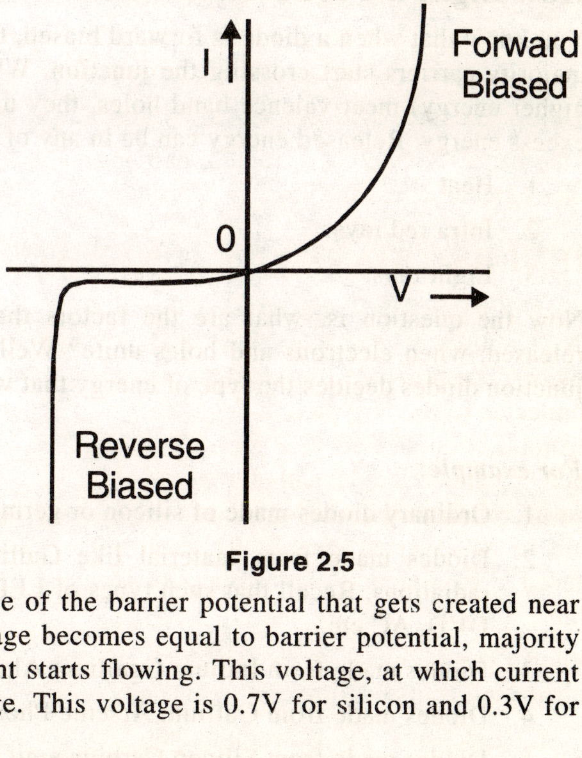

**Figure 2.5**

Refer V-I curve shown in figure 2.5 and note that under forward biased condition, current doesn't flow well through PN junction until voltage reaches certain value. This happens because of the barrier potential that gets created near the junction. When externally applied voltage becomes equal to barrier potential, majority carriers start crossing the barrier and current starts flowing. This voltage, at which current starts rapidly is called cut-in or knee voltage. This voltage is 0.7V for silicon and 0.3V for germanium.

## THE IDEAL DIODE

An ideal diode would be one, which would act as perfect conductor in forward biased condition and as a perfect insulator, when reversed biased. But in actual it doesn't happens so. A diode conducts well (but not like perfect conductor) when forward biased and poorly (not as bad as perfect insulator) when reversed biased.

## LIGHT EMITTING DIODES (LEDs)

Those diodes (PN junctions), which emit light when current passes through them are called light emitting diodes or LEDs in short. LEDs are frequently used as pilot lights in electronic appliances to indicate whether the device or a specific part of it is working or not. For example, when you switch on the computer, an LED on the front panel of its chassis starts glowing. This indicates that computer power is on. Similarly another LED connected to its hard disk glows, whenever read write operations are performed on the hard disk. LEDs look like tiny bulbs. Their physical appearance is shown in figure 2.6

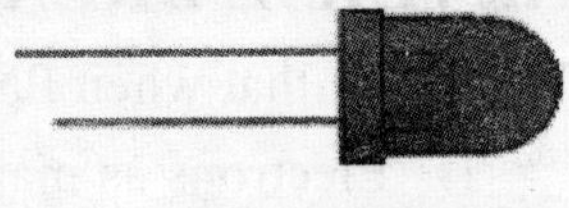

**Figure 2.6**

Diode electronics is enclosed in transparent bulb of special shape. This bulb concentrates the light in a particular direction so that intensity becomes high.

## How Light is Produced in LEDs

You know that when a diode is forward biased, the potential barrier is lowered and then the majority carriers start crossing the junction. When conduction band electrons (possessing higher energy) meet valence band holes, they unite. When they unite, electron releases its excess energy. Released energy can be in any of the following forms:

1. Heat
2. Infra red rays.
3. Light rays.

Now the question is, what are the factors that decide the type of energy that will be released, when electrons and holes unite? Well, the material used for construction of PN junction diodes decides the type of energy that will be released by diode.

***For example:***

1. Ordinary diodes made of silicon or germanium produces heat.
2. Diodes made from material like Gallium Aluminum Arsenide produce infrared radiations. Recall that such types of LEDs are used in remote controls of TV, VCR, DVD. AC etc.
3. Diodes made from Indium Ggallium Aluminum Phosphide etc. emit super red light.
4. Diodes made from Gallium Arsenic Phosphide etc. emit orange light.
5. Diodes made from Silicon Carbide emit cool white light.
6. Diodes made from Gallium Nitride emit green light.

## Characteristics of LED

Following are the characteristics of LEDs:

1. They work on very low voltage, as low as 1 to 2 volts and very low currents, as low as 5 to 10 mA.
2. They consume less power.
3. They do not require any heating time, so they work almost instantly.
4. They are small in size and light in weight.
5. They possess long life.

## PHOTODIODES

You know that when PN-junction is reversed biased, a reverse saturation current flows. This current flows because of minority carriers that get generated because of considerable amount of thermal agitation at room temperature, itself.

Recall that these minority carriers remain present in the form of electrons and holes in P and N regions respectively. The current that gets generated due to them is illustrated in V-I chart that has been plotted in figure 2.5.

Refer figure 2.5 and note that even by increasing the voltage, current doesn't change considerably in reverse bias. However if PN junction temperature is increased, more number of atoms get thermally agitated. Thus minority carriers increase in number and so does the reverse current. The same effect is produced if PN junction is illuminated. In such a situation even if the bias voltage is reduced to zero some current flows through the junction.

Photodiodes are basically PN junctions, which are reversed biased. They are encapsulated in a transparent case so that light radiations can reach the junction without any problem.

When the light rays strike the photodiode they excite its atoms. Thus they create few free electrons and holes. Electrons created near the junction, in P region are swept away in N region. Similarly holes created near the junction in N region are swept away in P region. Thus the movement of excess carriers near the junction produces photo current. This current adds to the current, which was previously there due to reverse bias.

If $I_0$ was the reverse current and $I_P$ is the photo current that has been generated due to the light falling on PN junction then total reverse current $I_r$ will be as follows:

$$I_r = I_0 + I_p$$

With zero light $I_p$ will be 0 and total current $I_r$ will be equal to reverse current $I_0$, called dark current.

Note that more the light that falls on photodiode, more is the current (Ip) produced.

V-I chart for photodiode is illustrated in figure 2.7.

V-I curve shown in figure 2.7 shows that reverse current is almost independent of the reverse voltage. It also shows that reverse current is proportional to the illumination. More the illumination more is the current.

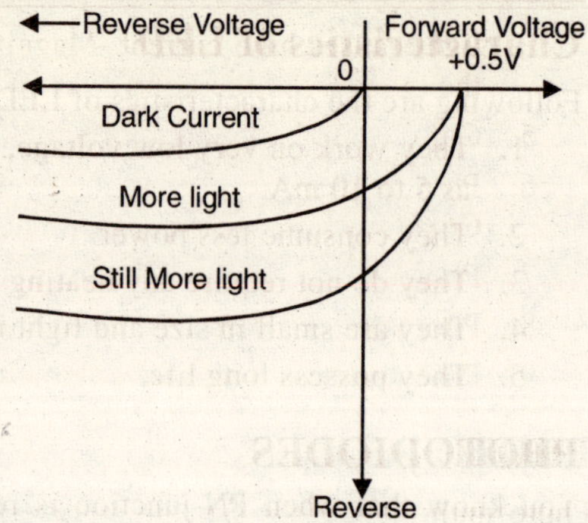


**Figure 2.7**

Note that unit of measurement for illumination is Lumens per square meter or lux. That means:

1 Lux =1 $L_m / m^2$

## Avalanche Photodiodes

Avalanche photodiodes are same as normal phodiodes but they are made to operate with much higher reverse bias. When reverse bias is very high the electrons are accelerated to such an extent that when they strike the atom in the material, they knock off other electrons of the atom setting them free. These ejected electrons then accelerate and knock off other electrons. Thus all of a sudden, the number of free electrons increases drastically. This makes the reverse current increase greatly. This phenomenon is called Avalanche Effect.

Operating photodiodes (avalanche photodiodes) in higher reverse bias condition increases the affectivity and responsivity of the diode because even a small radiation falling upon photodiode, results into larger current.

## Use of Photodiodes

Following are the uses of photodiodes:

1. Photodiodes are used in CD players. The light source (laser source) makes the light fall on compact disk. From there light gets reflected in either of the two modes i.e. high intensity or low intensity.

   The reflected light is made to fall on a photodiode, when the light falls on photodiode, current gets generated. Low intensity light generates low current while high intensity light generates high current. High intensity and low intensity currents are interpreted as 0 and 1 and thus data is read from the CD.

2. Photodiodes are used in smoke detectors. Smoke detectors are basically alarms that are used to alert nearby people in the event of fire.

   Smoke detectors basically contain a light source and a photodiode. In the absence of smoke the light doesn't fall on photodiode. Thus no current is generated.

   When fire takes place and smoke comes in the way of light rays, some of the light gets reflected by smoke particles. Reflected light falls on photodiode. This causes a current in the output. This current is then used to give an alarm.

3. Photodiodes are also used in the TV, VCR receivers that receive signals that are transmitted through their remote controls.
4. Photodiodes are also used in light intensity measuring equipments. Light is made to fall on photodiode. Low intensity lights causes small current while high intensity

light causes high current. Magnitude of the current produced is used for indicating the intensity of the light.

5. Color sensors make use of photodiodes.
6. Photodiodes are also used in bio-medical equipments like CT scan (now called Computed Topography) machine, blood gas monitors etc.
7. Photodiodes are also used in street light sensors.
8. Photodiodes are used in calibration detectors.
9. Photodiodes are also used in bar code readers.

## PHOTO MULTIPLIER TUBES

Photo multiplier tubes are very special vacuum tubes, constructed of glass or quartz. On front end they have photocathode and at back end they have electrical connections. Within the tube are housed photocathode, focusing electrode, many dynodes and an anode. Block diagram of photomultiplier tube is illustrated in figure 2.8.

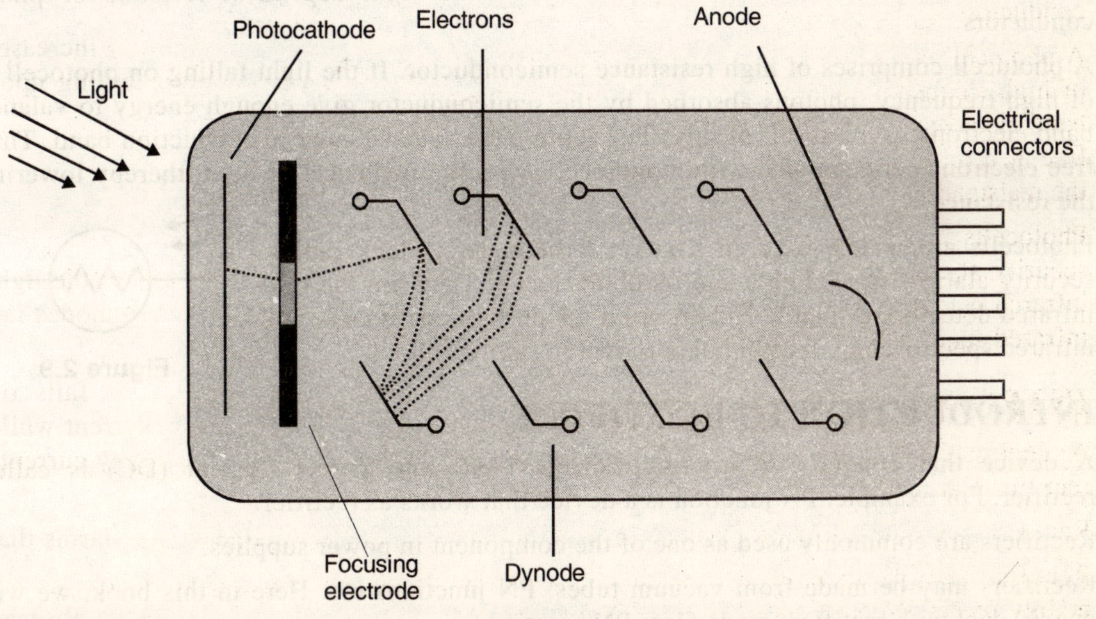


**Figure 2.8**

Photomultipliers are very sensitive to the radiations that fall into ultraviolet, visible and infrared region. They are mainly used for detecting the radiations.

Photomultiplier tube operates between 1 to 5 Kilo Volts. Each dynode in photomultiplier is maintained at higher positive voltage than the previous dynode. For example, the first dynode may be maintained at 1000 Volts. The second dynode may be at 1500 Volts, the third dynode may be at 2000 Volts and so on.

When the light/radiations fall on photocathode, photoelectrons are emitted. Focusing electrode focuses them towards first dynode. As they move towards it, they get accelerated due to the positive charge of the dynode. They hit the dynode with greater energy. This hitting brings out of more low energy electrons from the dynode. These electrons now move towards second dynode and get accelerated because of its positive electric field. Now these electrons hit the dynode and liberate more number of electrons from it.

The shape of dynode chain is kept such that electrons keep moving forward, with ever increasing number and ultimately reach the anode.

Accumulation of charge results in a sharp current pulse. This pulse indicates that high radiations had fallen on photocathode.

## PHOTOCELL

Photocell is basically a device for measuring the amount of light or electromagnetic radiations present in a space. Its electrical state is altered by the presence of light.

Photocell is a type of resistor. When the light strikes the cell, its resistivity decreases and it allows the current to flow more freely. When it is dark in the surrounding, its resistance decreases. This is the reason they are also called light dependent resistors or photo conductors.

A photocell comprises of high resistance semiconductor. If the light falling on photocell is of high frequency, photons absorbed by the semiconductor give enough energy to valence band electrons. As a result of this, they jump from valence band to conduction band. Thus free electrons get generated within photocell, with the incidence of light, thereby lowering the resistance.

Photocells are widely used in camera, light meters, clock radios, security alarms, street lights and outdoor clocks. They are the best infrared detectors available and are used for infrared astronomy and infrared spectroscopy. Its symbol is shown in figure 2.9.

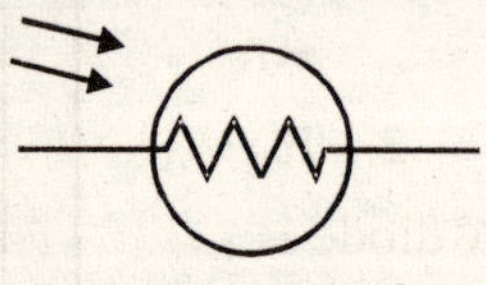

**Figure 2.9**

## INTRODUCTION TO RECTIFIER

A device that converts Alternating Current (AC) into Direct Current (DC) is called rectifier. For example, PN junction is a device that works as rectifier.

Rectifiers are commonly used as one of the component in power supplies.

Rectifiers may be made from vacuum tubes, PN junctions etc. Here in this book, we will mainly deal with rectifiers made from PN junctions.

## TYPE OF RECTIFIERS

Rectifiers can be broadly classified into following two categories:

1. Half wave rectifiers
2. Full wave rectifiers

Both types of rectifiers are described below.

## Half Wave Rectifiers

Half wave rectifier converts AC into DC. It allows either the positive or negative side of AC to pass through. The other half gets blocked by it. A half wave rectifier is shown in figure 2.10.

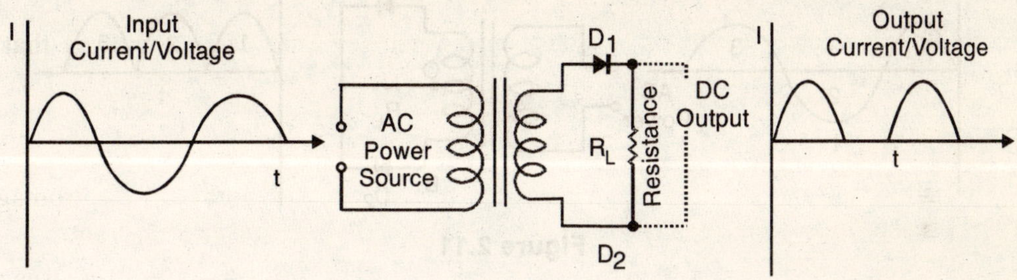


**Figure 2.10**

Refer figure 2.10 and note that the circuit shown there is a half wave rectifier. Transformer at input side works as AC source. It provides alternating current to the circuit. A transformer is used in the rectifier, with following objectives:

1. It steps up or steps down the voltage as per requirement. For example if 12 volt DC is required in the output. Then transformer can convert 220V normal power supply (or any other voltage available) to 12 Volt AC, which can then be converted into 12 volt DC.
2. It provides isolation from the power supply thus reduces the risk of electrical shock.

A diode is put in series with the resistance, across which output voltage is obtained.

AC voltage at the input varies from maximum positive voltage to minimum negative voltage. First half of the cycle, in which cycle remains positive, makes the diode forward biased. When the diode is forward biased, it conducts current. In this situation current is obtained in the output. Second half of the cycle, in which the voltage remains negative, makes the diode reversed biased. When the diode is reversed biased, it doesn't conduct the current. In this situation no current is obtained in the output.

The pattern of the current, which is obtained at the output, is shown in figure 2.10. Note that this current is not perfect DC but at least it is unidirectional.

## Full Wave Rectifier

As mentioned earlier, half wave rectifier works as a clipper. It utilizes only half the cycle of alternating current. It blocks the other half cycle. On contrary to this, full wave rectifier utilizes both the cycles of alternating current and makes them unidirectional. In this way it gives DC all throughout.

Full wave rectifiers are of two types. One is called center tap type rectifier. It makes use of two diodes. The other type is called bridge type and makes use of four diodes. Both types of rectifiers are described below.

## Center Tap Rectifier

Center tap rectifier is shown in figure 2.11.

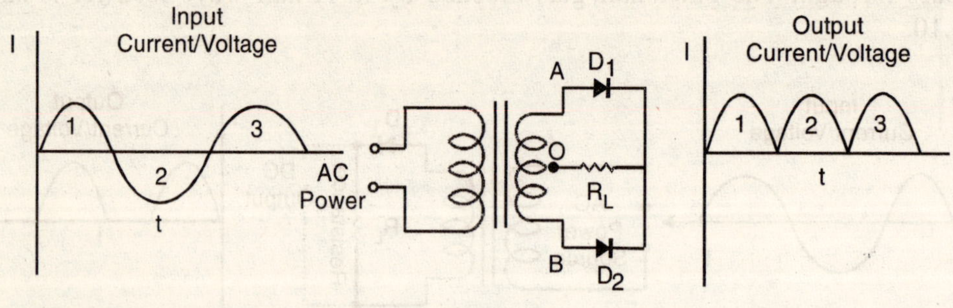


**Figure 2.11**

Note that center tap rectifier makes use of two diodes D1 and D2, with center tapped transformer.

During the positive half of the cycle, voltage at point A becomes positive with respect to center point O. This makes D1 diode forward biased. At the same time, voltage at point B becomes negative with respect to center point O. This makes D2 diode reversed biased. In this cycle, D1 being forward biased conducts while D2 being reversed biased doesn't conduct. As a result of this, current is obtained at the output due to D1 diode.

During the negative half of the cycle, voltage at point A becomes negative with respect to center point O. This make D1 diode reversed biased. At the same time, voltage at point B becomes positive with respect to center point O. This makes D2 diode forward biased. Thus in this cycle, D2 being forward biased conducts while D1 being reversed biased, doesn't conduct. As a result of this, current is obtained at the output due to D2 diode.

It is quite clear now that in full wave rectifier both the diodes put together produce current at output for both the halves of alternating current. This current is unidirectional, so it is DC.

Pattern of the current obtained at the output of full wave rectifier is clearly shown in figure 2.11.

Although efficiency of full wave center tap rectifier is generally found to be around 81% to 82%, but following are its drawbacks:

1. It is difficult to locate center tap on the secondary side of the transformer.
2. The DC output is small as each diode utilizes only one half of the transformer's secondary voltages.

## Bridge Rectifier

Bridge rectifier is modified form of center tap rectifier. In makes use of only four diodes instead of two. Four diodes are connected together to form a bridge, as shown in figure 2.12. This structure avoids the need for center-tapped transformer. Bridge rectifier is shown in figure 2.12.

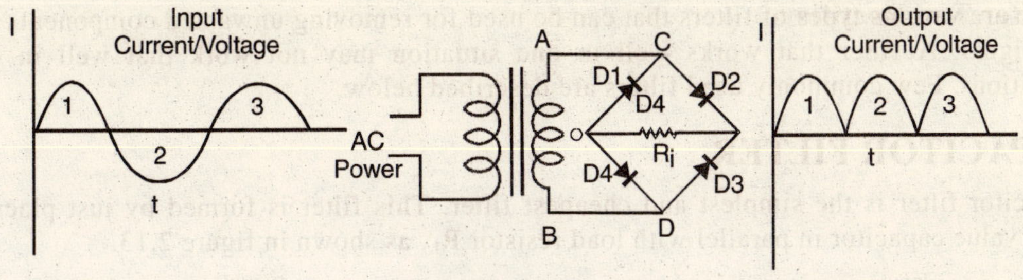


**Figure 2.12**

During the positive half cycle of the current, point A becomes positive and point B becomes negative. This makes point C positive in comparison to point O and point D negative in comparison to point O. Under this situation, D2 and D4 diodes become forward biased, while D1 and D3 diodes become reverse biased. During this cycle, D2 and D4 conduct while D1 and D3 don't. Thus for the positive half cycle of alternating current, unidirectional current is obtained at the output due to D2 and D4 diodes.

During the negative half cycle of the current, point A become negative and point B become positive. This makes point C negative with respect to point O and point D positive with respect to point O. Under this situation, D1 and D3 diodes become forward biased while D2 and D4 diodes, become reverse biased. During this cycle, D1 and D3 diodes conduct while D2 and D4 don't. Thus for the negative half cycle of AC, unidirectional current is obtained at the output due to D1 and D3 diodes.

Thus for both the halves of AC cycle, unidirectional current is obtained at the output.

The drawback with bridge rectifier is that it requires four diodes (instead of two) but it has following advantages:

1. It doesn't require tapped transformer.
2. The output is twice that of center-tap rectifier.
3. Peak inverse voltage is one half of the center tap rectifier.

## NEED FOR FILTERS

You know that unidirectional current that doesn't change its direction with time is called DC current. Different forms of DC current are illustrated in figure 1.5. Out of those three forms, first form is ideal form, which does not fluctuate with time. For example, the one obtained from a battery. If you compare the waveforms obtained from half wave and full wave rectifiers with ideal waveforms, you will conclude that although those waveforms are all DC but they are not ideal DC waveforms. They contain ripples. It is quite obvious that if somehow ripples are removed from such wave forms, they would get converted into ideal DC.

Filters are those electronic circuits, which operate upon given signal, perform signal processing functions on it and remove unwanted signal components from it. For example, an electronic filter of specified design, when added to rectifier will operate upon the fluctuating signal produced by the rectifier and convert it to almost ideal DC signal that has minimum fluctuations in it.

There are various types of filters that can be used for removing unwanted components from the signal. A filter that works well in one situation may not work that well in other conditions. Few commonly used filters are described below.

## CAPACITOR FILTER

Capacitor filter is the simplest and cheapest filter. This filter is formed by just placing a large value capacitor in parallel with load resistor $R_L$, as shown in figure 2.13.

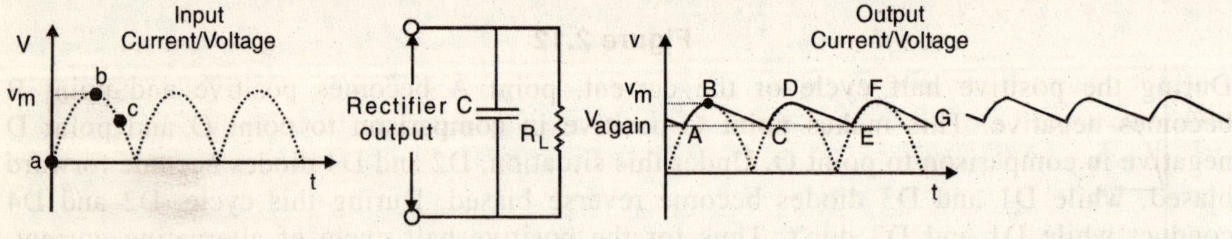


**Figure 2.13**

The fluctuating DC output of the rectifier is applied across the capacitor. In this arrangement capacitor filter works as follows:

1. When the rectifier voltage increases (refer portion ab in input side of figure 2.13), it charges the capacitor.
2. When the increasing voltage reaches its maximum i.e. $V_m$ , capacitor also gets charged to its maximum value $V_m$.
3. After touching the maximum value, rectifier voltage starts decreasing (Refer portion bc in input side of figure 2.13). Since now the capacitor is at a higher voltage than the source hence it sends the current into the load.
4. Back sent current reaches the diode also (of the rectifier) and makes it reversed biased.
5. Due to reverse bias, current can't flow from rectifier hence it stops flowing towards load $R_L$.
6. The capacitor continues to discharge until the source voltage becomes more than capacitor voltage i.e. $V_{again}$. (Refer point C in the output side of figure 2.13).
7. At this point, diode becomes forward biased and starts recharging the capacitor till it become charged to maximum voltage $V_m$
8. During the time capacitor is charging, rectifier supplies the current through the load.
9. After capacitor getting fully charged, it starts discharging and the cycle repeats.

It is quite clear from the process that current across load never falls to zero. It continuously remains available. Some times it is due to rectifier and some time it is due capacitor. The wave form BCDEFG, shown in output side of figure 2.13 illustrates the pattern of DC current that is obtained after putting capacitor filter. Note that output thus obtained is although fluctuating DC but not as fluctuating as the original signal. Due to the presence of the filter, it has been smoothed to a large extent.

Capacitor filters are low cost filters and are commonly used in transistors, radios, battery eliminators etc.

# CHOKE-INPUT LC FILTER

Choke-Input LC filter consists of an inductor and capacitor. Inductor is connected in series with the rectifier and a capacitor in parallel with load. Choke-Input LC filter is shown in figure 2.14

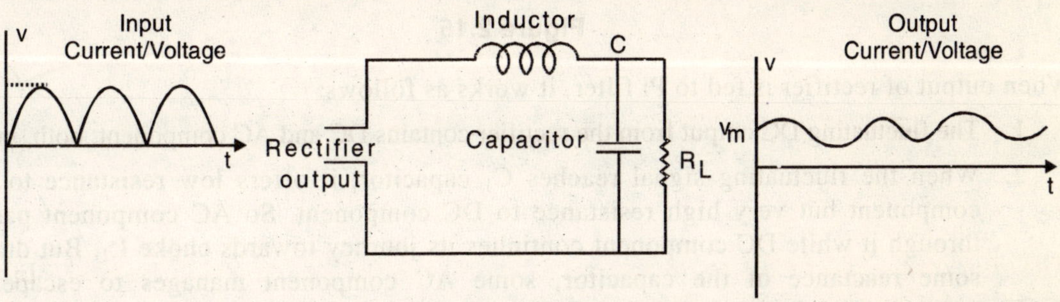


**Figure 2.14**

When fluctuating DC output from rectifier is connected to chock-input LC filter, following things happen:

1. Inductor allows the DC component of the input signal to pass through it and poses high resistance to AC component.
2. Whenever the current through the inductor changes, back electromotive force is produced. This back electromotive force prevents the current from changing its value. As a result of this, constant value current passes through inductor that has very few ripples in it.
3. Thus at point C, we have DC component and remaining AC components, which have managed to escape through choke due to its inductance.
4. Now the low reactance capacitor charges and discharges due to AC component (as explained in capacitor filter) and prevents DC component to pass through it.
5. Since DC component doesn't pass through capacitor hence it reaches the load $R_L$.

In this way, what we get at the output is much smoother DC current.

# Pi ( Π ) FILTER

Pi filter consists of two capacitors and one inductance arranged in the shape of Π. Pi filter is shown in figure 2.15.

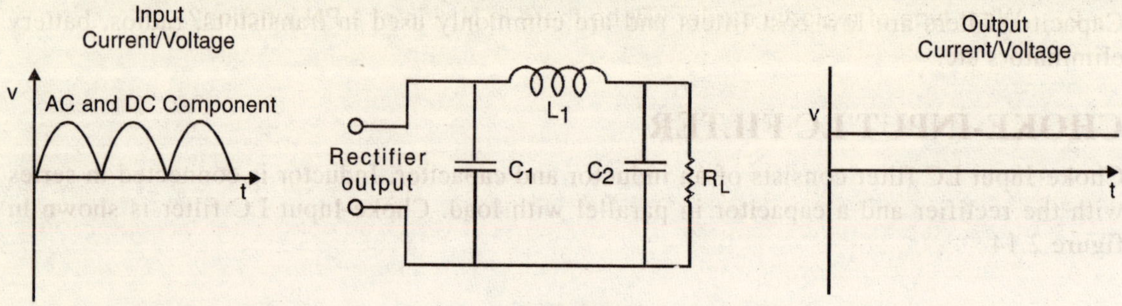


**Figure 2.15**

When output of rectifier is fed to Pi filter, it works as follows:

1. The fluctuating DC output from the rectifier contains DC and AC component, both, in it.
2. When the fluctuating signal reaches $C_1$ capacitor, it offers low resistance to AC component but very high resistance to DC component. So AC component passes through it while DC component continues its journey towards choke $L_1$. But due to some reactance of the capacitor, some AC component manages to escape $C_1$ capacitor and thus moves forward with DC component.
3. Thus, what inductor $L_1$ receives is the mixture of DC component and some AC component, which managed to escape $C_1$ capacitor.
4. Inductor $L_1$ offers low resistance to DC component but high resistance to AC component. Thus DC manages to move forward, while majority of AC component doesn't. Due to some reactance of $L_1$ a very small portion of AC component manages to escape the inductor.
5. Thus what capacitor $C_2$ receives is mixture of DC component and very little AC component that managed to escape the inductor.
6. The capacitor $C_2$ provides low resistance to AC component so it passes through it. It offers high resistance to DC component. Thus pure DC component moves towards load. Thus what $R_L$ receives is much smoother DC current.

## EXERCISES CHAPTER 2

### Short Type Questions

**A. Select most appropriate answers for following questions:**

1. What type of majority carrier will be there in P side of a PN junction?

(a) Holes (b) Electrons

(c) Fixed positively charge ions (d) None

2. What type of majority carrier will be there in N side of a PN junction?
   (a) Holes (b) Electrons
   (c) Fixed negative charged ions (d) None
3. Which of the following creates potential barrier in PN junction?
   (a) Minority carriers (b) Majority carriers
   (c) Both minority carriers and majority carriers (d) Fixed donors and acceptor ions.
4. Which of the following will make the PN junction forward biased?
   (a) Positive terminal connected to P side.
   (b) Negative terminal connected to N side.
   (c) Positive terminal connected to P side and negative terminal connected to N side.
   (d) Positive terminal connected to N side and negative terminal connected to P side.
5. Which of the following factor makes holes diffuse from P region to N region in PN junction?
   (a) Free electrons in N region attract them.
   (b) They are swept away due to potential difference.
   (c) There is greater concentration of holes in P region as compared to N region.
   (d) Holes have a tendency to move.
6. What will happen if a PN junction is reverse biased?
   (a) Potential barrier gets increased.
   (b) Potential barrier gets decreased.
   (c) Reduces the carrier current to zero.
   (d) Reduces the minority carrier to zero.
7. When will an avalanche break down occur in PN junction?
   (a) When forward bias will exceed certain limit.
   (b) When reverse bias will exceed certain limit.
   (c) When the polarity of the diode is changed.
   (d) When light is made to fall on PN junction.
8. When does LED emits light?
   (a) When light falls on it.
   (b) When current passes through it.
   (c) When avalanche break down occurs.
   (d) None of the above.
9. How many diodes will be required in full wave bridge rectifier?
   (a) 1 (b) 2 (c) 3 (4)

10. How many capacitors and inductors will be required in Pi filter?
    (a) 1 capacitor and 1 inductor.
    (b) 2 capacitors and 2 inductors.
    (c) 2 capacitors and 1 inductor.
    (d) 1 capacitor and 2 inductors.

## B. Fill in the blanks:

11. In a PN junction N sided contains .............as majority carriers and .........as minority carriers.
12. When a PN junction is forward biased, it. ............the width of the potential barrier.
13. A PN junction conducts when it is ...................................biased.
14. A reversed biased PN junction has positive terminal connected to.............side while negative terminal is connected to ....................................................side.
15. LED is acronym for.............. ................. .................
16. The PN junction, which conducts current, when light is made to fall on it, is called.....................
17. With no light falling on photodiode, the current in reversed biased conduction is referred to as.......................................
18. The rectifier, which converts only part of AC current into DC current is called............................rectifier.
19. The current that is obtained in domestic supply is an example of ..............current, while the current obtained from the battery is referred to as ..................current.
20. Out of capacitor, choke and Pi filters.................filter will give most smooth output.

## C. State, true or false for following statements:

21. A PN junction diode allows the current to flow only in one direction.
22. A rectifier converts DC current into AC current.
23. A reversed biased diode results into very high current.
24. Photodiodes are used in smoke detectors.
25. Photodiode emits light when current passes through it.
26. LEDs are special type of PN junctions.
27. The color of the light, emitted by LED, depends upon the intensity of the current that passes through it.
28. A capacitor offers very high resistance to AC but very low resistance to DC.
29. In choke-input LC filter, inductor blocks the AC component.
30. Bridge rectifier is a half wave rectifier.

## D. Answer the following questions in short:

31. In an unbiased PN junction, what makes electrons and holes cross the junction?
32. Which type of charged particles will be there in depletion region on N side of an unbiased PN junction?
33. Which type of majority carriers and minority carries will be there in P side of PN junction?
34. How will you connect battery terminals to make the PN junction reversed biased?
35. IN which type of biasing, current flows through PN junction?
36. Which types of junction is called LED?
37. Which type of PN junctions are called photodiodes?
38. Which electronic component is used in remote controls of a TV to send the signal?
39. Which electronic component is used for reading the data, written on CD?
40. Which two electronic components are used in Pi filter to block the fluctuating component in AC DC mixed signal?

### Detailed Answer Type Questions

## E. Answer the following questions in detail:

41. What is PN junction? What it is used for?
42. Initially when PN blocks of semiconductors are joined together, a current flows. What makes this current flow? Why after some time, this current stops?
43. Explain the following terms in relation to PN junction:
    (i) Majority carriers.
    (ii) Depletion region.
    (iii) Barrier height.
44. What do you understand by ideal diode?
45. What is reverse breakdown?
46. What is a full wave rectifier? Draw a circuit diagram for bridge rectifier?
47. Differentiate between center tap rectifier and bridge rectifier.
48. Differentiate between half wave rectifier and full wave rectifier.
49. What are electronic filters? Explain the functioning of photodiode.
50. Differentiate between LED and photodiode.
51. Explain the concept of photodiode and dark current.
52. Explain the functioning of Pi filter.

***

# BASIC ELECTRONICS

## CHAPTER 3

# *Transistors*

You are already familiar with PN junctions and know that they are formed by joining P and N type semiconductors smoothly.

You also know that P type material has holes as majority and electrons as minority carriers in it while N type material has electrons as majority and holes as minority carriers in it. When P type and N type materials are joined together to form a PN junction, electrons from N side and holes from P side cross the junction and thus a potential barrier gets created across the junction. This potential barrier stops further flow of carriers.

Such a PN junction works in very different way. When it is forward biased (positive terminal of battery connected to P side and negative terminal of the battery connected to N side), it conducts the current. On the other hand, when it is reversed biased (positive terminal connected to P side), it doesn't conduct.

Transistors are combination of two PN junctions, in which above mentioned characteristics are exploited for performing useful tasks. For example, transistors can be used for amplifying the signal. They can also be used as switch. In this chapter, we will describe transistors in detail.

## STRUCTURE OF A TRANSISTOR

A transistor is basically a combination of three differently doped semiconductor layers, in which one type of layer remains sandwiched between the two layers of other type. For example, when P type semiconductor is sandwiched between two N type semiconductor layers, an NPN transistor is formed. Note the name NPN clearly indicates that P layer has been sandwiched between two N type layers. Similarly, when N type semiconductor is sandwiched between two P type semiconductor layers, a PNP transistor is formed. Both types of transistor are illustrated in figure 3.1.

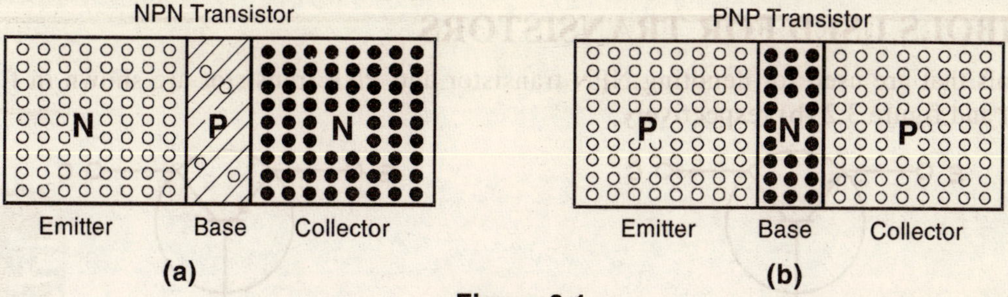


**Figure 3.1**

Refer figure 3.1(a) and note that in NPN transistor, first N region is called emitter, the sandwiched P region is called base and the third N region is called collector.

Similarly in PNP transistor, first P region is called emitter, sandwiched N region is called base and the third P region is called collector. Thus a transistor consists of three regions called emitter, base and collector.

Refer figure 3.1 and note that transistor has two PN junctions. One junction is formed between emitter and base. It is called emitter junction in short. Other junction is formed between base and collector. This junction is called collector junction in short.

## FUNCTION OF EMITTER, BASE AND COLLECTOR

In transistor all the regions works collectively and perform different functions. Their roles are described below:

1. The function of emitter is to emit electrons (in NPN transistor or holes in PNP junction) into the base.
2. The base acts as open door; it allows most of the electrons, coming into it to crossover to collector.
3. Collector collects all those electrons that are emitted by the emitter and are able to cross the base.

## PHYSICAL DETAILS OF TRANSISTOR

As mentioned above, transistor consists of three layers, out of which two layers are of same type but each layer of the transistor is different in its manufacturing and cannot be replaced by other. Following are few internal details of transistor layers:

1. The emitter is heavily doped.
2. Base is very lightly doped.
3. Width of the base is made very thin in comparison to emitter and collector.
4. The doping of the collector is between that of emitter and collector.
5. Length of the collector is kept much larger than that of emitter.

Note that although collector and emitter are always of the same type in a transistor but due to point (1), (4) and (5) they cannot be interchanged while using the transistor. If you do so, transistor may burn.

## SYMBOLS USED FOR TRANSISTORS

Symbols that are used for denoting NPN transistor and PNP transistor are shown in figure 3.2 (a) and figure 3.2 (b) respectively.

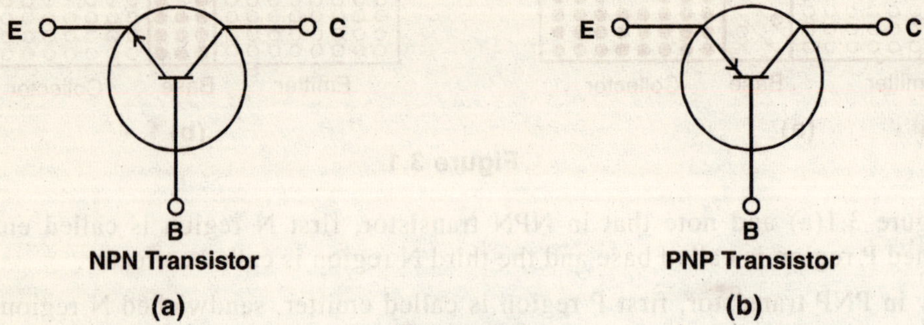


**Figure 3.2**

Refer figure 3.2 and note that emitters have an arrow. The arrow symbolizes emitter and the direction of the arrow represents the direction of the current i.e. electrons flow in opposite direction.

Both types of transistors are commonly used in electronic circuits. Their principle of operation is same. To keep the discussion simple, we will mainly deal with one type of transistor i.e. NPN. Similar analogy, with logical corrections can be applied for understanding the internal dynamics of PNP transistors.

## BIASING THE TRANSISTOR

As mentioned above, transistor comprises of two PN junctions. One is called emitter junction and another is called collector junction. Thus there are four ways in which transistors can be biased:

| Sr. No. | Emitter Junction | Collector Junction | Denoted as |
|---|---|---|---|
| 1 | Reversed biased | Reversed biased | RR |
| 2 | Reversed biased | Forward biased | RF |
| 3 | Forward biased | Forward biased | FF |
| 4 | Forward biased | Reversed biased | FR |

Now let's discuss each biasing condition in detail.

### RR Case

In this case, no junction conducts. Emitter doesn't emit electrons because emitter junction is reversed biased. Since there are no electrons hence collector has no electrons to collect. Thus collector current remains almost zero. As a net result the current doesn't flow through transistor in RR condition. In other words, you can say that in this situation transistor acts like an open switch.

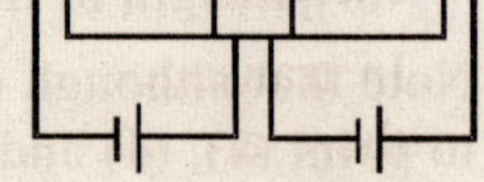


**Figure 3.3**

## RF Case

At first sight it seems that this condition can work as FR condition also because at both the ends, you have N type material. You may expect collector to emit electrons, which by crossing over the base region may be collected by emitter. But since both the N regions are doped differently and their dimensions differ hence their role cannot be interchanged. In fact this type of biasing may damage the transistor.

## FF Case

In this case both the junctions being forward biased will conduct. Thus the electrons emitted by emitter will crossover base region and pass through N region easily. Thus in this situation transistor will act as open switch.

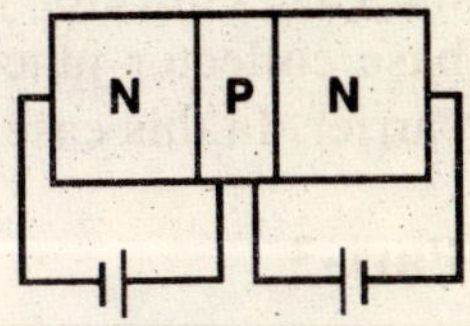


**Figure 3.5**

## FR Case

This is the most interesting and useful biasing case of transistor. This case is mostly used in day to day electronics. To understand the functioning of the transistor under FR biased condition, let's understand the functioning in parts.

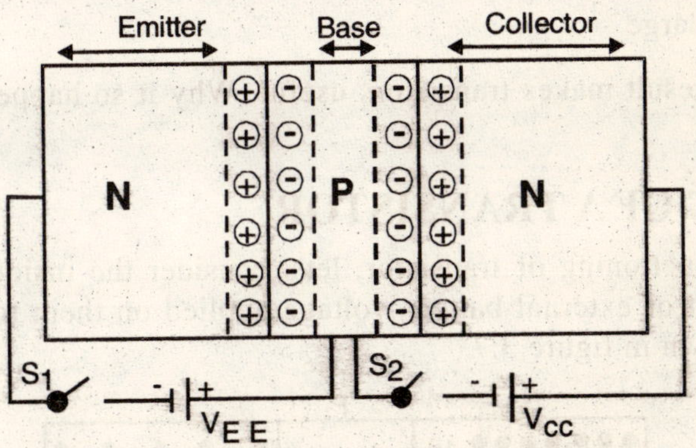


**Figure 3.6**

### Part-1

Refer figure 3.6 and assume that switch $S_1$ is closed and switch $S_2$ is open.

In this case emitter base junction will become forward biased. Since emitter region is heavily doped hence large current will pass through emitter region. At the same time some of the holes will diffuse from base to collector region and thus contribute to emitter current. Thus the total current flowing across collector junction $I_E$ will be sum of electron diffusion current and holes diffusion current.

In transistor, base region is very lightly doped in relation to emitter region. Due to this, there are very few holes in P region. As a result of this, very few electrons will combine with holes of base region and most of the electrons will move through base region. It can be expected that only 1% of the electrons will combine with holes, rest 99% will pass through base region. Thus base current $I_B$ will also be large and almost equal to $I_E$ (i.e. $I_B = I_E$). Since switch S2 is open hence collector current $I_C$ will be zero in this case.

**Part-2**

Now let's assume that switch $S_2$ is closed and $S_1$ is kept open. This situation will make the base collector junction reversed biased. Thus very little current $I_{CBO}$ will flow through base collector junction. Recall that this current will be due to thermally agitated minority carrier. In this case, since switch $S_1$ is open hence emitter current $I_E$ will be zero.

**Part-3**

Now consider that switches $S_1$ and $S_2$ are closed. In this case, obviously we would expect that case 1 and case 2 results should combine i.e. $I_E$ and $I_B$ should be large and $I_C$ should be very small. However the result is never so. In actual, following things happen:

1. $I_E$ remains large
2. $I_B$ becomes small
3. $I_C$ becomes large

In fact this peculiar result makes transistors useful. Why it so happens, is explained in the next section.

## FUNCTIONING OF A TRANSISTOR

To understand the functioning of transistor, let's consider the inside scenario of different regions and the effect of external battery voltage applied on them to bias the junctions. A complete view is shown in figure 3.7.

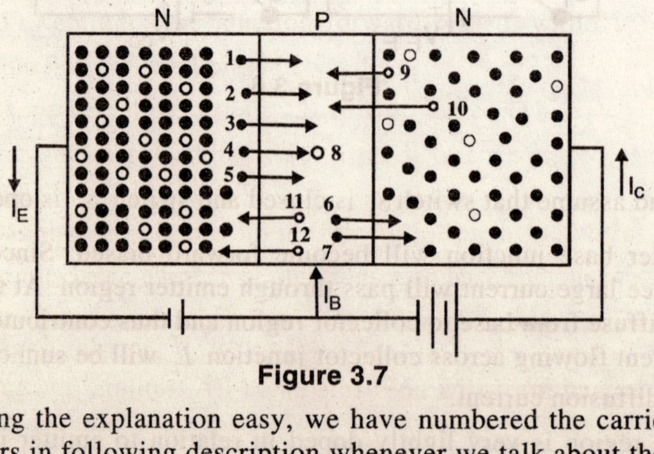


Figure 3.7

Note that for making the explanation easy, we have numbered the carriers. We will make use of these numbers in following description whenever we talk about the movement of the carriers. In the event of emitter base junction being forward biased and base collector junction being reversed biased, following activities occur:

1. Many electrons say 1,2,3,4,5 move from emitter, crossover the junction and enter base region.
2. Few holes, say 11 and 12 move from base to emitter.
3. Both the factors mentioned in point (1) and (2) constitute emitter current $I_E$. In this current major contribution comes from electrons, not by holes because base has been very lightly doped.

   The ratio of electron current to the total current (i.e. emitter current $I_E$) is generally 0.995. This ratio is called emitter injection ratio or emitter efficiency. It is denoted by $\gamma$ (gamma).
4. When electrons reach base, they become minority carriers (recall P region contain holes as majority and electrons as minority carriers). Since the base is lightly doped hence few electrons say 4 combines with holes 8.

   Since base in a transistor is very thin hence majority of emitter electrons, say 1, 2, 3, 5 crossover to the collector region. The ratio of electrons reaching the collector to the number of electrons emitted by emitter is generally 0.995. It is called base transportation factor and is denoted by symbol $\beta'$. Thus $\beta'=0.995$.
5. Now let's focus our attention to only those carriers whose movement is in between base and other two regions.
   (a) Consider emitter base junction to be an open circuit. Due to thermal agitation, few minority carriers like 9 and 10 move from collector region to base region. Similarly few minority carriers like 6 and 7 move from base to collector. They both put together form small leakage current, $I_{CBO}$.

   (b) Few electrons like 4 move from emitter to base region, They combine with holes of base region. Similarly few holes like 11 and 12 move from base to emitter region. These carriers, as mentioned earlier, form part of collector current.

   Note that the difference of these two currents (mentioned in point (a) and (b)) in base region makes base current $I_B$.

Now it is quite obvious that collector current $I_C$ will be less than emitter current $I_E$ for following two reasons.

(a) A part of collector current is due to the holes, moving from collector to base.

(b) Not all the electrons emitted by emitter reach collector, some of them combine with holes in base.

$$\gamma = \frac{ElectronCurrent}{I_E} \quad \text{OR} \quad I_E = \frac{EletronicCurrent}{\gamma}$$

$$\beta' = \frac{I_C}{ElectonCurrent} \quad \text{OR} \quad I_C = ElectronCurrent \times \beta'$$

Thus $I_C/I_E = \gamma \times \beta'$

Since $\gamma = 0.995$ and $\beta' = 0.995$

Hence $I_C/I_E = 0.99$

This ration is called alpha dc thus

$\alpha_{dc} = 0.99$

## RELATIONSHIP BETWEEN VARIOUS CURRENTS OF THE TRANSISTOR

You are now familiar with the internal carrier movements that take place within the transistor. Now let's analyze the relationship between different currents that flow in transistor circuits.

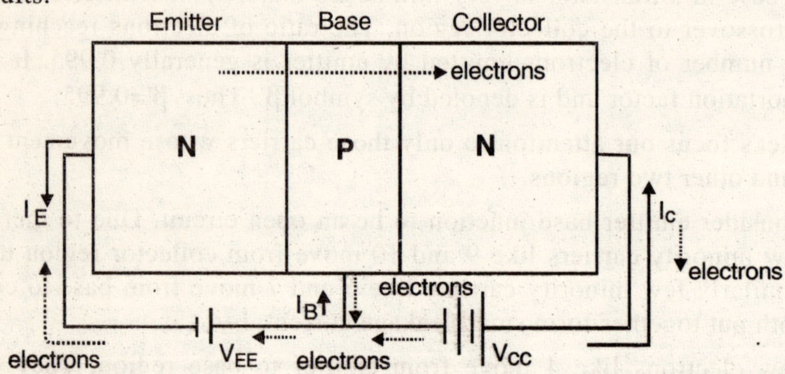


**Figure 3.8**

Refer figure 3.8 and note the following points.

1. $V_{EE}$ battery makes emitter base junction forward biased, while $V_{CC}$ battery makes base collector junction reversed biased.
2. $V_{EE}$ voltage is kept very low, usually 1 volt or lower, while $V_{CC}$ voltage is kept much higher than that e.g. 6 volts or so. This difference is necessary to make the current flow from emitter to collector.

When an arrangement, as shown in figure 3.8 is formed, following activities take place:

1. Emitter region emits large number of electrons into base region.
2. Few holes move from base to emitter. These holes recombine with electrons in emitter region. Thus for this short supply, electrons leave negative terminal of $V_{EE}$ battery and enter into emitter. All these electrons will form part of emitter current $I_E$ in the emitter section.

   Since the current flows in the opposite direction to the movement of electrons hence $I_E$ flows out of emitter terminal (as shown in figure 3.8).

3. All the electrons pushed by $V_{EE}$ will now travel through emitter terminal and enter into base. Here they will combine with few holes, and rest of them will travel to collector terminal. For each combination of electron with holes in base region, new holes are created in the base region. One electron for each new generation of hole is created, which travels out of base region. Thus as shown in figure 3.8, base current $I_B$ flows towards base terminal.

   Note that since, base region in transistor is deliberately kept thin and is very lightly doped hence, very few electrons combine with holes. As a result of this base current $I_B$ is very small in comparison to emitter current $I_E$.

4. Those electrons, which reach collector, experience attractive force due to battery $V_{CC}$. They travel through collector region and reach positive terminal of $V_{CC}$ battery. This forms collector current $I_C$. This current is obviously slightly less than $I_E$.

5. Negative terminal of $V_{CC}$ battery gives out as many electrons as enter its positive terminal. These electrons move towards positive terminal of $V_{EE}$ battery.

6. Electrons of $I_B$ current join with the electron shelled out by negative terminal of $V_{CC}$ battery and both of them enter the positive terminal of $V_{EE}$ battery.

In this way, as many electrons as were initially shelled out by negative terminal of $V_{EE}$ battery, reach its positive terminal. Thus the circuit gets completed. In this way the current flows through the transistor.

A very basic relationship that exists between these three current is, as mentioned below:

$$I_E = I_B + I_C$$

In other words, it can be said that emitter current $(I_E)$distributes itself into base current $(I_B)$ and collector current $(I_C)$.

Since the amount of current leaving the emitter is solely a function of emitter base bias voltage and collector receives most of the current hence a small change in emitter base voltage will have for greater effect on the magnitude of collector current than it will have on base current. This property of transistor makes it usable for amplifier application.

## BASIC PRINCIPLE BEHIND AMPLIFYING ACTION OF A TRANSISTOR

You are familiar with transistors and are aware of the following facts that when in a transistor, emitter base junction is forward biased and base-collector junction is reverse biased, following characteristic come into action:

1. Emitter base junction offers very low resistance to the current.
2. Base collector junction offers high resistance to the current.
3. Emitter current, $I_E$ is almost equal to collector current $I_C$.

Note that in spite of high resistance; almost same current is received at the output.

Power is defined as following:

$$P = I^2 R$$

So if a signal is supplied at the input end of the transistor then its power will be expressed as:

$$P_{in} = I_E^2 R_{in}$$

Similarly the power of the signal at the output will be expressed as:

$$P_{out} = I_C^2 R_{out}$$

Since $I_E^2 = I_C^2$ (almost) and $R_{out} > R_{in}$ hence $P_{out}$ will be greater than $P_{in}$. That means power gets amplified in transistor.

## TRANSISTOR AMPLIFIER

Basic transistor amplifier circuit is shown in figure 3.9.

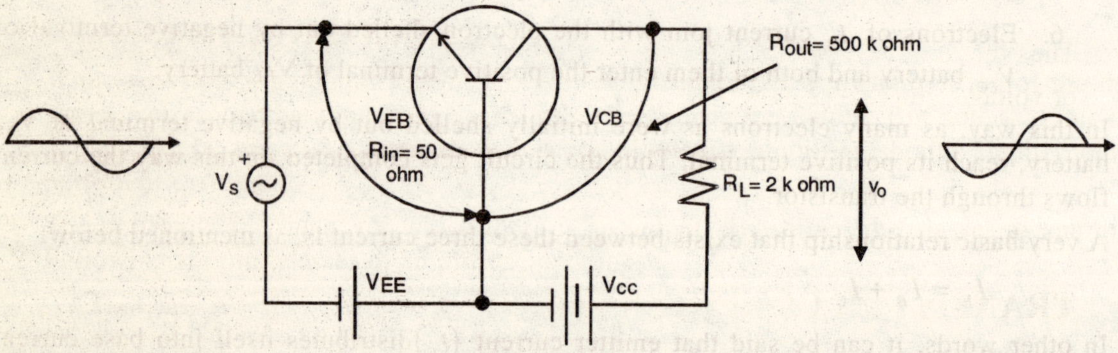


**Figure 3.9**

Note that in this circuit, base is common for input and output. Input signal is being provided between emitter and base, while output is being taken between base and collector. The battery $V_{EE}$ has made emitter base junction forward biased while other $V_{CC}$ has made collector terminal reverse biased.

An alternating signal $V_S$ has been put between emitter and base i.e. at the input end, which is to be amplified.

A load resistance $R_L$ has been put between collector and base, across which the amplification has to be observed.

To view the amplifying characteristics of transistor, let's assume the values of current, voltage and resistance. Say:

1. Emitter base junction is forward biased junction so it offers low resistance. Assume that resistance $R_{in}$ = 50 ohms.

2. Collector base junction is reversed biased junction, so it offers high resistance. Assume that resistance $R_{out} = 500$ Kilo ohms.
3. The alternating input signal that has to be amplified is 50 mV (rms).
4. The load resistance $R_L = 2$ Kilo ohms.

When alternating input is applied to the input emitter current, variation will take place as follows:

$$I_E = \frac{V_S}{R_{in}} = \frac{50\times10^{-3}}{50} = 1mA$$

Since collector current is almost the same as emitter current i.e. $I_E \simeq I_C = 1mA$

When $I_C$ will flow through $R_L$ a voltage will be formed across $R_L$. This voltage will be

$$\begin{aligned} V_O &= I_C R_L \\ &= (1\times10^{-3})(2\times10^{3}) \\ &= 2Volts \end{aligned}$$

Thus the ratio of the output voltage to input voltage $V_S$ will give the voltage amplification or voltage gain for the amplifier. Thus

$$A_V = \frac{V_O}{V_S} = \frac{2}{50\times10^{-3}} = \frac{2000}{50} = 40$$

That means the signal will have gain of 40.

## TRANSISTOR CONFIGURATIONS

You are already familiar with transistors and know that it has three electrodes, commonly known as emitter, base and collector. You also know that input and output activities, each require two electrodes. Thus one electrode in transistor circuit has to be common between input and output. For example, refer figure 3.9 and note that in that transistor circuit, input is being provided between emitter and base while output is being taken between base and collector. Thus base is common in input-output operations.

Now the question is, should base be always common in all the transistor circuits? Well, the answer is no. Transistor can be connected in any of the following configuration:

1. Common base configuration
2. Common emitter configuration
3. Common collector configuration

Be it any configuration, emitter base region has to be forward biased and base collector region has to be reverse biased. All the three configurations are shown in figure 3.10.

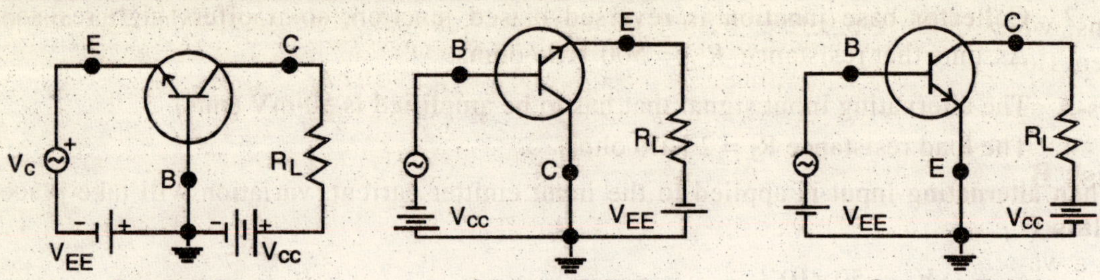


**Figure 3.10**

# QUIESCENT VOLTAGE AND CURRENT

When transistors are properly biased, DC voltage and current remain there in the circuit, even if the alternating signal is not applied. This voltage is called quiescent voltage and current is called quiescent current.

# CLASSES OF AMPLIFIER

While describing the amplification process, we assumed that for each portion of the input signal, corresponding amplifying signal has to appear in the output. But note that this may not always be the case. Often it may be required that amplifier should amplify only a portion of the signal.

The portion of the input, for which there is an output, determines the classes of amplifier. Following are four classes of amplifier operations:

1. Class A
2. Class B
3. Class AB
4. Class C

Details of all these classes are described below:

## Class A Amplifier Operation

In class A operations, transistor stays in active region throughout the AC cycle. Note that active region is that part, of the cycle, in which emitter base region remains forward biased and base collector region reverse biased. Thus transistor conducts.

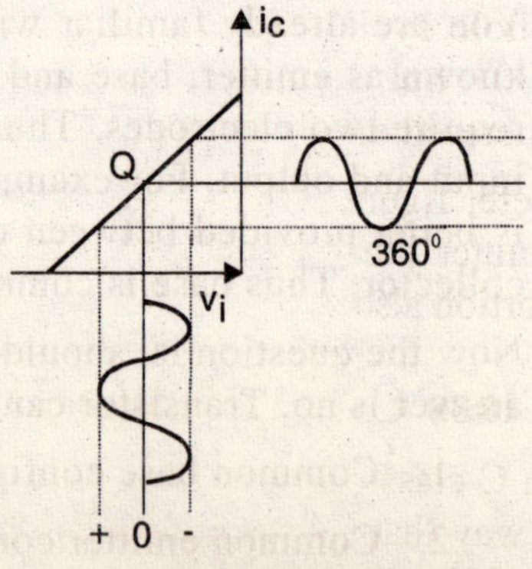


**Figure 3.11**

The biasing of the transistor is kept such that for each portion of input signal there is current in the output. That means current flows for $360^0$. This characteristic of class A operated amplifier is shown in figure 3.9.

Refer figure 3.11 and note that output signal remains $180^0$ out of phase with input. That means, when input is positive, output is negative, but with same amplitude. The output current flows for complete direction as shown in figure 3.11

Class-A operated amplifiers are used in audio systems, radars etc.

## Class B Amplifier Operations

In class B operated amplifiers, biasing is maintained in such a way that transistor stays in active region, only for half of the cycle. That means current flows only for half of the cycle. That means current flows only for $180^0$. This characteristic of class B operated amplifier is shown in figure 3.12.

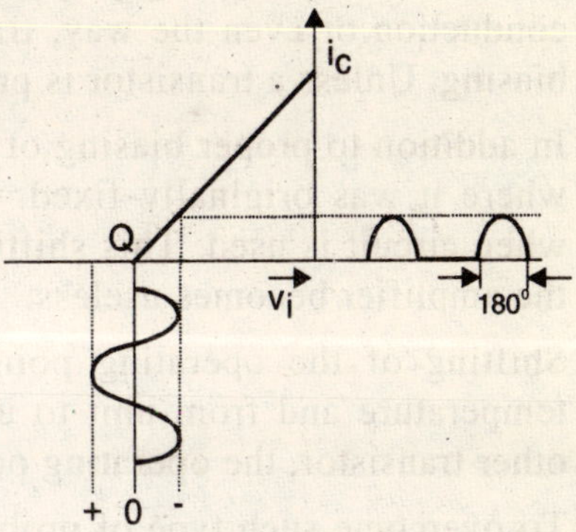

**Figure 3.12**

Refer figure 3.12 and note that emitter junction has been so biased that when signal turns negative, transistor becomes forward biased. Thus conduction takes place and current is received in output. But when signal turns positive, emitter junction doesn't conduct so for all positive portions of the signal there appears no current in the output. Thus broken amplified signal is obtained in the output.

Class B operated amplifiers are used in all those audio systems, which give high power output. They are also used in transmitters.

## Class AB amplifier Operations

In AB class operated amplifiers, biasing is maintained in such a way that emitter junction remains positively biased for more than half the cycle, but less than the whole cycle. Thus the output current flows for more than $180^0$ but less $360^0$. This characteristic of class AB operated amplifier is shown in figure 3.13.

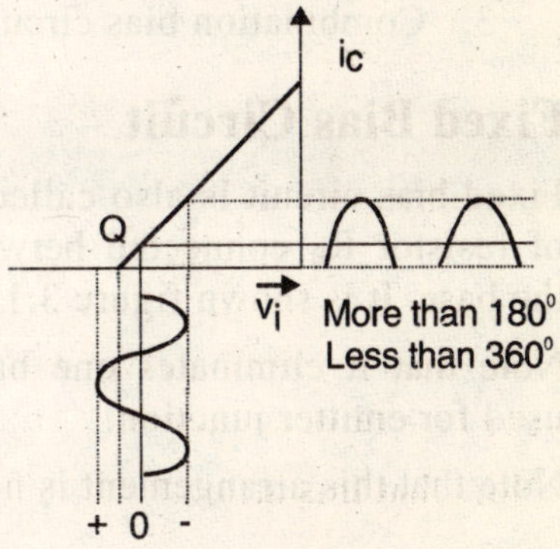

**Figure 3.13**

Refer figure 3.13 and note that emitter junction is so biased emitter junction not only conducts for negative cycle of input signal but partially for positive portion also.

## Class C Amplifier Operations

In C class operated amplifiers, biasing is maintained in such a way that emitter junction is positively biased for less than half the cycle. Thus output current remains zero for more than half the cycle. This characteristic of class AB operated amplifiers is shown in figure 3.14.

C amplifiers are used as radio-frequency amplifiers in transmitters.

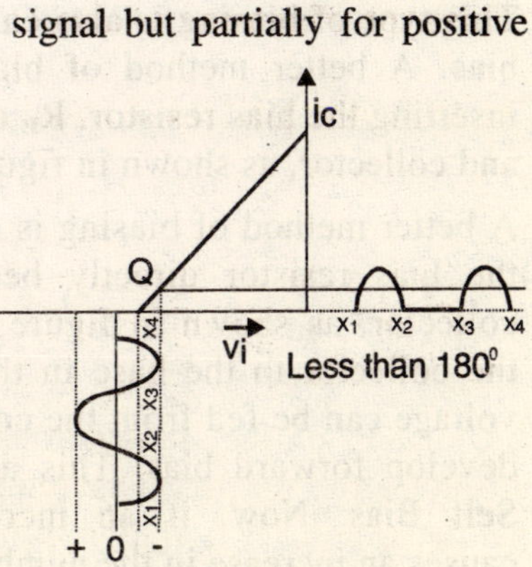

**Figure 3.14**

# TRANSISTOR BIASING AND STABILIZATION OF OPERATING POINT

You know that biasing plays an important role in transistor circuits. Conduction or non conduction or even the way, transistor will conduct (amplify the signal) depends upon its biasing. Unless a transistor is properly biased, desired output will not be received.

In addition to proper biasing of the transistor, one has to also ensure that it remains biased, where it was originally fixed. Unfortunately, in transistor circuits operating point shifts, when circuit is used. This shifting obviously derives the transistor in unwanted region and the amplifier becomes useless.

Shifting of the operating point takes place because transistor parameters change with temperature and from unit to unit. Thus if a transistor in a circuit is replaced by similar other transistor, the operating point shifts.

To overcome such type of problems, biasing circuits are used. Following are few common biasing circuits:

1. Fixed bias circuit
2. Self bias circuit
3. Combination bias circuit

## Fixed Bias Circuit

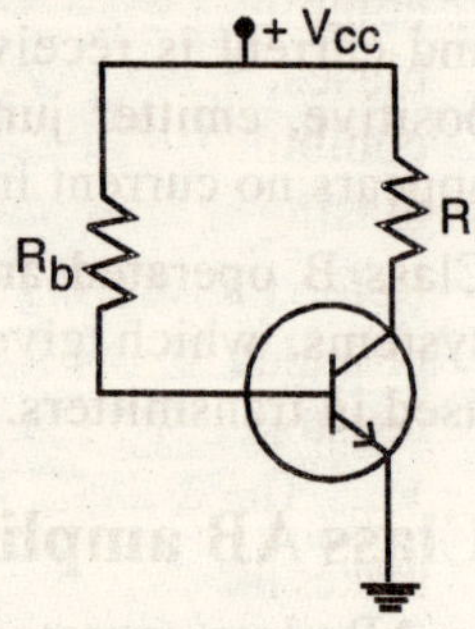

Figure 3.15

Fixed bias circuit is also called base current bias circuit. In consists of resistor $R_B$ connected between the collector supply voltage and the base. It is shown figure 3.15.

Note that it eliminates one battery $V_{EE}$ that was previously being used for emitter junction.

Note that this arrangement is not thermally very stable.

## Self Bias Circuit

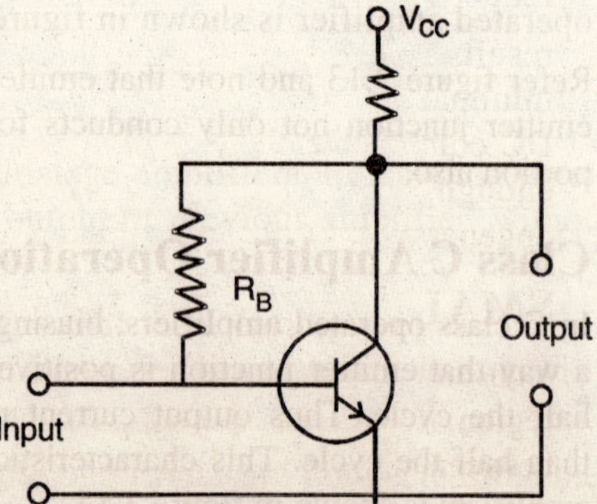

Figure 3.16

This type of biasing is also called collector-to-base bias. A better method of biasing is obtained by inserting the bias resistor, $R_B$ directly between base and collector, as shown in figure 3.16.

A better method of biasing is obtained by inserting the bias resistor directly between the base and collector, as shown in figure 3.16. By connecting the collector to the base in this manner, feedback voltage can be fed from the coilector to the base to develop forward bias. This arrangement is called Self Bias. Now, if an increase in temperature causes an increase in the number of charge carriers, which increases collector current, the collector voltage ($V_C$) will fall because of the increase of voltage produced across the load resistor ($R_B$). This drop in $V_C$ will be fed back to the base and will result in a decrease in the base current. The decrease in base current will oppose the original increase in collector current and tend to stabilize it. The exact opposite effect is produced when the collector current decreases. Note that opposite effect will be produce if collector current decreases.

This type of biasing has following drawbacks:

1. It is partially effective, so it is only used where moderate changes in temperature are expected.
2. It reduces amplification, since the signal on collector also affects the base voltage. This is because input signals and output signals are always $180^0$out of phase.

## Combination Bias Circuit

As the name indicates, this biasing is a combination of both the previously mentioned biases. This is the reason, why it is also called combination of fixed and self bias. Such a biasing is shown in figure 3.17.

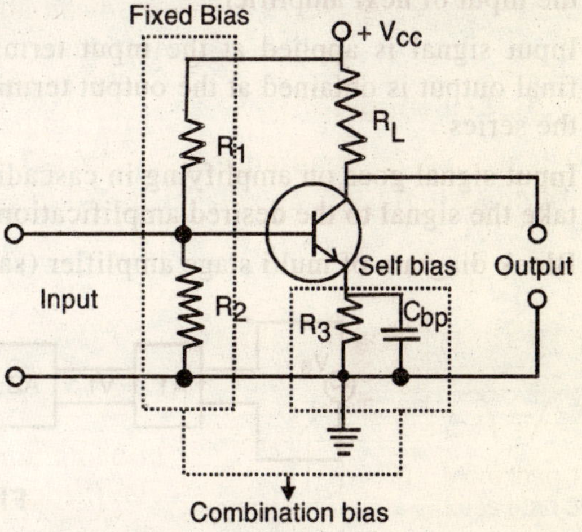


**Figure 3.17**

Refer this figure and note that fixed biasing is provided by voltage-divider network made by resistors R1, R2 and collector supply voltage $V_{CC}$. The DC current flowing through this network makes the base positive with respect to emitter.

Self biasing arrangement is made by introducing resistor R3 in series with emitter. When emitter current $I_E$ increases, the voltage drop across R3 will also increase. This will reduce $V_C$. This reaction to an increase in $I_E$ by R3 is another form of degeneration, which results in less output from the amplifier. But to provide DC thermal stability and provide minimal AC degeneration a capacitor $C_{bp}$ is placed across R3. Recall that capacitor blocks DC component of the signal, while it allows AC component to pass through. If $C_{bp}$ is large enough, rapid signal variations will not charge its charge materially and thus no degeneration of signal will occur.

# SMALL SIGNAL AMPLIFIERS

You are already familiar with transistors and know that when it is properly biased in active region, it works as an amplifier. In transistor amplifier circuit, when alternating voltage is applied between the emitter and base. Fluctuations are made in collector current. Thus an amplified output signal is obtained when the current flows through output load resistance $R_L$.

When the input signal is so weak that it produces small variations in collector current in comparison to its quiescent value, the amplifiers are called small signal amplifiers or voltage amplifiers.

Such types of amplifiers are used as first stage amplifiers in radio receivers, TV receivers, Tape recorders, measuring instruments etc.

Small signal amplifier that is used for amplifying AC signal is shown in figure 3.18.

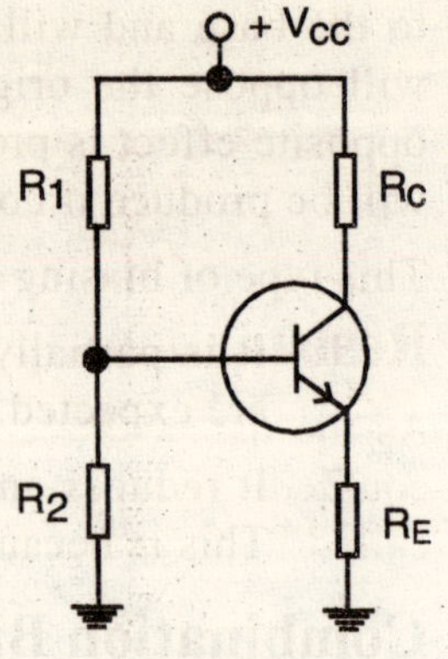


**Figure 3.18**

# MULTI STAGE AMPLIFIERS

Practically speaking the voltage or current gain obtained from a single amplifier is not sufficient for most of the applications. More amplified voltage or current is required for performing the task to the satisfactory level. Thus multiple amplifiers are connected one after the other, in such a way that output of one amplifier is fed to the input of next amplifier.

Input signal is applied at the input terminal of first amplifier and final output is obtained at the output terminal of the last amplifier of the series.

Input signal goes on amplifying in cascading fashion. Thus multiple amplifiers put together take the signal to the desired amplification level.

Block diagram of multi stage amplifier (say 4 stages) is shown in figure 3.19.

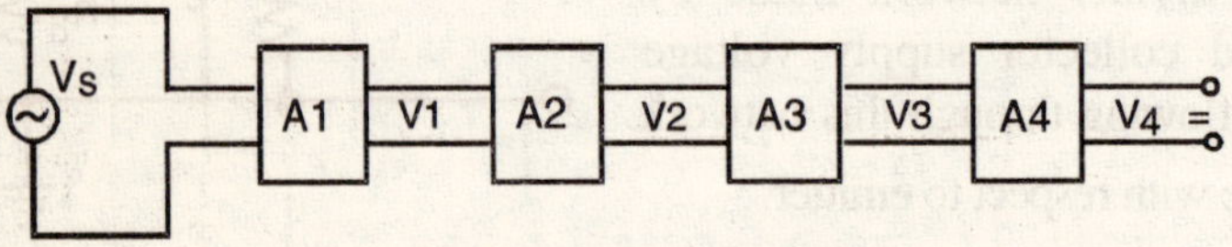


**Figure 3.19**

In multistage amplifiers, total gain is the product of the gain of the individual stages.

**Gain (A) = (Gain A1) (Gain A2) (Gain A3) (Gain A4)**

# COUPLING

Now the question is how the two amplifiers of multistage amplifiers are coupled together, so that the next amplifier in the series receives the output of previous amplifier, without or with minimal loss?

One of the major requirements in this arrangement is that the DC output of one amplifier should not be allowed to go to the input of another amplifier because if it is done, it will disturb the biasing condition of next amplifier.

Another point to be noted here is that when multiple amplifiers are connected together in above mentioned fashion, all previous amplifiers put together form a part of load impedance (resistance) for the next amplifier of the series. Thus the performance of multistage amplifier depends upon the method that has been used for coupling the amplifiers.

Following are the three common methods that are used for coupling the amplifiers of multistage amplifiers:

1. Resistance-Capacitance Coupling
2. Transformer Coupling
3. Direct Coupling

## Resistance Capacitance Coupling

As the name indicates, in resistance capacitance coupling, two stages of the amplifier are coupled, using resistance and capacitance. This is the reason, why this scheme is also called RC coupled scheme. RC coupled multistage amplifier is shown in figure 3.20.

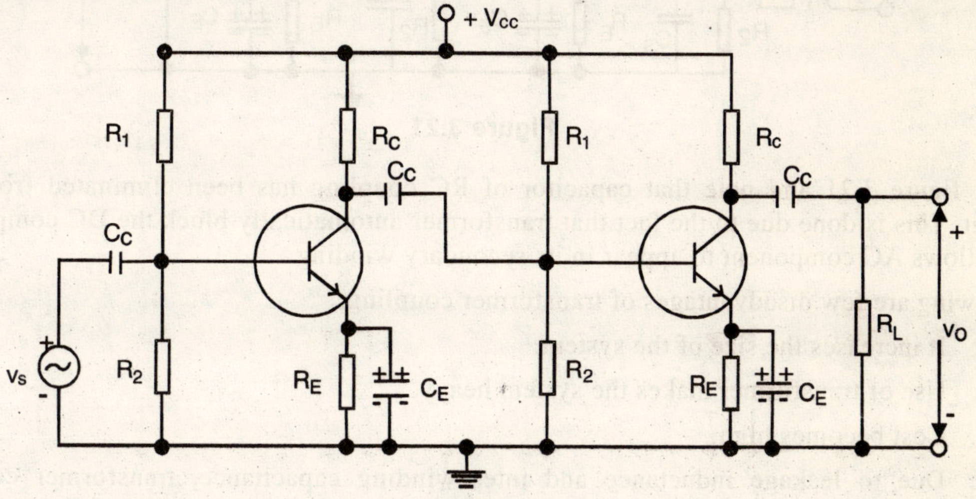


**Figure 3.20**

In this type of coupling, signal developed across RC of first stage is linked to the base of the second stage through a capacitor $C_C$

The coupling capacitor blocks DC component of the output signal (from first stage) and permits AC component to appear at input of the second stage. In this way, biasing of the second stage is not disturbed.

Since capacitor scheme has its own independence hence some voltage loss always occurs across coupling capacitor.

RC coupling scheme is mostly used in audio small signal amplifiers that are used in tape recorders, CD player, public address systems radio receivers etc.

## Transformer Coupling

In this type of coupling, transformer is used to transfer the AC output of one stage to the input of next stage. The resistor $R_C$ of RC coupled amplifier is replaced by the primary winding of the transformer. Wire between the voltage divider and the base of the second amplifier is replaced by the secondary winding of the transformer.

A transformers coupled amplifier is shown in figure 3.21.

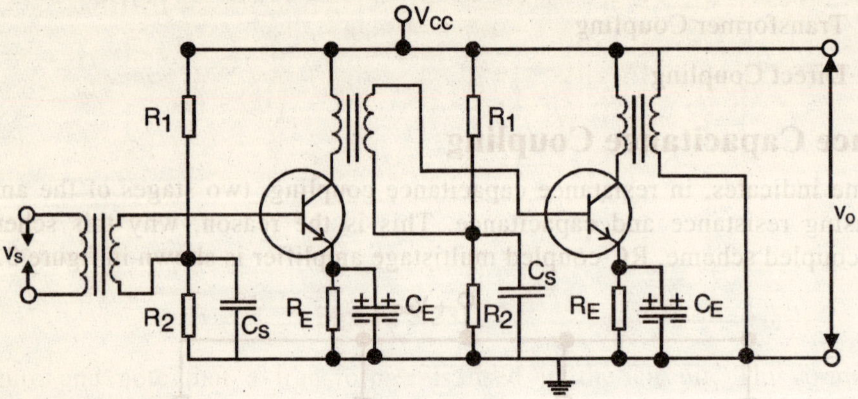


**Figure 3.21**

Refer figure 3.21 and note that capacitor of RC coupling has been eliminated from the circuit. This is done due to the fact that transformer automatically block the DC component and allows AC component to appear in its secondary winding.

Following are few disadvantages of transformer coupling:

1. It increases the size of the system.
2. Use of transformer makes the system heavy.
3. Cost becomes high.
4. Due to leakage inductance and inter winding capacitance, transformer coupled amplifier doesn't amplify the signal of different frequencies equally well.

Following are the advantages of transformer coupled amplifiers:

1. Power loss is saved to a large extent.
2. It provides proper impedance matching.

Transformer coupled amplifiers are widely used for amplifying radio frequency signals.

## Direct Coupling

As the name indicates, in direct coupling the output of one stage amplifier is connected to the input of next stage amplifier, using simple connecting wire.

This type of coupling is used in the amplifiers that are required to amplify the signals of very low frequencies. For example, in thermocouple.

It is quite obvious that when the signal frequency is below 10 Hz, coupling capacitors and bypassing capacitors cannot be used. Capacitors will tend to block such frequencies hence amplifier will not work.

Since no bypassing capacitor is used in the output hence DC component of previous stage amplifier reaches input of next stage amplifier. This DC voltage is taken into account, while biasing the transistor of next stage amplifier. A direct coupled amplifier is shown in figure 3.22

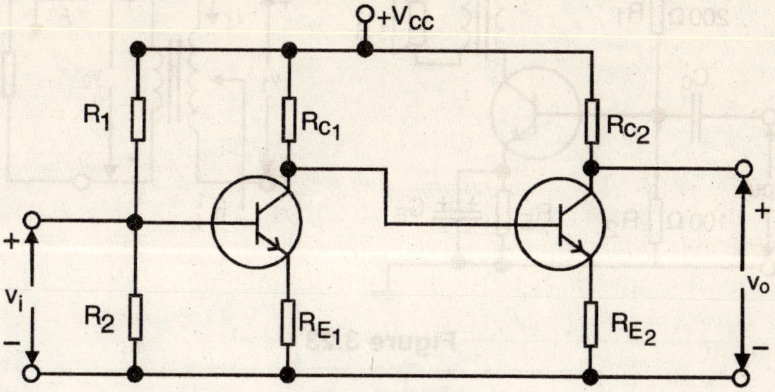


**Figure 3.22**

Following is the disadvantage of direct coupled amplifier:

1. Transistor parameters change with temperature. This cause collector voltage and current to change. Due to direct coupling this change appears at final output.

## POWER AMPLIFIER

Power amplifiers are basically the amplifiers that deliver power at their output.

To understand the utility of power amplifier, consider the case of audio system. The output of an audio system is obviously the sound that we get from its speakers.

Speaker receives electronic signal from its back assembly. This electric signal moves the diaphragm of the speaker. The movement of the diaphragm causes sound. It is quite obvious electric signal should be powerful enough to cause the movement of the diaphragm.

Note that voltage amplifiers amplify the voltage. Even the largely amplified voltages are weak in power. They do not posses sufficient energy to vibrate the diaphragm of the speakers. Thus in all such applications and systems, a different type of amplifier is required, which can provide sufficient power to make the speaker work. Practically speaking, in such type of systems, multi stage amplifier is used for amplifying the voltage and in the last stage amplified voltage is fed as input to the power amplifier. Power amplifier provides enough power at its output terminal. Speaker is connected to the output terminal of power amplifier, and the power thus generated gets transferred to speaker and diaphragm of the speaker vibrates to produce sound.

Now the question is, does power amplifier amplify power? Well, the answer is no. It takes DC power from the power supply connected to the output circuit and converts it to powerful AC signal. But in this process, the type of AC power at output terminal is controlled by the input signal. In other words you can say that power amplifier typically works as DC-to-AC power converter. A power amplifier is shown in figure 3.23.

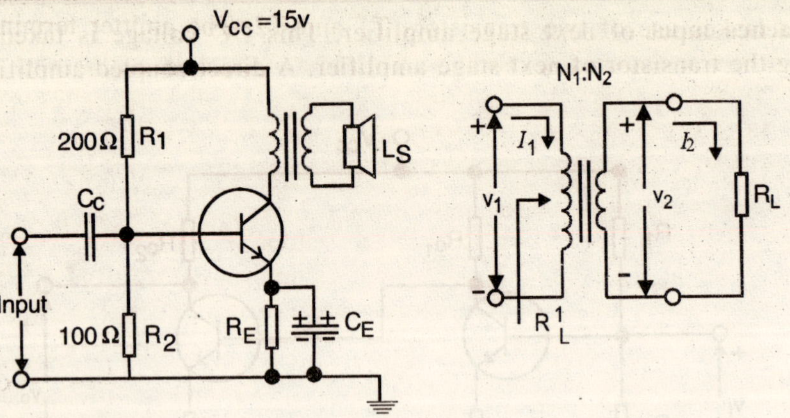


**Figure 3.23**

Refer this figure and note that a transformer is used in the circuit. This transformer performs two functions:

1. It provides impedance matching mechanism between the amplifier and the speaker. Due to impedance matching, maximum power is transferred from the amplifier to the speaker.
2. It provides enough power signal to operate the speaker.

Note that in almost all the electronic systems, the last stage amplifier is power amplifier.

## PUSH-PULL AMPLIFIER

Push-pull amplifier is basically a power amplifier. It is shown in figure 3.24.

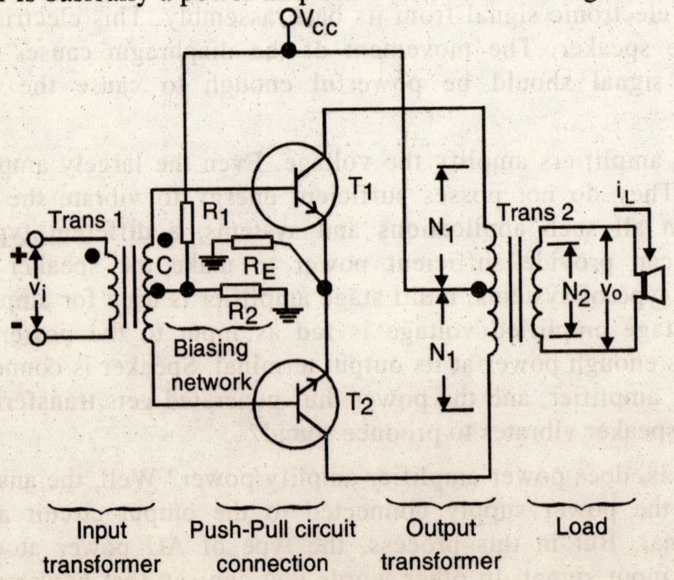


**Figure 3.24**

Note that it makes use of two transistors T1 and T2. The emitter terminals of both the transistor are connected together. Push pull amplifier makes use of two transformers Trans1 and Trans2. Trans1 is placed on the input side of the amplifier. Note that it is a center tapped type transformer. Its center taping in the secondary is used for biasing the transistor in active region. Due to the center tapping it provides opposite polarity signals to the two transistors T1 and T2.

The other transformer, Trans2 is placed at the output. This transformer is also center tap type but center tapping mechanism is in primary side of the transformer. Collector terminal of both the transistors, T1 and T2 are connected to the power supply $V_{CC}$ through the primary of output-end transformer.

Speaker in the circuit forms load resistance. Turn ratio of the transformer is so chosen that load resistance of the speaker matches the output impedance of the transformer (which is the basic requirement of power amplifier for transferring the maximum power to the load). Resistance $R_1$, $R_2$ and $R_E$ provide biasing mechanism for the transistors.

## Functioning Of Push-Pull Amplifier

When an arrangement as shown in figure 3.24 is made, push-pull amplifier works as following:

1. When sinusoidal signal is introduced at the primary winding of Trans1. The voltage is automatically developed in the secondary winding of the transformer.
2. Point C is center tap point. Since this point has been grounded through R2 resistor, two different potentials of same magnitude but of opposite polarity are formed on either side of it. When upper side potential will be positive the lower side potential will be negative and when upper side potential will be negative the lower side potential will be positive. Thus emitter of both the transistors will always remain opposite in polarity.
3. As a result of oppositely phased biasing, the quiescent collector current will flow in opposite direction, in the output section of these transistor, which is center tapped primary winding of output transformer Trans2. As result of this, two opposite fluxes of magnitude will be developed in the magnetic core of the transformer. If both of them are equal in magnitude the net result will be zero.
4. When an AC signal is applied to transistors, transistor T1 generates $I_{C_1}$ current and transistor T2 generates $I_{C_2}$ current. These currents will also be alternating in nature and will vary above and below $I_{C_Q}$, which is quiescent current.

   Since transistor T1 and T2 are $180^0$ out of phase with each other hence when $I_{C1}$ current will be positive then $I_{C2}$ will be negative. Similarly, when $I_{C1}$ will be negative, $I_{C_2}$ will be positive.

   Nature of these two currents is shown in figure in 3.25 (a) and 3.25 (b).

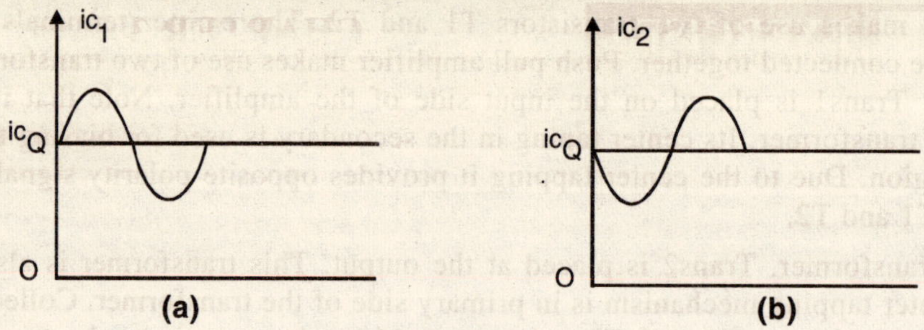


**Figure 3.25**

5. The next flux produced by these current will also be in opposite directions. The net flux produced in the core will the same as that produced by the difference of the current that will flow in the primary winding of transformer, Trans2 i.e. $I_{C_1} - I_{C_2}$.

   Note that $-I_{C_2}$ will be the variation of the signal around $-I_{C_Q}$. Thus $-I_{C_2}$ will be as shown in figure 3.26

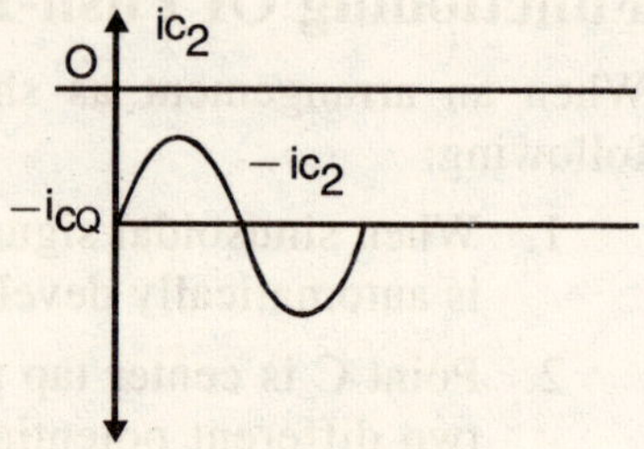


**Figure 3.26**

6. Now the net effect will be due to the current shown in figure 3.25 (a) and current shown in figure 3.26.

   The net effect is as shown in figure 3.27.

7. Thus the current, as shown in figure 3.27 will flow in the primary winding of transformer, Tans2. This will cause same voltage fluctuations in secondary winding of transformer,Trans2 and the AC power will be delivered to the load.

Refer figure 3.25 (a) and (b) and note that first figure is for transistor T1 and second figure is for transistor T2. The current curves clearly state that when the current in one transistor is being pushed up, the current in other is being pulled down. This is the reason, why this amplifier is called push-pull amplifier.

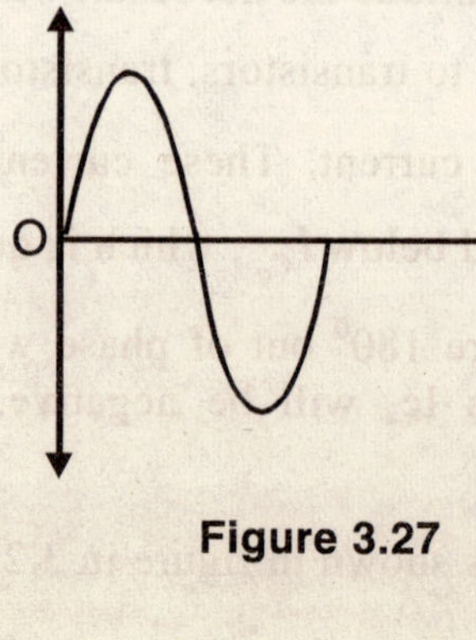


**Figure 3.27**

# EXERCISES CHAPTER 3

## Short Type Questions

**A. Select most appropriate answers for following questions:**

1. How many PN junctions will be required to form a transistor?
   (a) 1 (b) 2
   (c) 3 (d) 4
2. What type of biasing would be required for a transistor to work as amplifier?
   (a) FF (b) FB
   (c) BB (d) BF
3. Which of the following doesn't hold good for the transistor?
   (a) Emitter is heavily doped.
   (b) Base is very thin in comparison to emitter and collector.
   (c) Size of the collector is kept small.
   (d) Collector is doped more than base but less than emitter.
4. In which class of operation, an amplifier produces full wave at the output?
   (a) A (b) B
   (c) AB (d) C
5. Which of the following is not a valid configuration?
   (a) Common emitter (b) Common base
   (c) Common collector (d) Common current
6. Which of the following is not a valid biasing method?
   (a) Fixed biasing (b) Variable biasing
   (c) Self biasing (d) Combination biasing
7. Which of the following is not a valid coupling method for the amplifier of multistage amplifiers?
   (a) RC coupling (b) Direct coupling
   (c) Indirect coupling (d)Transformer coupling
8. Which of the following is used in the last section of multi stage amplifier of an audio system?
   (a) Power amplifier (b) Voltage amplifier
   (c) Current amplifier (d) Small signal amplifier
9. Into which category does Push-Pull amplifier fall?
   (a) Power amplifier (b) Voltage amplifier
   (c) Current amplifier (d) Small signal amplifier

10. How many transformers are used in Push- Pull amplifier?
(a) None (b) One (c) Two (d) Three

## B. Fill in the blanks:

11. In a transistor ..........................................................................is most heavily doped.
12. ...............................................is the thinnest of all the three sections of transistor.
13. The arrow shown in transistor symbol represents ................................................
14. To make the transistor work as amplifier, base collector junction is ........................ biased.
15. Out of three currents $I_C, I_B$ and $I_E$ in a transistor....................will be the largest.
16. In ......................class of operation of amplifier, only one half portion of input signal appears in the output.
17. Combination biasing circuits make use of..............biasing circuit and.......................biasing circuit, both.
18. Small signal amplifiers are basically ..................................................amplifiers.
19. For very low frequency signals............................coupling method is used in multi stage amplifiers.
20. Push-Pull amplifier is an example of ....................................................amplifier.

## C. State, true or false for following statements:

21. In a transistor, base section is most heavily doped portion.
22. PNP transistor would consist of 2 N types and 1 P type region.
23. To make the transistor work as amplifier, collector junction is forward biased while collector junction is reversed biased.
24. While transistor works as amplifier, emitter current is almost equal to collector current.
25. In class AB operations, the signal received at the output is always less than 3600 but more than 1800.
26. A circuit, in which input is provided between emitter and base and the output is received between base and collector will be a common base configuration.
27. A transistor biased as FF will act as an open switch.
28. Transformer coupling suits well for high frequency signals.
29. Self biasing method is the best biasing method.
30. In Push Pull amplifier, both the transistors produce currents, which are 3600 out of phase, with each other.

## D. Answer the following questions in short:

31. Name the three regions of a transistor?
32. Which region of the transistor is made largest in dimension?

33. Which current out of emitter current, base current and collector current is the smallest?
34. How the two junctions of a transistor are biased so that it work as amplifier?
35. How will you define alpha factor of a transistor?
36. How will you define beta factor of a transistor?
37. What does the equation $I_E = I_B + I_C$ signify for a transistor?
38. Name any three methods for coupling the amplifiers of multi stage transformers?
39. In which class of amplifier operations, the signal received at the output is for less than half the cycle of input signal?
40. Which type of amplifier will be required at the last stage of multistage amplifier to produce sound through speaker?

**Detailed Answer Type Questions**

## E. Answer the following questions in detail:

41. What is transistor? Describe its structure in detail.
42. Differentiate between NPN and PNP transistors. Draw the symbols that are used to denote them.
43. Draw a structure of NPN transistor and describe why $I_E = I_B + I_C$
44. Describe the basic principle behind amplifying action of transistor.
45. Draw a circuit for transistor amplifier?
46. Draw circuit diagrams to shown common base, common emitter and common collector configurations.
47. What do you understand by class of amplifier? How many classes of amplifier operations are there? Name them.
48. Differentiate between class A and class B operations of amplifier.
49. What is fixed bias circuit? Why it is required? Draw its circuit diagram.
50. Draw a circuit diagram for push pull amplifier.
51. Describe any two coupling methods for multi stage amplifiers.
52. Why power amplifiers are required?

****

# BASIC ELECTRONICS

## CHAPTER 4

# *Introduction To Integrated Circuits*

You are now familiar with electronic components like resistors, capacitor, diode and transistors. You know that transistor was invented in the year 1947.

With the advent of transistor, electronics started the journey of miniaturization. With the use of transistors, electronic circuits became small and their reliability became higher. Soon after their invention they were frequently used in most of the electronic equipments like radios, record players, audio systems, tape recorders, calculators etc. But soon after that when the complexity of the circuits grew, problems started arising.

For all the electronic equipments that were to be made, assembly workers had to make all the circuits by hand, thus soldering each electronic component at right place and connecting them with metallic path or wires. A single mistake here or there stopped the flow of current and thus the circuit failed.

Engineers soon realized that assembling numerous electronic components manually for complex circuit of the equipments like, computer, etc. was impossible. Another problem was of size. In spite of transistor being a small component, its use in bigger and complex circuits could not reduce the size of the circuit to the extent where speed of operations could fall into appreciable limits. For example, computers that were manufactured, using transistors were very large and their speed of operations was too slow.

For all these reasons, further miniaturization was required, in which many circuits including all the electronic components and wires could be packed into a single small unit.

## BIRTH OF INTEGRATED CIRCUIT

As it happens with most of the inventions, two people came out with the idea of integrating the circuits, at almost the same time.

In summer of 1958 Jack Kilby joined Texas Instruments. He was given the project to build smaller electronic circuits. One day, in July he was sitting alone in his office because he was not supposed to take leave, as he was new. The office was deserted and he had a lot of time to think. All of a sudden an idea came to his mind that not only the transistor, but all parts of the circuit could be made from semiconductors. If it could be done then the entire circuit could be built, using a single crystal, making it very small and very easy to produce. By February 1959, he was ready with the first integrated circuit. This circuit was named Solid Circuit. Its size was of pencil point. Soon it was sent for patent.

In January 1959, Robert Noyce, working at Fairchild Semiconductor Company came out with the same idea. While Kilby had framed the details of making the individual components from semiconductor, Noyce went a step ahead and framed the idea of also connecting them together within the same unit.

Fairchild named this circuit "Unitary Circuit". On April 1961, patent office awarded the first patent for an integrated circuit to Robert Noyce. Jack Kibly did not get the patent.

## WHAT IS AN INTEGRATED CIRCUIT

An Integrated circuit is a small electronic device made from semiconductor that has numerous electronic circuits packed into it.

Integrated circuit consists of a single crystal chip of silicon, containing both active and passive electronic components like resistors, capacitors, diodes, transistors and their inter connections in them. Circuit is often called a chip.

It is packed in hermetically sealed case, with its leads extending from it in the form of pins for input, output and power supply. External appearance of integrated circuit is shown in figure 4.1

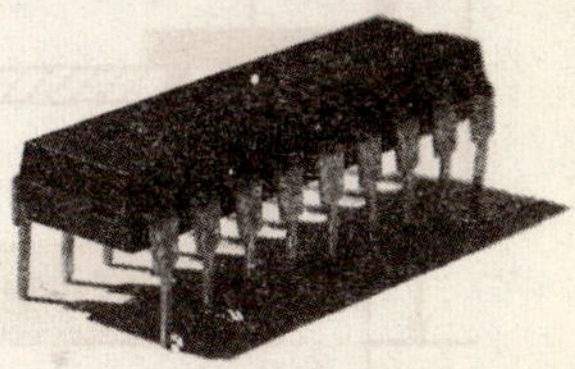

**Figure 4.1**

## LEVELS OF COMPLEXITY OF ICs

Originally when IC came into existence, only few electronic circuits were packed on a single chip. But as the time passed, technology improved and it became possible to pack more number of transistors into the same space.

IC technology started from Small Scale Integration (called SSI in short) and traveled a long path. Currently it has reached the stage of Ultra Large Scale Integration. Progress of integrated circuit technology is summarized below.

1. Small Scale Integration (SSI): Using this technology, about 100 electronic components were packed within a single chip.
2. Medium Scale Integration SMSI): After Small Scale Integration, came Medium Scale Integration technology. Using this technology about 100 to 3000 electronic components were packed in a single chip
3. Large Scale Integration (LSI): Large Scale Integration technology is more advanced than Medium Scale Integrations technology. It allows more number of electronic components to be packed in single chip. Using this technology about 3,000 to 1,00,000 electronic component can be packed in a single chip.
4. Very Large Scale Integration (VLSI): Most of the modern electronic chips make use, of VLSI technology. Using this technology, it is possible to pack around 1,000,000 to 10,000,000 electronic components in a single chip.

5. Ultra Large Scale Integration (ULSI): The line of demarcation between VLSI and ULSI is very vague. The technology, using which over 10,000,000 electronic components are packed into a single chip is called ULSI technology. For example, Intel Pentium processors make use of ULSI technology.

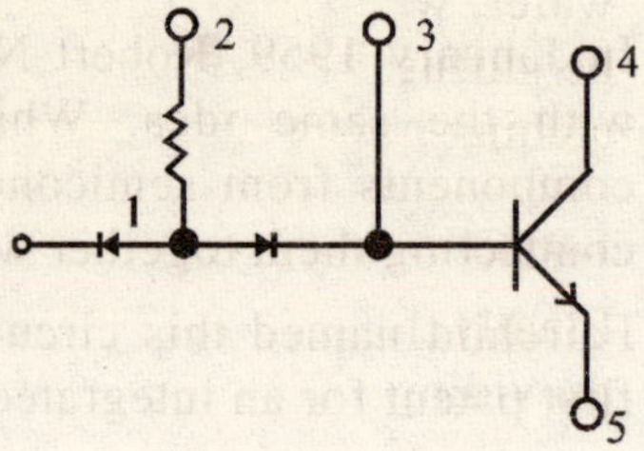

Figure 4.2 (a)

## BASIC INTERNAL STRUCTURE OF IC

Basic internal structure of integrated circuit is illustrated in figure 4.2. Refer this figure and note that integrated circuit is made from four different layers. Each layer has its own importance and a specific role to play.

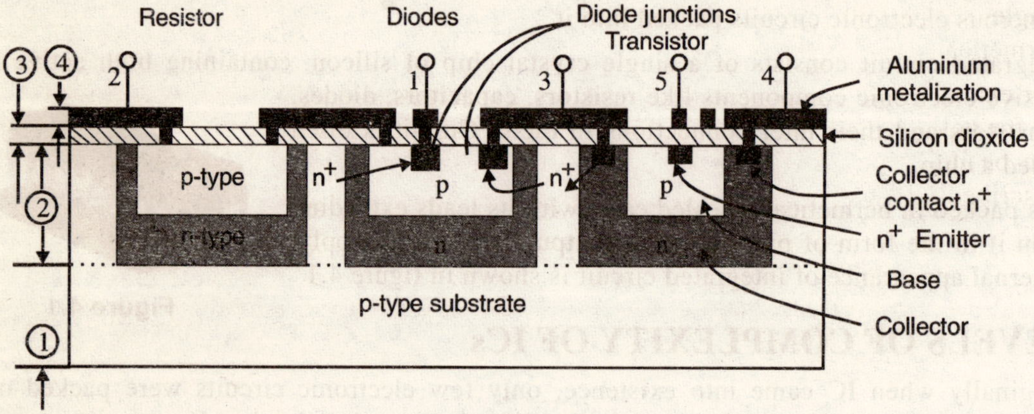

Figure 4.2 (b)

Details of each layer are given below.

### Layer-1

First layer of IC lies at the bottom. It is made from P type silicon and serves as substrate, on which IC is built. This layer is 6 mils thick.

### Layer-2

1. The second layer is thinner than first layer and comprises of N type material. This layer is grown as single crystal extension of first layer.
2. All electronic components, such as resistors, capacitors, diode and transistors are built within this layer, using a series of diffusion steps.
3. For diffusion, p-type and n-type impurities are used. The impurities are diffused at selective places, with very high precision. The selective diffusion of impurities is accomplished by using $SiO_2$ as a barrier, which protects portion of the wafer against impurity penetration.

### Layer-3

Third layer is a layer of $SiO_2$ (mentioned above). It provides protection to the area of the wafer, where impurities are not to be diffused and opening to the areas where they have to be diffused. Openings are created by a process called etching. The process by which etching is done is called Photolithography. Photolithography is explained in this chapter.

### Layer-4

Finally fourth layer is added on top. The layer is made of aluminum and is used for providing all necessary connections.

## BASIC MONOLITHIC CIRCUIT

Fabrication of an IC is basically a process, in which a real working circuit is carved into single semiconductor material, in the form shown in figure 4.2 and is packed into hermetically sealed unit.

For example, if a voltage amplifier circuit is transformed into the form shown in figure 4.2, IC for voltage amplifier will be made. Wherever in circuit, voltage amplifier is required IC can be inserted there. Obtaining monolith circuit is a multi step process these steps are described below:

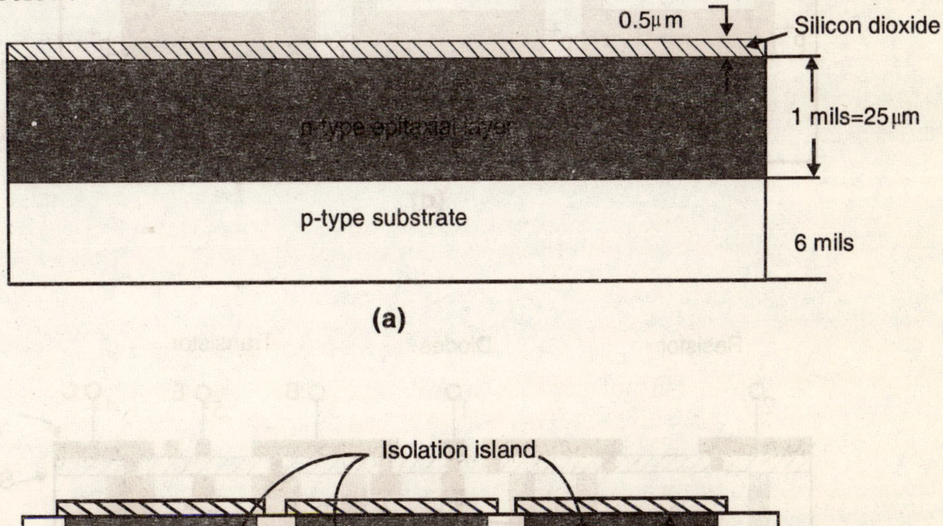


**(a)**

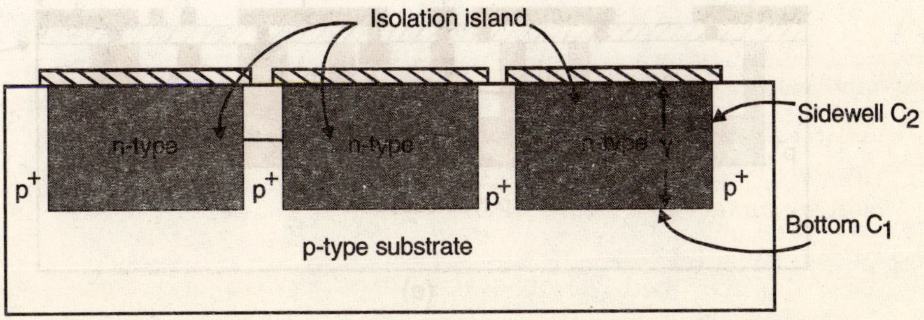


**(b)**

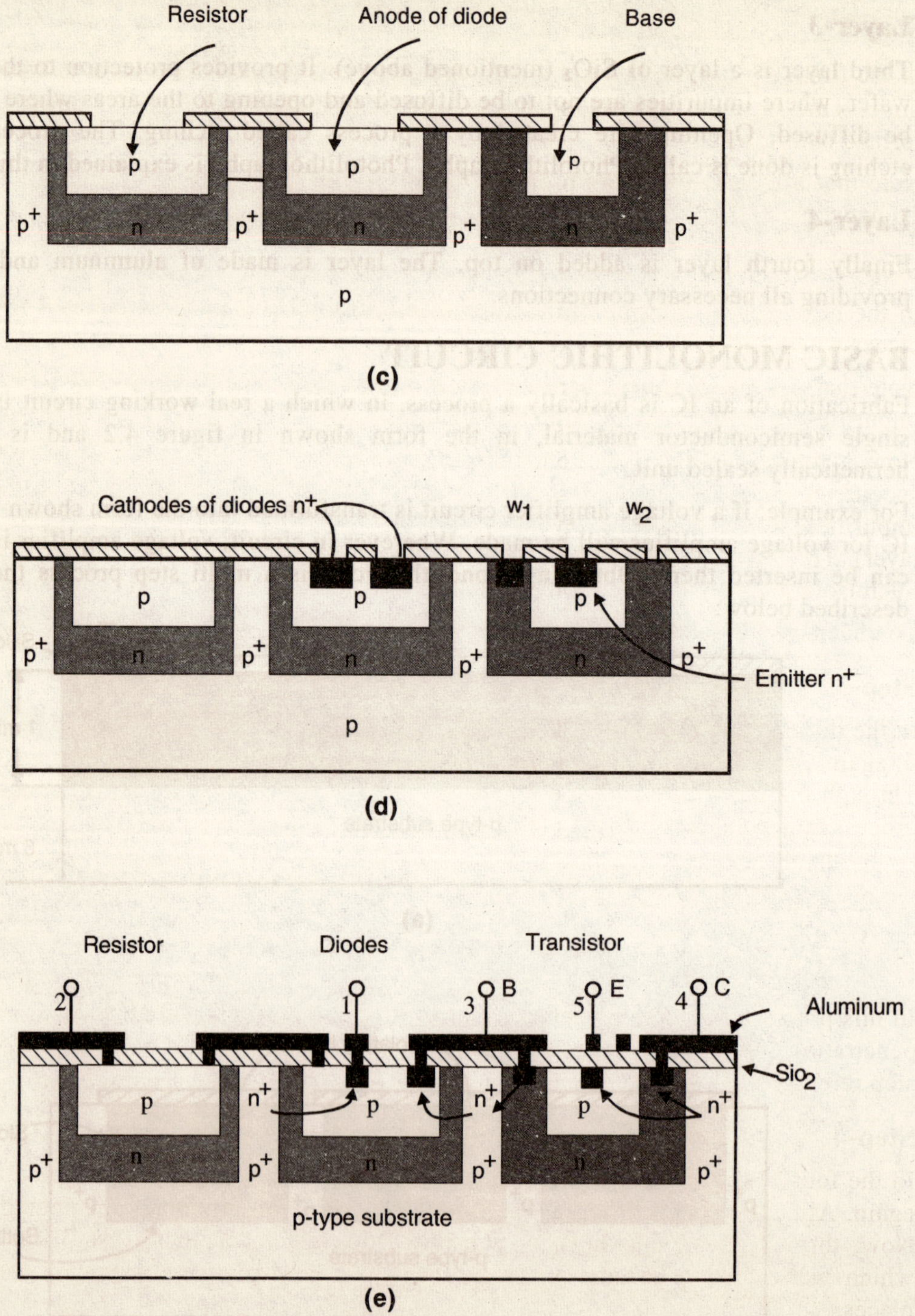
Resistor
Anode of diode
Base
p
n
p+
(c)
Cathodes of diodes n+
w1
w2
Emitter n+
(d)
Resistor
Diodes
Transistor
2
1
3 B
5 E
4 C
Aluminum
Sio2
n+
p-type substrate
(e)

**Figure 4.3**

### Step-1

In the first step, n-type layer is grown over, p-type substrate. This layer is typically 25 µm thick. This layer is then polished and cleaned. After this a layer of 0.5 µm of $SiO_2$ is formed over the entire wafer.

Note that $SiO_2$ possesses the property of preventing the diffusion of impurities through it. To understand the process of this step, refer figure 4.3(a).

### Step-2

In the next step, using photolithographic etching process, $SiO_2$ is removed from all those places where impurity is to be diffused into the wafer.

Note that the places from where $SiO_2$ is not to be removed, serves as mask or shield. Thus impurities cannot diffuse through this section.

Now a process called isolation diffusion is carried out. This process is carried out at specific temperature and for specific duration. Both these factors make sure that p-type impurities penetrate n-type layer (i.e. second layer) and reach p-type material (i.e. first layer). This creates isolated regions in the wafer. These regions are shown in figure 4.3(b). Such isolation provides electronic isolation between different circuit components.

To understand the activities of this step refer figure 4.3 (b).

### Step-3

In the third step, a new layer of $SiO_2$ is laid over the wafer (which was end product of step-2) again. After this, photolithographic process is used again to create openings in the wafer. Through these openings, p-type impurities are made to diffuse in the isolated islands. This process thus creates following elements in the wafer, at all required places:

1. Transistor base region.
2. Capacitors
3. Resistors
4. Anodes of diodes

In this process the depth of penetration of p-type impurities is controlled so as to keep the penetration shallow and keep it away from substrate. To understand the activities of this step refer figure 4.3 (c).

### Step-4

In the fourth step, a layer of $SiO_2$ is laid over the wafer (which was end product of step-3) again. After this, photolithographic process is used again to create openings in $SiO_2$ layer. Now, through these openings, n-type impurities are made to diffused in p-type region, which were created in step-3. This process creates following elements at all required places:

1. Transistor emitters
2. Cathode region of diode
3. Junction capacitors

After this, additional windows are created into n region for making provision for interconnection of components.

Refer figure 4.3 (d) to understand the activities of this step.

### Step-5

Note that by the end of step-4, all the electronic components required for the circuit have been carved in the wafer. Now they need to be interconnected as per requirement.

To make these connections, a new layer of $Sio_2$ is laid over the wafer and required openings are made using photolithographic process.

The inter connections are made by first laying thin quoting of aluminum over the entire wafer, using vacuum deposition process.

Now using photoresist technique, aluminum is etched away from all undesired places and leaving it deposited over those areas, where conduction has to take place. Refer figure 4.3(e) to understand the activities of this step.

In actual manufacturing process, many circuits of same type are carved together on a single wafer. After step-5, the wafer is fine cut, using a diamond tipped tool and separated into individual chip.

Each chip is then mounted on ceramic wafer and package leads are connected to the integrated circuit by stitch bounding of a 1-mil aluminum or gold wire from the terminal pad on the circuit to the package leads, which ultimately connects IC with external world. This is shown in figure 4.4.

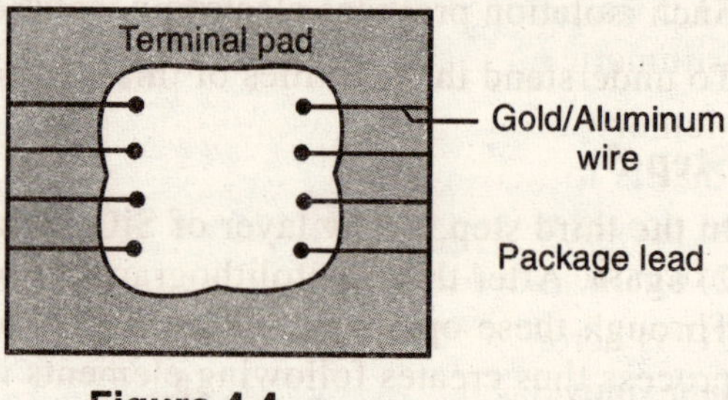


Figure 4.4

## CLASSIFICATION OF ICS

Integrated circuits can be classified into following classes:

1. Digital Integrated circuits
2. Analog Integrated circuits
3. Hybrid Integrated circuits

### Digital Integrated Circuits

Digital integrated circuits can contain millions of electronic circuits, such as logic gates, flip-flops, multiplexers and other complex circuits in them. Extremely small size of the IC promotes high speed, low power dissipation and reduced manufacturing cost. Microprocessors, memory chips etc. are the best example of digital integrated circuits. A brief introduction of microprocessor is given later in this chapter.

### Analog Integrated Circuits

Analog integrated circuits operate upon continuously varying signals. Sensors, power management circuits, operational amplifiers are examples of analog integrated circuits. They perform functions like amplification, active filtering, demodulation, mixing etc.

## Hybrid Integrated Circuits

As the name indicates, hybrid integrated circuits operate upon digital and analog signals, both. Chips for analog to digital conversion (A/D), digital to analog conversion (D/A) etc. are examples of hybrid integrated circuits. In such ICs CMOS technology is generally used for digital operation and bipolar transistors are used for analog performance.

## MICROPROCESSOR

A microprocessor is basically a digital integrated circuit, which can be programmed. Its main function is to perform arithmetic and logical operations. It is used in all those electronic devices, which are designed for performing mathematical functions and decision-making operations.

Internally the microprocessor consists of three main units called ALU, CU and Memory. Block diagram of microprocessor is shown in figure 4.5.

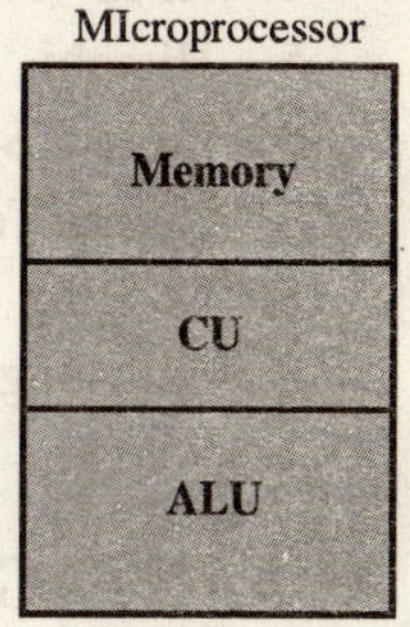


**Figure 4.5**

ALU is acronym for Arithmetic and Logic Unit. This unit performs arithmetic and logic operations.

Memory of microprocessor is basically a storage space. It holds data (that is to be processed) and instructions (according to which data is to be processed) temporary. It also holds the results that are received after processing the data.

CU is acronym for Control Unit. This unit of microprocessor controls the data movements. For example, activities like, transfer of data and instructions from memory to ALU and result from ALU to memory etc. are performed by CPU.

The first microprocessor was introduced in 1971 by a famous chip manufacturing company of USA, called Intel. The chip was identified as Intel 4004. It was not a very powerful chip. It could only perform addition and subtraction functions. But on the other hand, it was a remarkable step that complete circuit was a single small chip.

As the time passed, technology advanced and more sophisticated microprocessors came into existence. Today microprocessors like Pentium 4 "Prescott" are available. It contains around 125,000,000 transistors packed in it. It operates at a speed of 3.60 Hz. It is capable of reading 64 bits of data simultaneously from the memory and performs 7000 Million instructions per second (MIPS). It suits well for multimedia operations also. Such microprocessors are mainly used in computers.

## FEW PROMINENT ICS

There are many famous companies and many famous ICs. Some of them are described below:

1. The 555. It is multivibrator circuit.
2. The 741. It is an operational amplifier circuit.

3. The 7400. It is TTL logic circuit.
4. Intel Pentium series. They are microprocessors, designed and developed by Intel.
5. Celeron series. They are scaled down versions of Pentium series processors, manufactured by Intel Company.
6. Intel Mobile series microprocessor. They are basically energy efficient processors. Because of their small size, they are mainly used in laptops and palmtop computers.

## IC MANUFACTURES

Following are few famous manufactures of ICs:

1. IBM
2. Fair child semiconductor
3. Intel
4. AMD (Advanced Micro Devices)
5. Hitachi
6. Lattice Semiconductor
7. Linear Technology
8. Toshiba
9. Samsung Electronics
10. Zilog

## ADVANTAGES OF INTEGRATED CIRCUITS

Following are the advantages of integrated circuits:

1. Small in size
2. Very light in weight.
3. Very low cost.
4. High reliability.
5. Low power consumption.
6. Very suitable for small signal operations.
7. Easily replaceable in the circuit.

## DISADVANTAGES OF INTEGRATED CIRCUITS

Following are the disadvantages of integrated circuit:

1. Coils or inductors cannot be fabricated.
2. They operate upon low voltages only.
3. They can handle only limited amount of power.
4. They cannot withstand excessive heat.
5. They cannot be repaired. They can only be replaced by a new piece.

## EXERCISES CHAPTER 4

### Short Type Questions

**A. Select most appropriate answers for following questions:**

1. Who developed first IC?
   (a) C.V Raman (b) J.S Kilby (c) Robert Noyce (d) AMilkon
2. Which of the following is the base material for Integrated Circuit?
   (a) Metal (b) Insulator (c) Semiconductor (d) Gold
3. Which layer of IC is called substrate?
   (a) First (b) Second (c) Third (d) Fourth
4. Which material is used for masking during the fabrication of IC?
   (a) Silicon (b) $Sio_2$ (c) Aluminum (d) Gold
5. Which process is used for creating window in wafer of and diffusing impurities in the created space?
   (a) Photoemission (b) Photo Synthesis
   (c) Photolithography (d) Photo electric effect

**B. Fill in the blanks:**

6. IC is acronym for.................................................................................................
7. ICs can be classified into three classes. They are .................. ....................and .....................................integrated circuit.
8. .......................is an example of digital integrated circuit.
9. The IC, which would perform analog to digital conversion will be termed as .......................integrated circuit.
10. ............., .................and............ are the three main units of microprocessor.

**C. State, true or false for following statements:**

11. Robert Noyce's work in the area of IC development was awarded first patent.
12. Last layer of an IC is a metallic layer comprising of aluminum.
13. Monolith ICs are fabricated from ceramic base.
14. MSI is acronym for Mass Scale Integration.
15. Modern ICs are the result of ULSI technology.

**D. Answer the following questions in short:**

16. Write the full form of SSI. Using this technology, how many electronic components can be packed in a single chip?
17. What is the biggest advantage of converting a normal circuit into IC?

18. Who discovered integrated circuits?
19. Which material is used for creating substrate in IC fabrication?
20. Name any three electronic components that can be fabricated in monolith IC.

## Detailed Answer Type Questions

**E. Answer the following questions in detail:**

21. What is IC and what are its advantages over normal circuit?
22. Describe the basic internal structure of IC.
23. Describe the process of IC fabrication, using suitable diagram.
24. Describe the classification of integrated circuits.
25. What is microprocessor? What it is used for? Draw a block diagram of microprocessor.
26. What do you understand by monolith circuit design?
27. Describe the following:
    (a) Second layer of IC.
    (b) Fourth layer of IC.
    (c) Role of $SiO_2$ in IC fabrication.
    (d) Advantages of IC.
    (e) ALU of a microprocessor.
28. How are modern microprocessors different from older microprocessors?
29. How is LSI different from VLSI?
30. Name any five ICs and five IC manufactures. Name at least two microprocessors.

****

# COMPUTERS

# CHAPTER 5

## *Introduction to Computer*

Calculations have always been part of man's life. Be it his personal life or professional work in office, he has to perform calculations. Without calculations he cannot lead a normal life.

It has always been man's desire to perform all the calculations correctly, effortlessly and at very fast speed. For this, he has been continuously evolving new mechanisms and building better tools for calculations.

Ever since ancient time, a technological race has been set up between the complexity of the problems and calculation capabilities of calculating devices.

In this race both the runners have tried their best to stay ahead of each other. Whenever complexity of the calculations takes stride, a machine with more capabilities and power comes into existence. But at the same time, with the advent of new machines, complexity grows multi fold to place new challenges before the machine.

Thus in this race, many calculating devices came into existence and became obsolete, within short period of time.

The race has not yet finished, but to meet the challenges of contemporary complexity, computer in its most powerful form is available for use. There is hardly any sphere of life left today, in which computers are not being used. But unattended complexities are still demanding for more capable machines.

In this chapter, we will take a close look on the development of computer, which has taken place over past so many years.

### HISTORY OF COMPUTER

Computers of today are quite powerful, sophisticated and very fast. They were not so in their earlier stages. In fact they have traveled a long distance from primitiveness to sophistication. They took many years to reach the present stage.

Ancient people used pebbles for counting and made scratches on the wall to keep track of their belongings. For example, ten scratches on the wall represented ten goats that they had.

After that two new scratches nearby used to be clear indication of two new additions made in the fleet.

As they became bit wise, they made use of the objects nearby them and made counting and calculating devices. For example, long time ago they collected wires, wood and beads from here and there and made the first counting device, called Abacus. As the time passed, many computing devices got developed. After lot of advancements and improvements calculating devices reached the stage of modern computer. Ever since the birth of abacus, calculating devices have been making continuous progress. A brief historic development of these devices is given below.

## Abacus

Around 1000 years before the birth of Jesus Christ the Mesopotamians quite unknowingly laid the foundation of computer era. They made a device of stones sand and pebbles, called Abacus. It was mainly used for counting.

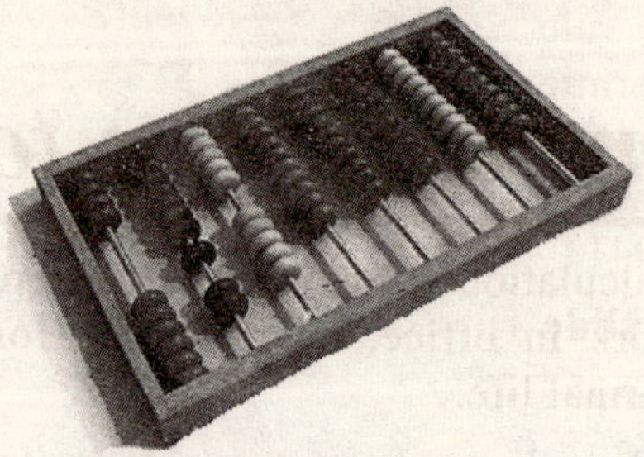

**Figure 5.1**

Later reforms were made in Abacus. Straight wires were fixed in a wooden frame and small beads were put in the wires. Beads could be moved along the length of the wire. Mathematical calculations like addition, subtraction, multiplication and division could be performed using new Abacus. Abacus is shown in figure 5.1.

Abacus is still being used today. Children are taught counting, using Abacus in schools. Many shopkeepers in Asia and "Chinatown" in North America, still make use of it for calculating the total payable amount for the articles sold by them.

## Napier's Bones

John Napier was a Scottish mathematician who lived from 1550 to 1670. He worked for more than 20 years to develop the theory of tables, which became quite famous, with the name Logarithmic Tables.

Towards the end of his life, John Napier devised set of rods, called Bones, because they were constituted from bones. Digits were printed on them. Napier's rods essentially transformed the process of complex multiplication and division into much simpler tasks of addition and subtraction.

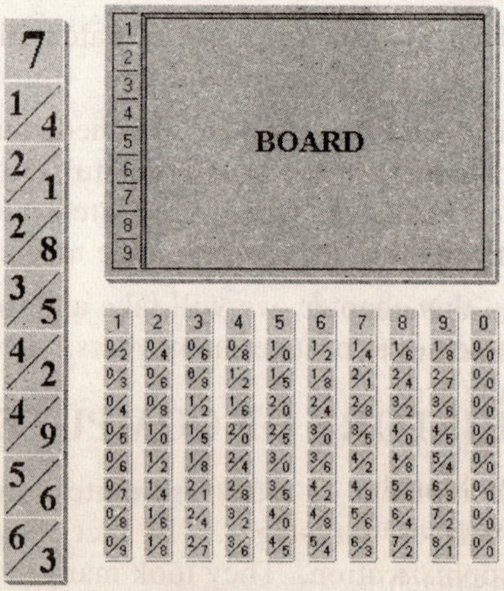


**Figure 5.2**

Napier's Bones are shown in figure 5.2. Refer this figure and note that each rod contains ten squares. The first square has the serial number of the rod written in it. Remaining nine squares are bisected by a diagonal, leaving aside the first square. The $n^{th}$ square in it contains the result of multiplication of rod number by n. So that the upper triangle of the square contains the most significant figure and the lowest triangle contains the least significance figure. For example, the third square in fifth rod will have 15 in it.

Now let's multiply 786 x 3. The 3[rd] square in 6[th] rod contains 18. So 18 will be at unit place of the product. Similarly, third square of the 8[th] rod contains 24. Thus 24 will be at 10[th] position. Third square of 7[th] rod contains 21. Thus 21 will be at 100[th] position. Thus we get:

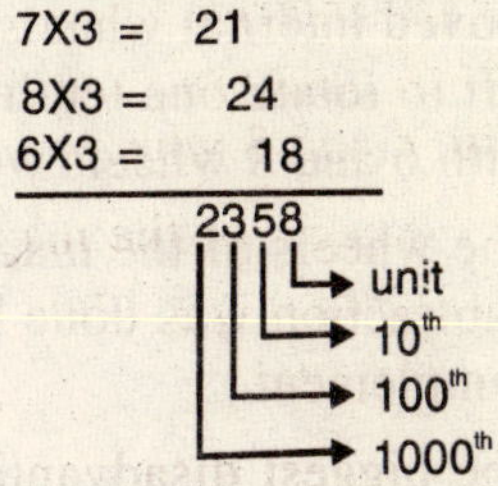


## Slide Rule

Slide rule was once-upon-a-time favorite mechanical device that was quite often used for performing complex and complicated calculations at very fast speed. Using slide rule mathematical, statistical and engineering calculations are performed with ease. For doing all these calculations, it makes use of logarithmic tables that were invented by John Napier in 1941.

The **slide rule** consists of a pair of finely divided fixed scales. Most often, outer pair of scales is fixed and inner scale is movable. A sliding window called cursor is put around the scales.

In slide rule, all the calculations are done by sliding the scales and the cursor. This is the reason why it drives its name slide rule.

Before the advent of pocket calculator, slide rules were quite popular. From 1950 to 1960, they were seen as the best available tool for calculations. When computers became popular and pocket calculator became cheap, their use became limited.

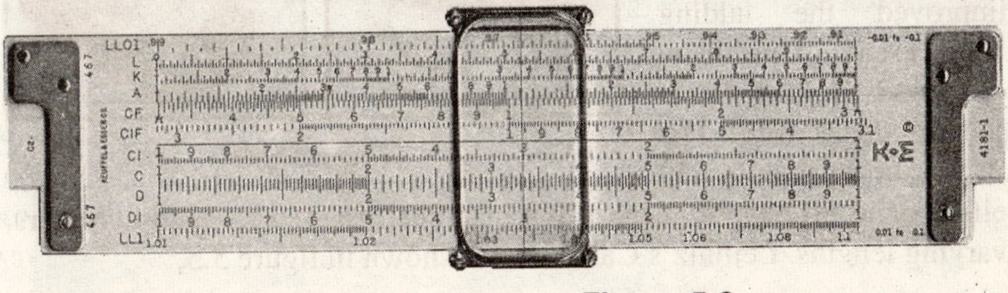

**Figure 5.3**

## Pascal's Calculator

Blaise Pascal was a French mathematician, physist and theologian. He was the son of tax collector. His father had to do lot of additions and subtractions, in which he quite often made errors.

At the age of 18 in the year 1840 Pascal started developing a mechanical device, using which arithmetic operations like addition and subtractions could be done very easily. The first operating model was introduced in 1942.

This machine was called Arithmetic Machine. Later Pascal created 50 more machines, over next 10 years.

Later this machine was called, Pascal's Adding Machine. It is shown in figure 5.4.

Pascal's adding machine was built on a brass rectangular box where, a set of notched dials moved internal wheels, in such a way that a full rotation of a wheel caused the wheel, at the left to rotate one tenth. Although the first prototype contained only 5 wheels but later, units with 6 and 8 wheels were also made.

The wheels of the machine moved only clock wise and were designed to perform additions. Subtraction was done by applying a cumbersome technique, based on addition of the nine's complement.

The biggest disadvantage of this machine was that operations like multiplication, division etc. could not be done through it.

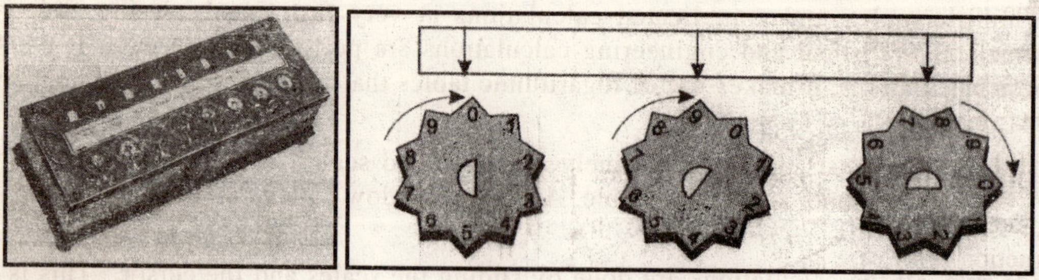

**Figure 5.4**

## Leibniz's Calculator

A German mathematician, whose name was Gottfried Leibniz, improved the adding machine, developed by Pascal. This new machine, which he constructed in year 1671, was able to multiply and divide as well. The concept of multiplication was to repeatedly add the same number to itself. Instead of using wheels in his machine, he used stepped-cylinders, each with nine teeth of varying lengths. Leibniz's Calculator is shown in figure 5.5.

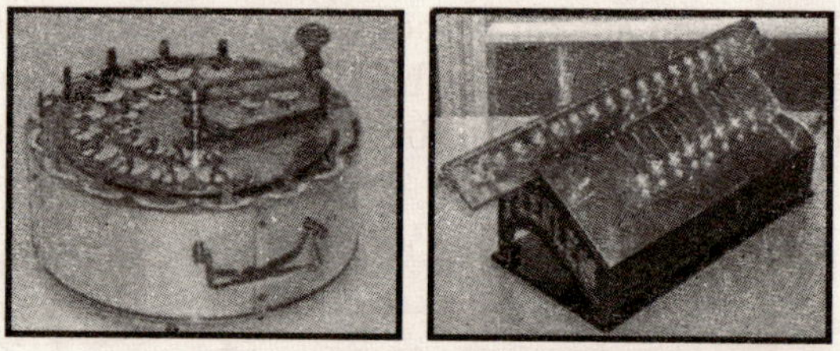

**Figure 5.5**

## Jacquard's Loom

A French weaver Joseph Jacquard invented an automated loom in 1804. This loom used punched cards to produce patterns and designs on cloth. The weaving action of this machine was totally governed by the absence or presence of holes in the cards. In other words, you can say that the entire operation was under a program's control. With this, historic invention of punched cards, the era of storing and retrieving information started that greatly influenced the later inventions and advancements of computing devices.

**Figure 5.6**

## Babbage's Difference Engine

Figure 5.7

In year 1822, Charles Babbage, a professor of mathematics developed a machine called Difference Engine. This machine was able to calculate logarithms up to high precision degree. It had capability to calculate different functional values and could also calculate polynomials by finite difference technique.

## Babbage's Analytical Engine

Figure 5.8

Charles Babbage started working on a device named Anaiytical Engine in year 1833. This Analytical Engine was in real terms the ancestor of modern computer. The Analytical Engine was able to perform all four arithmetic operations. It was conceptualized around processor, storage area, memory and input/output devices. The two revolutionary inventions in Analytical Engine were comparison and modification of stored information. This was the first invention, which enabled machine to compare quantities and then decide which of the instruction sequence to follow. Due to lack of technology at that time, it was not possible for Babbage to build such an engine. Its design remained conceptual. His great invention of Difference Engine and Analytical Engine earned Charles Babbage the title Father of Modern Computers - a suitable tribute to him.

## Hollerith's Machine

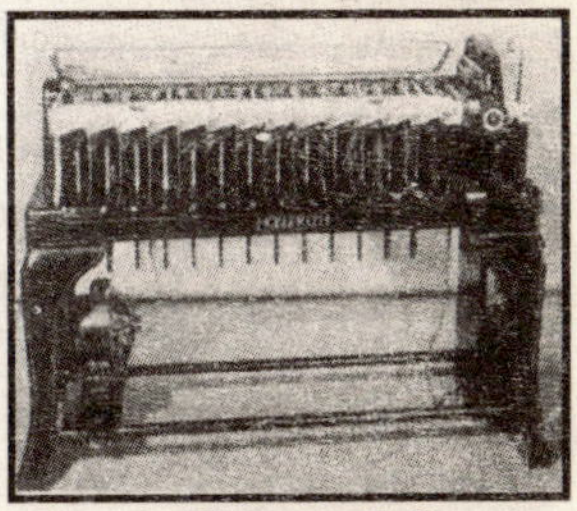

Figure 5.9

The dream of Charles Babbage was fabricated by an American, named Herman Hollerith (1869-1926). In year 1887 he fabricated the first electromechanical punched card tabulator that used punched cards for input, output and instructions. This machine was used by American department of census to compile their 1880 census data and was able to complete compilation in 3 years, which earlier used to take around 10 years.

## Mark-1

Figure 5.10

Prof Howard Aiken (1900-1973) of U.S.A constructed an electromechanical computer named Mark-1 in year 1943. Mark-1 had capability to multiply two 10-digit numbers in 5 seconds, which was a record at that time. Mark-1 was the first machine, which could perform according to pre programmed instructions automatically without any manual interference. This was the first operational general-purpose computer.

## ENIAC

Figure 5.11

The defense department of USA was supposed to design advanced weapons, missiles and aircrafts. For this, they had to do lot of complex calculations. This process was not only time consuming but was error prone also. To do this work efficiently, in 1946, scientists of Pennsylvania University designed a new computer.

This computer was 20 feet long and 40 feet high. It consumed as much electric power as was sufficient to light a small city. But this machine was much faster than any earlier machine. It was capable of doing more than 5000 additions, subtractions or 350 multiplications and divisions in a second. This machine was called ENIAC. **ENIAC** is acronym for **Electronic Numerical Integrator and Computer.** The biggest problem with this computer was that all the instructions were hardwired within the computer. Thus making modifications in instructions was a very difficult task. This problem was removed from the computers, which came after this computer.

Figure 5.12

## EDVAC

EDVAC is acronym for **Electronic Discreet Variable Automatic Computer**. It was designed and created in 1950 in Moore school of America. It was capable of taking instructions from external resources. Data and instructions, both could be provided to it, using punched cards. Thus problem of hardwired instructions, which was there in ENIAC, was solved in this computer. Its performance was much better than earlier computers.

## Modern Computer

In the year 1981, IBM developed a modern computer. It was small in size but very powerful. Its functioning was very simple. This computer was called **Personal Computer** or **PC**. It became popular very soon. After 1981 lot of improvements were made in PC. These improvements made PC very powerful and easy to operate. Personal Computer's functioning has now become so simple that even a child can operate it.

Figure 5.13

# GENERATIONS OF COMPUTER

Computer being an electronic device, its size, capabilities and power at any time are greatly influenced by the existing state of electronics of that time.

If electronic components and devices of any time were big, slow and unreliable then computers of that time were also big, slow and unreliable. On the other hand fast, reliable and smaller electronic components made the computers small, reliable and fast in operations.

You know that technological developments of electronics passed through few major phases. For example, there was a time when vacuum tubes were quite popular. They were freely used in majority of electronic equipments. After some time, they were replaced by transistors. Later Integrated Circuits came and set a new path for electronic developments.

Computers of specific electronic era are often referred to as a generation. Various generations of computer that came into existence, are described below.

## First Generation Computers (1942-1955)

Figure 5.14

First generation computers were voluminous in size. **They were made using vacuum tubes**. Vacuum tube was a fragile glass device that could control and amplify electronic signals.

First generation computers used the idea of execution of stored programs. Following were the main drawbacks of first generation computers:

1. Their size was very large.
2. Speed of instruction-execution was very slow.
3. They generated lot of heat.
4. Switching through vacuum tubes was not reliable.
5. Heavy air conditioning was required.
6. Only scientists and engineers were able to use these computers and that too in limited areas due to their high cost.
7. They needed a constant maintenance, which was very expensive.

Some of first generation computers are ENIAC, EDSAC and UNIVAC I etc.

## Second Generation Computers (1955-1964)

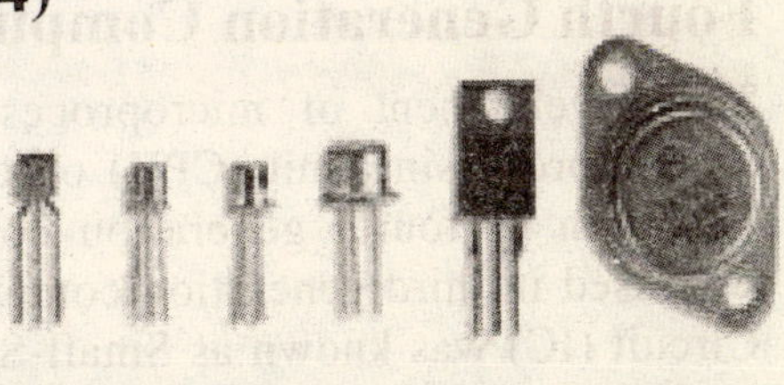

Figure 5.15

Transistor was invented as a smaller and more reliable successor of the vacuum tube. Although transistor was invented in 1948, its application in computers started in the year 1958. The transistor was more reliable, less expensive, and smaller than the smallest vacuum tube. It consumed very little power and produced much less heat.

A transistor is also called a solid-state device. It is made of a material whose principal part is silicon element. It is also called semiconductor device. It performs all the functions of a vacuum tube i.e. switches circuit on and off at a very high speed. **Second generation computer made use of transistors.**

Apart from the above listed advantages, the second generation computers used better devices such as card readers, printers etc. This generation is also important from the point of view of language development and use. First generation computers used machine level language or low-level language i.e. program was written directly in binary digits, which was very cumbersome and tedious. Later the development of Assembly Language helped the programmers to some extent. The high level language FORTRAN was also developed for this generation of computers. This development brought the computers very close to man and made it a very popular machine. Some of the names of second generation computers are IBM series, UNIVAC III, CDC 1400 series, Honeywell etc.

## Third Generation Computers (1964-75)

The development of integrated circuit was another milestone in the history of computer science. It was micro electronic technology, in which it was possible to integrate large number of circuit elements into very small surface area (less than 5 mm square) of silicon known as Chip. This generation was based on Integrated Circuits (IC) technology and the **computers that were designed with the use of integrated circuits were called third generation computers.**

Figure 5.16

These computers had better storage devices and improved input / output devices, such as Visual Display Unit (VDU), Magnetic Ink Character Reader (MICR) and high-speed line printers. These computers were 10,000 times faster than the first generation computers and arithmetic and logical operations could be performed in nano seconds (1 nano second=$10^{-9}$ seconds). It was this time, around which major advancement in the field of hardware as well as software took place. Many useful high-level languages were developed. These computers used operating system. The use of faster input and output devices enhanced the capabilities of these computers and made multi-programming possible, in which a number of programs could be executed simultaneously by a single processor. Some of the popular computers of this generation are IBM 360 series, 370 series, UNIVAC, HOG 9000 series etc. Some of them are still in use.

## Fourth Generation Computers (1975 Onwards)

Figure 5.17

The development of microprocessor chip that contains entire central processing unit (CPU) on a single silicon chip led to the invention of fourth generation computers. The technology that was used in third generation computers in the form of Integrated Circuit (IC) was known as Small-Scale Integration (SSI). This IC contained about ten to twenty transistors on it. Later, the Medium Scale Integration (MSI) was developed, in which it was possible to assemble up to hundred transistors on a single chip. After this, came the era of Large-Scale Integration (LSI) in which, it was possible to integrate 30,000 transistors on a single chip. Efforts were made and now the era of VLSI (Very Large Scale Integration) has come, in which millions of transistors can be assembled on single chip. **Current generation computers are fourth generation computers and are making use of VLSI chip.** This has made the size of current computer very small. Fourth generation of computers can be called microcomputer generation. The input/output devices used with fourth generation computers are quite advanced. These devices are optical readers, which can transfer the contents of CD within few microseconds to the computers. This generation's computers are compact, faster, and more reliable.

## Fifth Generation Computers

Till fourth generation computers, the main stress was on hardware technology. Though these computers are faster, more accurate, reliable, smaller in size and very cheap, still they lack thinking power. **The fifth generation computers will have thinking power and capability to make decisions like human beings,** and may prove better than man in certain aspects. They will be more useful in the field of knowledge processing rather than in data processing.

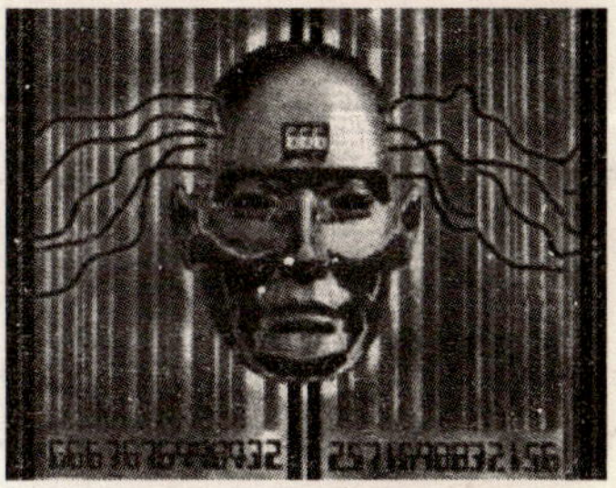

**Figure 1.18**

In fact fifth generation computers are dreams of Japanese scientists. They have started work in this direction few years ago. The concept of Artificial Intelligence (AI) is being used in these computers. These computers will have Knowledge Information Processing System (KIPS) rather than the present Data/Logic Information Processing System. Japan has chosen the PROLOG (Programming in logic) language as its operating software and plans to have the final machine, which will talk with human beings, see and deliver pictures and hear man's natural language.

## TYPES OF COMPUTERS

You know that the device, which computes is called computer. For computation, computer takes some input, performs computation on it and gives result.

The type of input that computer accepts, decides its nature and capabilities and the type of computation that it will be able to perform. For example, some of the computers accept input in the form of signals. On the other hand, majority of computers take input in the form of digits. Some computers accept input in both the forms i.e. in the form of digits and signals. On the basis of input type, computers have been classified into following three categories:

1. Analog computers
2. Digital Computers
3. Hybrid computers

A brief description of all these types of computers is given below:

## Analog Computers

Let's first understand what is signal. Signals can be defined as continuous flow of physical identity. Temperature, pressure, water level, speed, voltage, current etc. are few examples of signals.

**Those computers, which operate upon signals** (by accepting temperature, pressure, speed etc. as input) **are called analog computers.** Speedometer, thermometer, barometer, voltmeter, ammeter etc. are few examples of analog computers.

To understand the concept of analog computers, let's take the example of speedometer, which is used in scooters and cars. It makes use of cable. One end of the cable remains in contact with the wheel of the vehicle. The other end of the cable is inserted in the speedometer. When the wheel rotates, cable also rotates. Cable's rotation is thus passed to the internal assembly of the speedometer, which in turn rotates the needle on a calibrated scale. Thus when the vehicle moves, its speed gets displayed on the speedometer's scale.

The slide rule is hand-operated analog computer for doing arithmetic operations like multiplication, division etc.

## Digital Computers

In computer terminology, alphabets, numbers, special characters etc. are called digits. Those computers, which accept digits (like name, age, e-mail address etc.) as input, are called digital computers. Computers that we see in offices, shops, hotels etc. are digital computers.

## Hybrid Computers

Those computers, which accept both types of inputs i.e. signals and digits are called hybrid computers. For example, computers placed in ICU (Intensive Care Unit) are examples of hybrid computers. They accept heartbeats of the patient (signals) as input. They also accept instructions that need to be carried out (digits). Similarly computers placed in atomic reactors (taking radioactivity level and instructions as input) and mines (which take carbon monoxide level and instructions as input) are other examples of hybrid computers.

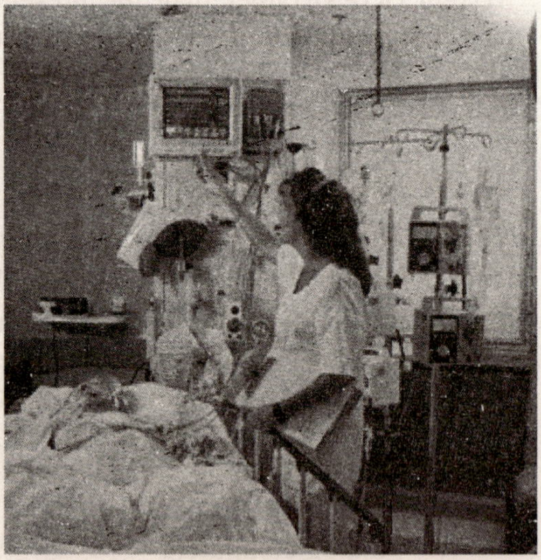

Figure 5.19

Note that in this book, we will mainly deal with digital computers. All the aspects, explained now onwards would mainly apply to digital computers unless stated otherwise.

## EXERCISES CHAPTER 5

### Short Type Questions

**A. Select most appropriate answers for following questions:**

1. What do you call that calculating device, which makes use of beads for calculations?
   (a) Napier's Bones (b) Slide rule
   (c) Pascal's adding machine (d) Abacus
2. Which of the following computing device was invented by the same person, who discovered logarithmic tables?
   (a) Napier's Bones (b) Slide rule
   (c) Pascal's adding machine (d) Abacus
3. Name the electronic component, which was used in first generation computers.
   (a) Transistor (b) Vacuum tube
   (c) PCB (d) IC
4. Which of the following device was the first device to make use of punched codes?
   (a) Abacus (b) Jacquard's loom
   (c) Second generation computers (d) Modern computers
5. Which of the following computer accepts signal and digits as input?
   (a) Analog computers (b) Digital computers
   (c) General computers (d) Hybrid computers

**B. Fill in the blanks:**

6. .................................. generation computers made use of VLSI chips.
7. Fifth generation computers will have..........................................................power.
8. Slide rule is an example of..........................................................................computer.
9. Computers that we use in computer lab are....................................computers.
10. Computers that are used in mines, ICUs etc. are examples of...................computer.

**C. State, true or false for following statements:**

11. Abacus is first generation computer.
12. Pascal's adding machine could perform multiplication and division also.
13. ENIAC computer made use of vacuum tubes.
14. Speedometer of a vehicle is an example of digital computer.
15. Integrated circuits were first introduced in third generation computers.
16. ENIAC computer came after EDVAC computer.

17. Vacuum tubes came into existence before transistor.
18. Fourth generation computers are smaller in size and faster than third generation computers.

**D. Answer the following questions in short:**

19. Which computing device was invented by Blaise Pascal?
20. Modern computers belong to which generation?
21. Name the mathematician who invented the calculating device called Napier's Bones.
22. Name any two types of signals that could be given as input to analog computers.
23. Which type of input is given to digital computers?
24. Which type of input is given to hybrid computers?
25. Which word will characterize fifth generation computer?
26. Which calculating machine made use of punched cards and which was the non-calculating device, from which the concept of punched card was taken?
27. Give an example of digital computer.

## Detailed Answer Type Questions

**E. Answer the following questions in detail:**

28. Describe any five historic calculating devices, in the order, in which they came into existence.
29. What is ENIAC? Describe its characteristics.
30. What is vacuum tube? What is its relationship with computers?
31. How were second-generation computers different from third generation computers?
32. Write a short note on fifth generation computer.
33. What are analog computers? Give at least two examples of analog computers.
34. What are digital computers? Give an example of digital computer.
35. How are hybrid computers different from analog and digital computers? Give example of hybrid computers.
36. Write a short note on generations of computer.

****

# COMPUTERS

# CHAPTER 6

## *Computer And Its Devices*

Computer is electronic equipment. It is mainly used for automating manual tasks. It comprises of many units. These units work in coordination with each other and perform the given task. A block diagram of computer is illustrated in figure 6.1.

Refer this figure and note that computer comprises of following units:

1. Input unit
2. Output Unit
3. Secondary Storage Unit
4. Memory Unit
5. Arithmetic and Logical Unit
6. Control Unit

Physically these units exist in the form of electronic devices in computer. Each unit performs its own function in co-ordination with other devices and thus computer performs the given task. What is the role of each unit, and in what form do they physically exist, is explained below.

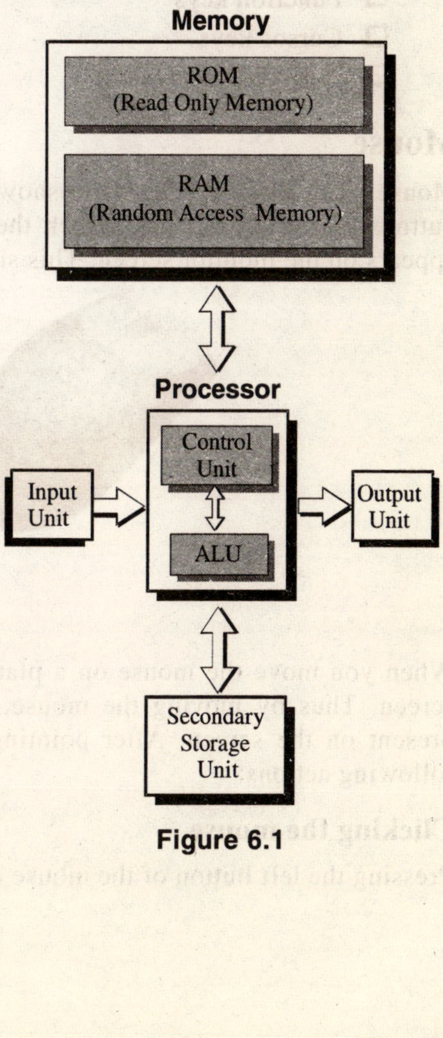


**Figure 6.1**

### INPUT UNIT

The role of input unit in computer is to provide means for supplying data or instructions to the computer. In other words, you can say that what you supply to the computer for processing is supplied through input unit. For example, if you wish to calculate simple interest on Rs. 10,000, @ 7% for 5 years then the data 10,000, 7, 5% etc. and the instructions for calculating the simple interest will have to be given to the computer through input unit.

All those devices, using which input is given to the computer are termed as input units. Following are few commonly used input units of computer.

## Keyboard

Keyboard consists of many keys. Keys have different, alphabets, numbers, characters or words written on them. Keyboard is illustrated in figure 6.2.

When you press an alphabet or number key, the character written on the key gets typed at current position.

**Figure 6.2**

### Keys

Keys of the keyboard are broadly classified into following categories:

- Alphabet keys
- Number keys
- Function keys
- Cursor keys
- Action character keys

## Mouse

Mouse is an input device. It is shown in figure 6.3. It looks like a real mouse with few buttons placed on its back. When the mouse is connected to the computer, an arrow sign appears on the monitor screen. This sign is called Mouse Pointer.

**Figure 6.3**

When you move the mouse on a plane surface, mouse pointer also moves on the monitor screen. Thus by moving the mouse, you can make the mouse pointer point to anything present on the screen. After pointing the mouse on any desired thing, you can perform following actions:

### Clicking the mouse

Pressing the left button of the mouse and releasing it is called clicking the mouse.

### Double clicking the mouse

Pressing the left button of the mouse twice, very quickly and then leaving it, is called double clicking the mouse.

### Dragging the mouse

Moving the mouse, with its left button pressed is called dragging the mouse.

### Dropping

Moving the finger away from the mouse after dragging it is called dropping.

### Right clicking the mouse

Pressing the right button of the mouse and releasing it is called right clicking the mouse.

## OUTPUT UNIT

The role of output unit is to show the result of processing. In other words you can say that computer displays all the results on its output unit. For example, if you are calculating simple interest on computer then it will display the interest amount on its output unit. Following are few commonly used output units of computer.

## VDU

VDU is acronym for Visual Display Unit. It looks like portable TV. VDU is illustrated in figure 6.4. It is also called monitor. It primarily displays following articles:

- ❑ The input that you supply through keyboard.
- ❑ Result of processing (output)

**Figure 6.4**

## Printer

Printer is an output device. It prints the output on paper. Different types of printers print differently. Their quality and speed of printing differ from each other. Following are different types of printers, which are being commonly used these days:

1. Dot Matrix Printer
2. Inkjet Printer
3. Laser Printer

### Dot Matrix Printers

These printers are very versatile printers and drive their name, "Dot Matrix" due to the fact that they print the characters with the help of dots. Pins of the printer print these dots. The dots are printed so close to each other that the gap between them hardly remains visible. Viewer gets an impression of full continuity. These printers are versatile printers and are capable of printing not only text, but up to some extent graphics also. They are relatively fast printers and their speed varies from 300 Character Per Second (CPS) to 600 CPS.

### Inkjet Printers

For printing the characters and drawings, inkjet printers make use of colored inks. They use cyan, magenta, yellow and black inks. Using these four basic colors, inkjet printers generate all other colors. These printers are very popular due to their high quality and low cost.

### Laser Printers

These are high quality, high speed and high cost printers. In such type of printers, printing is done, using laser beam and black magnetic powder, called toner. Speed of these printers can range from 10 pages a minute to about 200 pages per minute. They cost much higher than dot matrix and inkjet printers. All the three types of printer are illustrated in figure 6.5.

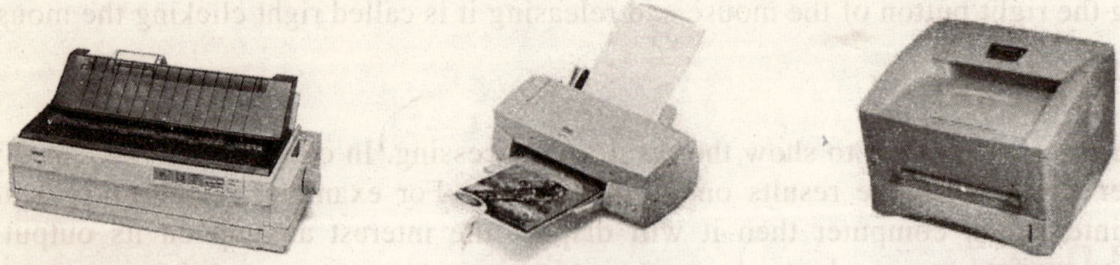

Dot Matrix Printer Inkjet Printer Laser Printer

Figure 6.5

## SECONDARY STORAGE UNITS

Secondary storage units are those devices, which hold the data, instructions, information etc. permanently within computer. Whatever is written on them remains there till the time someone removes it deliberately from there. This is the reason, why they are also called permanent storage units.

Whatever is required for longer duration is stored in secondary storage unit. Later, whenever stored item is required, it is read from there. Unlike input and output units, they allow users to perform both, input and output operations. This is the reason why they are also called Input/Output units. Following are few commonly used secondary storage devices of computer.

### Floppy Drive

Floppy drive is secondary storage device. It is basically an assembly, in which you put the floppy, close its door and then perform read / write operations on the floppy. For analogy, you can think of tape recorder's cassette housing assembly, in which you put the cassette, close the door and then perform read / write operations on the cassette. Similarly you either write the data on the floppy or read the data from the floppy. Remember that, whatever you write on floppy gets permanently stored there, until you delete it yourself or mishandle the floppy. Floppy and floppy drive are illustrated in figure 6.6.

**Floppy** **Floppy Drive**

**(a)** **(b)**

**Figure 6.6**

Floppy is basically a data storage medium, on which you record the data or programs to read them later. Floppy consist of flexible plastic, coated with magnetic material Iron oxide. This plastic is enclosed in a thick black jacket, which protects it from heat, dust, sun etc. Most commonly used floppies are of 3 ½" size.

## Hard Disk

Hard disk is another secondary storage unit, which stores the data permanently in computer. Externally, it looks like a box and remains fixed inside the computer. Its internal composition is shown in figure 6.7.

Refer this figure and note that hard disk consists of multiple dishes, on which the data gets stored. These dishes are called platters. Each platter has a corresponding read write head.

Read write heads write the data or instructions on the platters. Thus they get permanently stored on the platter.

Note that floppy and hard disk both are used for storing data but the capacity of the hard disk is generally many times more than that of a floppy. Floppy is a removable storage medium while hard disk is a fixed medium.

## CD-Writer

CD writer is another Input/Output device, which stores data, instructions, results etc. permanently on medium called CD-R and CD-R/W.

### CD-R

CD-R is acronym for Compact Disk-Recordable. It resembles a compact disk, on which data can be written. Data is written on CD-R through a device called CD Writer. CD-R looks like CD but its internal composition remains different. Data can only be written once on CD-R, not again and again.

### CD-R/W

CD-R/W is acronym for Compact Disk Rewriteable. It looks similar to CD-R but on CD-R/W, data can be written multiple times. If you wish, you can write data on it, read data from it and erase the data as many times as you want. CD-Writer and CD-R are illustrated in figure 6.8.

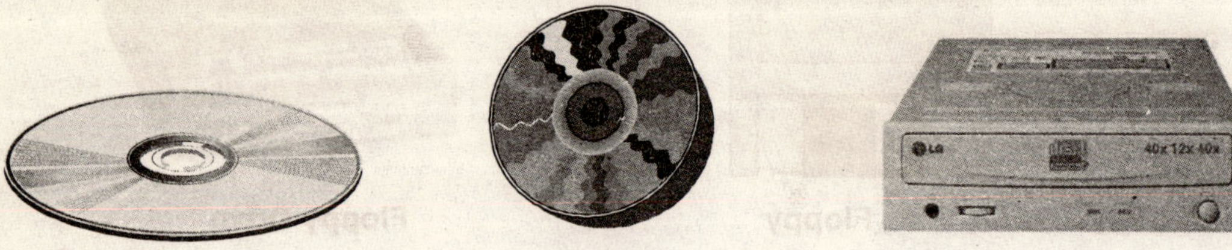

CD-R CD-R/W CD-Writer

Figure 6.8

# MEMORY

Those units of computer, which hold the data, instructions, information etc. either temporarily or permanently are called memory. Different types of memories exist in computer. Refer figure 6.1 and note that two types of memories viz. Random Access Memory and Read Only Memory are quite essential for computer's functioning. A brief introduction of these memories is given below.

## Random Access Memory (RAM)

Random Access Memory (called RAM in short) is the memory where computer writes data, instruction, results etc. temporarily. It keeps them in this memory till the time it requires them. As soon as their requirement is over, it removes them from there.

To understand its role, assume that two numbers, 100 and 200, are to be added. When these number and adding instructions are provided to the computer, it will hold them in random access memory, till the time result is printed. For performing the addition, it will fetch the numbers and addition instruction from random access memory.

After calculating the result, it will store the result in the memory. At the end, after printing/displaying the result, it will remove the numbers, instructions and result from the memory, so as to make the memory free for other tasks.

Random access memory is also called temporary memory. It keeps contents stored in it, till the time power is on. As soon as the power goes off, contents in it get erased. Note that this is a type of memory; in which computer performs both read and write operations.

## Read Only Memory (ROM)

Read Only Memory (called ROM in short) is special type of memory, in which computer doesn't write anything. As the name indicates, it only reads contents from it.

ROM physically exists in the form of an electronic chip. Contents are written into it, using a special device called ROM programmer and the chip is placed within the computer circuitry.

Now the question is what is written in ROM? Well, instructions (called boot strap program) that load the operating system into computer's memory are written into it. When you switch on the computer, it reads contents of ROM and executes them. Thus the operating system gets loaded into memory and computer becomes active.

## Virtual Memory

Virtual memory is a specifically allocated space on secondary storage device, which works in coordination with RAM.

Virtual memory makes it possible to work with programs that have size greater than the size of RAM. For example, if the size of RAM is 128 MB then virtual memory makes it possible to execute programs that exceed 128 MB (say 150 MB, 175 MB or 200 MB etc.).

This is made possible by dividing the large program into many smaller segments. Those segments, which are immediately required are maintained in RAM and those, which would be required after some time are placed in virtual memory. Thus a program, larger than RAM size becomes ready for execution.

Computer starts the execution of the program by executing the first segment present in RAM.

Whenever a segment present in virtual memory is required, a segment, which is not immediately required in RAM is written in the virtual memory and a free space is created in RAM. Immediately required segment is read from virtual memory and written in RAM and then executed. Whenever segment written in virtual memory is required again, some other segment from RAM is written in virtual memory and this segment is placed there. In this way swapping of segments keeps continuing.

This process is done so fast that user never realizes that these read write operations are going on in the background. He never gets a feel of memory limitation. All this is made possible through virtual memory.

## ALU

Refer figure 6.1 and note that there exists a unit in computer, called ALU. ALU is acronym for Arithmetic and Logic Unit. It performs arithmetic and logical calculations in computer. Calculations like, addition, subtraction, multiplication, division etc. fall into the category of arithmetic calculations. Operations like a>b, c>=d etc. are logical operations. In computer terminology, the activity of performing arithmetic and logical calculations is called processing. In other words, you can say that in computer ALU performs data processing.

## CONTROL UNIT

While computer functions, data moves from one unit to another unit. For example, when you give input to computer, data and instructions move from input unit to memory. When data is processed, it moves from memory to ALU. Similarly when processing is over, results move from ALU to memory and from memory to output unit. Now the question is which part of computer organizes data movement within the computer? Well, control unit is the main unit, which performs traffic police like task. It ensures movement of right data from right source, to right destination at right time so that all the units of computer perform their functions in coordination with each other.

## CENTRAL PROCESSING UNIT

Often ALU and Control Unit, put together are called Central Processing Unit or CPU in short.

## CLASSIFICATION OF COMPUTERS

Computers can be classified into various categories, on the basis of their sizes. When we say, size of the computer, it refers to multiple factors like, size of the memory, number of terminals that can be connected to the computer, storage capacity of the hard disk, type of processor used in the computer etc. Computers are generally classified into following categories:

1. Microcomputers
2. Mini computers
3. Mainframe computers
4. Supercomputers

## Micro Computers

Microcomputer is the smallest category of computers, in which single microchip is used for two basic units i.e. ALU and Control Unit. This microchip is often referred to as microprocessor. Microcomputers are further classified into following categories:

1. Personal computers
2. Laptop computers

### Personal Computers (PC)

Computers that you commonly see these days are examples of Personal Computers. In short, they are called PC. Their outward appearance is shown in figure 6.9. They mainly comprise of four units i.e. CPU box called chaises, keyboard, VDU and mouse. Only one person can work on them at a time, so they are called single user computer. Originally they were designed and manufactured by IBM. Later, as they became popular, other manufacturers also started manufacturing similar type of computers called IBM clones.

**Figure 6.9**

### Laptop Computers

Laptop computers also fall into the category of microcomputers. They possess same capabilities and resources that of Personal Computers. A laptop is shown in figure 6.10.

**Figure 6.10**

In their outward appearance, Laptop computers look like a briefcase. Upper panel of the briefcase comprises of LCD screen. Lower panel contains keyboard and mouse arrangement. Other components and circuitry remain packed inside the briefcase. They are deliberately designed to be light in weight. They run on chargeable battery, which resides inside. Due to their size, weight and independence from power (for two to three hours) people carry them while traveling in car, train or aeroplane and work on them.

## Minicomputers

Minicomputers are relatively larger and faster computers. Apart from single user environment, they also support multi user environment. They are generally used for automating those applications, which are large in size, require fast processing capabilities and demand for resource sharing among multiple users. Main characteristics of minicomputers are described below.

1. They are built, using high performance and high capability processors.
2. Memory size in such type of computers is generally very large.
3. They support multiple terminals, which may range from 2 to 16.
4. Large capacity disks are used in multiple numbers so that the data and programs of all the users could be put on-line.
5. They provide facility to connect multiple printers.
6. They possess the capability of performing computer network related major operations.

## Mainframes

Mainframes fall into next higher category of computers. Their internal architecture and circuitry remains different from minicomputers. They use specially designed proprietary circuits instead of just single microprocessor as their CPU. Their circuitry promotes higher connectivity, faster throughputs and large data processing capabilities.

Due to proprietary circuitry, mainframes become quite big in size and provide facility for connecting minicomputers and microcomputers with them. Few characteristics of mainframes are listed below.

1. They possess the capability of addressing larger memory sizes than that of minicomputers.
2. They also support larger capacity disks like optical disks etc.
3. Their terminal connectivity can go as high as 256 or beyond.
4. They possess large number of application libraries that provide great help to developers in developing useful applications.

DEC 20, IBM 370 etc. are the examples of few mainframes. IBM (USA) is the largest manufacturer of Mainframe computers.

**Figure 6.11**

## Supercomputers

Figure 6.12

Supercomputer is the largest category of computers. They use multiple CPUs for processing the data and executing the instructions. While performing a task through Supercomputer, the complete task gets divided into multiple independent tasks. Each CPU takes up individual task and completes it in parallel. This methodology of processing is called parallel processing.

Due to parallel processing, Supercomputers become capable of processing large volumes of data at very fast speed. They process multi million to few trillion commands per second. Applications like weather forecasting, nuclear weapon design, projection system modeling, aero modeling etc., which require tremendous processing capabilities, can be very well executed on Supercomputers. CRAY, XMP 24 and NEC 500 are few examples of Supercomputers.

# FEW IMPORTANT DEFINITIONS

You are now familiar with computer, its devices and its nature. Following are few important definitions related to computer:

## Hardware

The physical part of computer that can be seen or touched is called hardware. For example, keyboard, mouse, memory, ALU etc. are examples of hardware.

## Software

Computer hardware cannot work on its own. It requires instructions for functioning. Instructions that make the computer function are called software.

## Program

All those instructions, which are written for performing a specific task are called program. For example, if you write 25 instructions for accepting principle amount, rate of interest and the year of deposit for calculating compound interest then these 25 instructions put together will be called program.

## Operating System

All those programs, which put together activate the computer, so that it becomes capable of performing input, output and processing functions, are called operating system. Without operating system, computer neither understands the instructions given by you nor does it perform them. MS-DOS, Windows, Linux, Unix etc. are few popular operating systems.

## Booting

The process of loading the operating system in computer's memory is called booting. Only after booting the computer, it becomes capable of accepting the data or commands and processing them.

## DIFFERENCE BETWEEN CALCULATOR AND COMPUTER

Calculator and computer, both are digital equipments. Both of them are used for performing calculations. They are commonly used as assisting tools but there is a great difference between them. Few important differences are listed below:

1. Calculator is purely a numeric device, while computer works with numbers and alphabets also. Calculator cannot process alphabetic data such as name, address, city etc. while computer can do it.
2. Calculator cannot repeat previous operations, while computer can do. For example, operations done for calculating the percentage of marks for one student cannot be repeated for other students, in calculator. But they can be repeated in computer to calculate the percentage of all the students of other classes of the college.
3. Result of calculation obtained in calculator cannot be stored permanently, while they can be done in computer.
4. Capabilities of calculator always remain fixed. Calculators cannot perform different types of tasks.

   Computers are quite versatile. By executing different programs on computer, different types of task can be performed. For example, using same computer, letters can be written to friends and relatives, movies can be viewed, games can be played, reports can be printed on printer etc.
5. Power and capabilities of computer can always be enhanced by adding new devices to it. But no additional devices can be attached with calculator.
6. Only the functions, for which the buttons are available on keypad, can be performed in calculator. By writing a new program in computer, new types of operations can always be performed.

## EXERCISES CHAPTER 6

### Short Type Questions

**A. Select most appropriate answers for following questions:**

1. Which of the following is used as input device in computer?
   (a) Keyboard (b) VDU
   (c) Printer (d) Hard disk
2. Which of the following is used as an output device in computer?
   (a) Keyboard (b) Mouse
   (c) Printer (d) Hard disk
3. Which of the following doesn't store data permanently?
   (a) Floppy (b) Hard disk
   (c) RAM (d) CD
4. Which of the following unit performs calculations in computer?
   (a) ALU (b) Control Unit
   (c) RAM (d) ROM
5. Which of the following makes it possible to execute bigger programs than memory size?
   (a) Random Access Memory (b) Real Only Memory
   (c) Cache Memory (d) Virtual Memory

**B. Fill in the blanks:**

6. RAM is acronym for ......................... ......................... .........................
7. ROM is acronym for......................... ......................... .........................
8. ALU is acronym for......................... ......................... .........................
9. CU is acronym for......................... ......................... .........................
10. CD is acronym for......................... ......................... .........................

**C. State, true or false for following statements:**

11. Those computers, which make use of microprocessor, are called microcomputers.
12. Laptop is an example of microcomputer.
13. Personal computer is an example of minicomputers.
14. Minicomputers are larger than microcomputers but smaller than mainframes.
15. Mainframe is the largest category of computer.
16. CRAY, NEC 500 are two mini computers.

17. Hard disk stores more data than floppy.
18. Floppy and CD both are removable media.
19. To drag the mouse you have to press its left button twice.
20. VDU and printer both are examples, of output devices.

## D. Answer the following questions in short:

21. Name any two input devices of computer.
22. Name any two output devices of computer.
23. Name any two storage devices of computer.
24. Name three storage mediums that are used for storing data in computer.
25. Name any three types of computers.
26. Name any three types of memories used in computer.
27. Name any two types of computers that fall into the category of microcomputers.
28. Name that category of computer, which is bigger than minicomputers and smaller than super computers.
29. Which types of computers are used in weather forecasting applications?
30. What do you call the physical part computer?

## Detailed Answer Type Questions

## E. Answer the following questions in detail:

31. Draw a block diagram of computer and describe the role of each unit, shown in it.
32. Keyboard and mouse, both, are input devices. What is the main difference in the processes of providing inputs to the computer through them?
33. VDU and printer, both, are output devices, what is the main difference between their outputs?
34. Differentiate between CD-R and CD-R/W.
35. What is hard disk? Describe in short. Describe advantages of floppy over hard disk.
36. What is random access memory? What is its role in computers?
37. What is read only memory? Describe its role in computers, taking a suitable example.
38. What is virtual memory? What are its advantages? Describe its role in computers.
39. Personal computer and laptop, both are microcomputers. Describe at least five differences between them.
40. What is super computer? Name any three super computers. Describe their characteristics.

****

# COMPUTERS

# CHAPTER 7

## *Operating Systems*

A computer system is basically combination of hardware and software. For its functioning, it requites different types of hardware devices, electronic components and various types of software. Operating system is one of the software, which computer uses for its internal functioning. What operating system is, what are its functions and what types of operating systems are commonly available for use are the issues that are described in this chapter.

### OPERATING SYSTEM BASICS

Operating System is an essential software that is required for a computer to become operational. It provides functionality to computer hardware so that electro-mechanical components of it perform read, write and processing functions as human being do.

Without operating system, computer cannot work. In the absence of operating system, neither input devices will be able to provide data to the computer nor memory will be able to store anything it nor processor will be able to process the data nor will output devices be able to show the result. Managing computer's resources and making them work in coordination with each other is the responsibility of operating system.

Any instruction given by the user to the computer to perform a function is actually carried out by operating system. It is the operating system, which on receiving instructions from the user invokes all internal units to perform their duties and workout the result. Details of operating system are described below.

### THE PURPOSE OF OPERATING SYSTEM

Operating system is essential software, purpose of which is to activate the computer and:

1. perform internal management functions
2. provide services.

Internal management functions are the functions that have to be essentially performed to make the computer work. Services are bunch of commands and utilities that operating system provides to its users to have better control over computer. Block diagram, showing the basic structure of operating system is shown in figure 7.1. It also depicts its purpose.

# MANAGEMENT FUNCTIONS OF OPERATING SYSTEM

Essential functions that an operating system performs are:

- Process management functions
- Memory management functions
- Information management functions
- Device management functions

What these functions are, where do they fit in data processing cycle, is explained below.

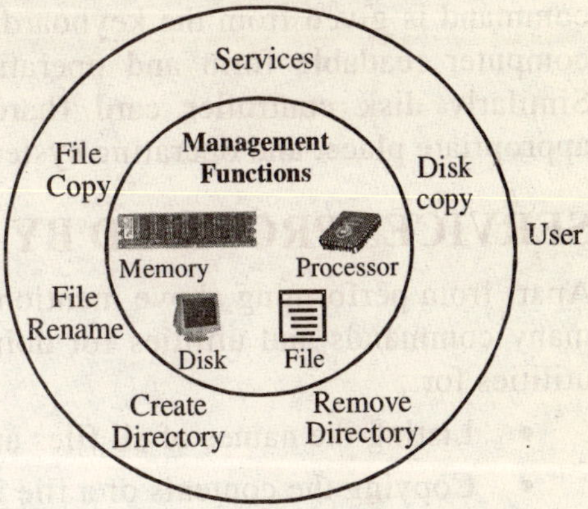


**Figure 7.1**

## Process Management Functions

Programs that are to be executed and their data, initially reside within computer's memory. For executing them they are taken to the processor, arithmetic and logical operations are performed within the processor, result of processing are transferred back to the memory and processor is set free for performing other tasks.

All these processor related operations are called process management functions. It is quite obvious that these functions cannot be carried out on their own. Some software is required to perform these functions. Operating system is the software, which performs these process management functions.

## Memory Management Functions

All the programs that are to be executed and the result that are to be printed, have to be essentially present in memory. When any such item comes to memory, it is accommodated in some free portion of the memory. If there is no free space, swapping procedure is started.

On the other hand, if any item present in memory is no more required then memory is reclaimed and free space is created. All these functions are memory management functions and are performed by operating system.

## Information Management Functions

Permanent storage of information within computer is an essential requirement of any computer environment. For this, data/information is written on the secondary storage devices, in systematic way. In this process, all arrangements are made to ensure that new data/information being written on the device doesn't overwrite previously written data/information and afterwards it is easily retrievable in its, original form. All those activities, which are carried out in computers for this objective are called information management functions and are carried out by operating system.

## Device Management System

Keyboard, monitor, floppy drive, hard disk etc. are few devices that are part of almost all the computers. In computers, their functionality is partially controlled by hardware and partially controlled by software.

Operating system is the software that controls these devices. For example, when a command is given from the keyboard, keyboard controller card (hardware) converts it into computer readable form and operating system (software) interprets it and executes it. Similarly disk controller card (hardware) places the read/write head of the disk at appropriate place, and operating system performs the writing function.

## SERVICES PROVIDED BY OPERATING SYSTEM

Apart from performing above mentioned management functions, operating system provides many commands and utilities for doing housekeeping tasks. For example, it may provide utilities for:

- Listing the names of all files and directories, present on a disk.
- Copying the contents of a file into another.
- Compressing and uncompressing the files.
- De-fragmenting the disk.

## TYPES OF OPERATING SYSTEMS

Various types of operating systems are available for use. For example, MS-DOS, Windows, Linux, Unix etc. are the names, of few popular operating systems.

Although all of them perform above-mentioned management functions but they differ from each other in terms of capabilities, user interface and services that they extend to their users. Operating systems can be broadly classified into following three categories:

1. Single user operating systems
2. Multi user operating systems
3. Network operating systems

A detailed description of all the types of operating systems is given below.

### Single User Operating Systems

Single user operating systems are those operating systems, which allow only one user to work on the system at a time. In no way, multiple users can work on the system simultaneously. Concept of single user operating system is illustrated in figure 7.2.

Figure 7.2

Single user operating systems can be broadly categorized into two categories:

1. Single user single task operating systems.
2. Single user multitasking operating systems

Single user single task operating systems allow user to execute one program at a time. For example, if he is preparing a worksheet on computer, he cannot print a report on the printer. As soon as he invokes a program, computer gets dedicated to that task only. CP/M, MS-DOS etc. are few examples of single user single task operating systems.

Single user multitasking operating systems allows user to execute more than one program at a time. For example, while he is preparing a worksheet on computer, he can print a report on the printer and play music too. Practically there is no limit from operating system's side for the number of tasks that can be executed simultaneously. Obviously physical size of the memory (hardware available) puts limitation on the number. Windows-Me, Windows-XP, Macintosh, OS/2 etc. are few examples of single user multitasking operating systems.

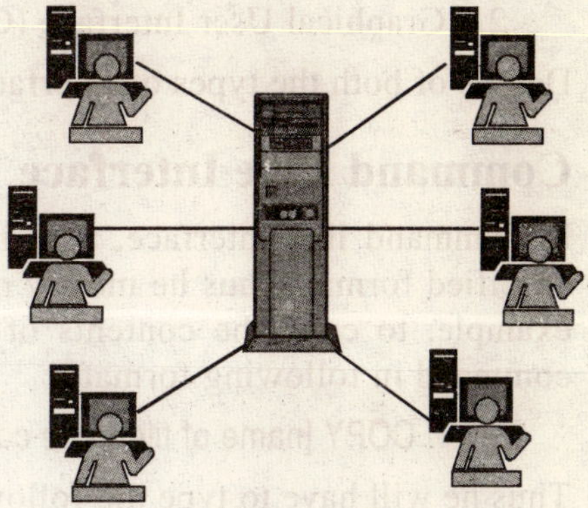

**Figure 7.3**

## Multi-User Operating Systems

Multi-User operating systems are larger and more complex operating systems. They allow multiple users to work on the system simultaneously. Concept of multi-user operating system is illustrated in figure 7.3.

Following are the features of multi-user operating systems, which are not provided in single user operating systems.

- Time sharing (CPU devotes time to all the users in round robin fashion).
- Tight security features.
- Resource sharing among users.
- System administrator privileges.

Linux, Unix, VMS etc. are few examples of multi-user operating systems.

## Network Operating System

When two or more computers are connected to each other with the objective of sharing their resources, they form a computer network. A network operating system is an operating system, which includes special functions, protocols and device drives for connecting the computers of the network. When a computer is booted with network operating system, all computers get connected together and the network becomes operational. When network becomes operational:

- Security features get implemented
- Devices of one computer can be used by other network users.
- Files can be shared among them.
- They can communicate with each other, sending mail and messages.

Windows-NT, Windows server 2000, Windows server 3000, Novell Netware, IBM AIX, Red Hat Linux etc. are few examples of network operating systems.

# PROVIDING USER INTERFACE

While working on computer, user has to interact with computer. For example, if a program is to be executed, user has to communicate to the computer that this is the program, which is to be executed.

Note that different operating systems provide different ways of interacting with computer. In computer terminology, the method of interaction is often referred to as user interface. User interfaces are basically of two types:

1. Command Line Interface (CLT).
2. Graphical User Interface (GUI).

Details of both the types of interfaces are given below:

## Command Line Interface

In command line interface, user interacts with the computer by typing the commands in specified format. Thus he mostly makes use of keyboard to interact with the computer. For example, to copy the contents of data.dat file in data.bak file, he will have to type the command in following format:

COPY [name of file to be copied] [name of file in which contents have to be copied]

Thus he will have to type the following command:

COPY data.dat data.bak

Note that in such type of communications, user has to remember the name and format of the commands. Spelling mistakes and deviations in format lead to errors and the task is not performed.

MS-DOS, Unix, Linux etc. are few operating systems, which provide command line interface. Visual appearance of MS-DOS's command line interface is illustrated in figure 7.4.

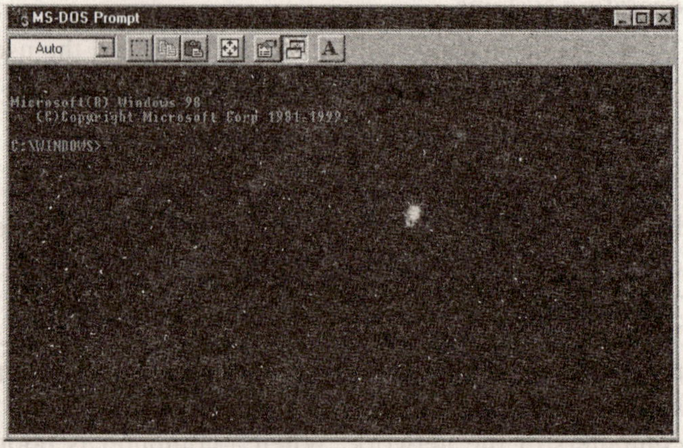


Figure 7.4

## Graphical User Interface

In Graphical User Interface (GUI), user interacts with the computer through graphical items, such as icons, menus dialog boxes etc. As and when required, these items automatically appear on the monitors and user has to simply select his choice from the displayed items, by clicking the mouse on desired option. Thus in this type of communications, user mostly makes use of mouse to interact with computer.

To understand the concept of GUI, refer, menus shown in figure 7.5. To invoke de fragmentation program, user need not remember the command name and command format, as he had to do in command line interface. For this, he simply need to click the mouse on the corresponding option present in the menu.

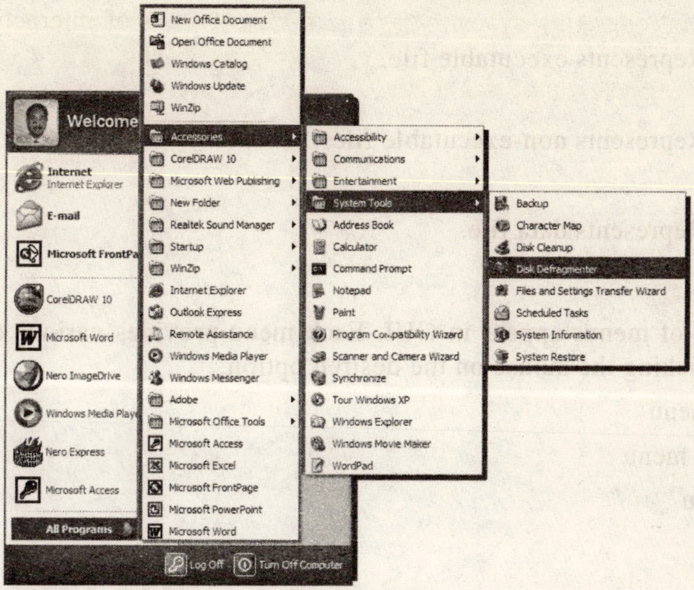


**Figure 7.5**

Windows, Macintosh operating system, OS/2, some versions of Linux, Unix etc. provide graphical user interface to interact with computer.

# GUI Components

Graphical user interface makes use of following GUI components to make the user interaction easy and interesting:

## Program Icons

All those icons that represent programs are called program icons. Some of the program icons are illustrated in figure 7.6(a). To run the program, you need to double click the mouse on its program icon. For example, to execute Date/Time program, you will have to double click the mouse on its icon.

## Objects Icons

All those icons that represent objects like, devices, files, folders etc. are called object icons. Some of the object icons are illustrated below.

Represents Floppy drive.

Represents hard disk.

Represents a folder.

Represents executable file.

Represents non-executable files.

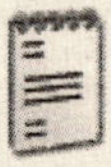

Represents data file.

### Menus

Following types of menus appear in GUI. Each menu provides various options. An option is selected by clicking the mouse on the desired option.

1. Popup menu
2. Shortcut menu
3. Bar menu

### List Boxes

List boxes display various choices within a box (names, numbers, symbols etc.). An item of choice is selected by clicking the mouse on it. A list box is shown in figure 7.6(d).

### Drop Down List Box

Drop down list box is another type of list box that looks like a text box with a down pointing arrow button on its right hand side. When you click the mouse on the arrow, a list of options drops down. To select any option from this list, you need to click the mouse on it. A drop down list is shown in figure 7.6 (d).

### Check Boxes

Small square shaped buttons (could be one or more) form check boxes. Labels written near the buttons give an indication of the type of data / information that will get selected, when you click the mouse on it. When you select a check box, a mark gets placed within the check box. If you wish, you could select more than one check box from the group. Check boxes are illustrated in figure 7.6 (c).

### Radio Buttons

Group of at least two or more rounded buttons form radio buttons. Labels written near buttons give an indication of the type of data/information that will get selected when you click the mouse on it. Only one radio button, from the group can be selected. When you select a button, a black dot appears within the selected button. Radio buttons are illustrated in figure 7.6 (c).

### Spinner

A spinner consists of two arrow buttons and a text box, with some value written into it. One arrow points up and the other points down. When you click the mouse on up arrow, the value in the text box increases. When you click it on down arrow, the value in the text box decreases. A typical spinner is illustrated in figure 7.6 (b).

### Text box

Text box consists of rectangular blank area, in which contents are typed. A typical text box is illustrated in figure 7.6 (d). Note that the label written beside the text box gives the idea of the contents that are to be typed in text box. For example, the label "File name" indicates that the name of the file is to be typed in this text box.

### Buttons

Rectangular labels, with some text written on them form buttons. Few buttons are illustrated in figure 7.6 (d). When you click the mouse on the button, an action takes place. The text written on the buttons, generally gives an indication of the action, which will take place, when you click the mouse on it. For example, when you click the mouse on "Open" button, the file mentioned in the dialog box gets opened.

### Toolbars

A row of small buttons, with icons in it forms a toolbar. Each button performs a specific task when you click the mouse on it. A typical toolbar is illustrated in figure 7.6 (a).

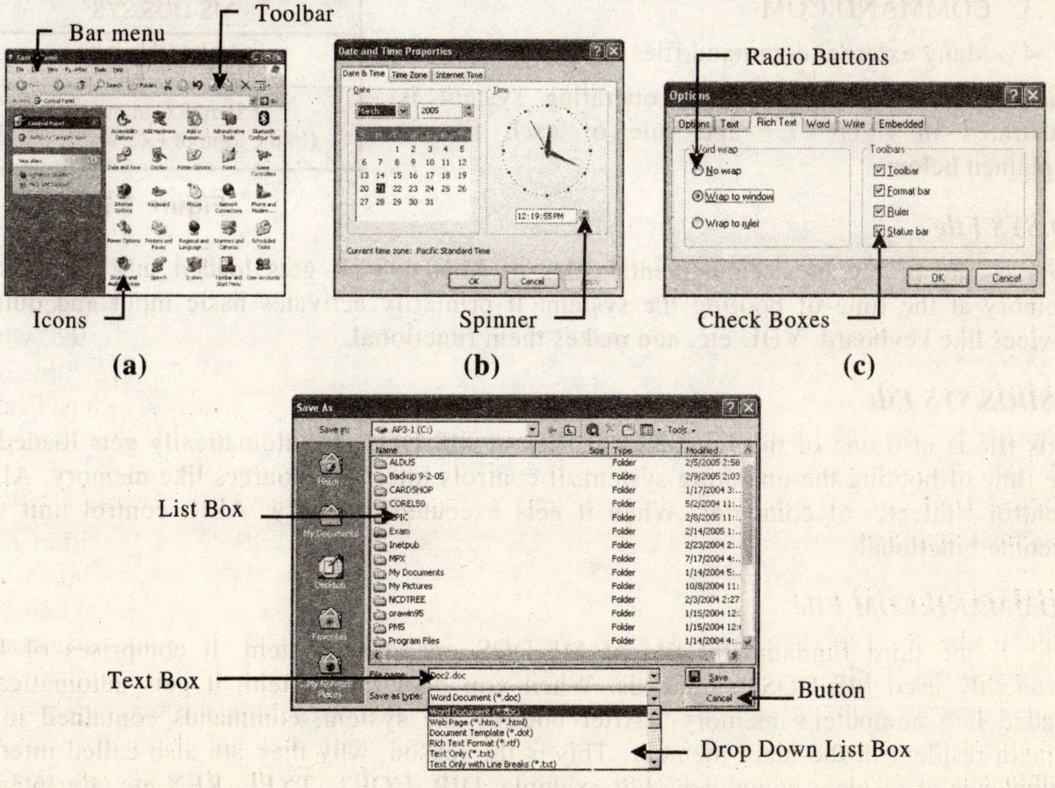


**Figure 7.6**

## MS-DOS OPERATING SYSTEM

MS-DOS is one of the most popular, powerful and useful operating system. It was designed and developed in the initial days of Personal Computers (PC) by Microsoft Corporation of USA. Due to its versatility and ease of operations, it became quite popular, within short span of time.

MS-DOS is single user, single task, character user interface (CUI) based operating system. To execute any command in MS-DOS, you need to know the command and its format. Any mistake in its spelling or format leads to error in command's execution.

MS-DOS not only activates computer resources and controls them but it also provides many commands for performing day-to-day tasks.

## ARCHITECTURE OF MS-DOS

MS-DOS is a modular operating system and comprises of many files. It utilizes these files as and when required. Following files put together constitute MS-DOS operating system.

1. IO.SYS
2. MSDOS.SYS
3. COMMAND.COM
4. Many external command files

Basic architecture of MS-DOS operating system is illustrated in figure 7.7 and role of each file is explained below.

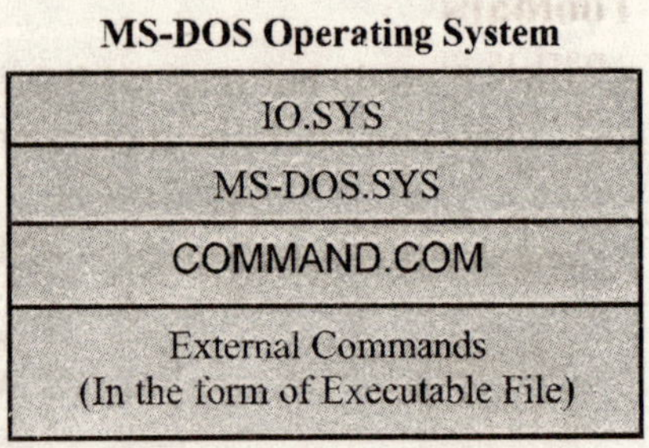


Figure 7.7

### *IO.SYS File*

This is one of the most fundamental files of MS-DOS. It gets loaded into computer's memory at the time of booting the system. It primarily activates basic input and output devices like keyboard, VDU etc. and makes them functional.

### *MSDOS.SYS File*

This file is also one of the fundamental files of MS-DOS. It automatically gets loaded at the time of booting the operating system. It controls internal resources like memory, ALU, Control Unit etc. of computer. When it gets executed, memory, ALU, control unit etc. become functional.

### *COMMAND.COM File*

This is the third fundamental file of MS-DOS operating system. It comprises of few frequently used MS-DOS commands. When you boot the system, it gets automatically loaded into computer's memory. After booting the system, commands contained in it, remain resident in the main memory. This is the reason, why they are also called internal commands or resident commands. For example, DIR, COPY, TYPE, REN etc. are internal commands of MS-DOS and reside in COMMAND.COM file.

To execute an internal command, you need to type the command from the keyboard in its recommended format. When you do so, it is directly read from the memory and executed. Role and function of commonly used internal commands are described later in this chapter.

### External Commands

MS-DOS comprises of many external commands. Each command exists in the form of executable file and resides on the disk. For example, FORMAT, XCOPY etc. are the examples of external commands. They exist in the form of FORMAT.COM, XCOPY.EXE files respectively.

To execute an external command, you need to type the name of the command from the keyboard and press Enter key. When you do so, computer reads the file from the disk, loads it into memory and executes it. When its execution is complete, it is removed from the memory. Role and functions of few external commands are described later in this chapter.

## File Naming Convention

In MS-DOS, file name comprises of two parts. First part is called primary name and second part is called extension. Both the parts remain separated from each other by a dot. Primary name can be at most 8 characters long and extension can be at most 3 characters long. The name can either be written in upper case or lower case letters but operating system stores all the file names in upper case only. As per rules, mathematical signs like +,-,*,/ etc. can not be used in constituting the file name. Following is the general format on MS-DOS file name:

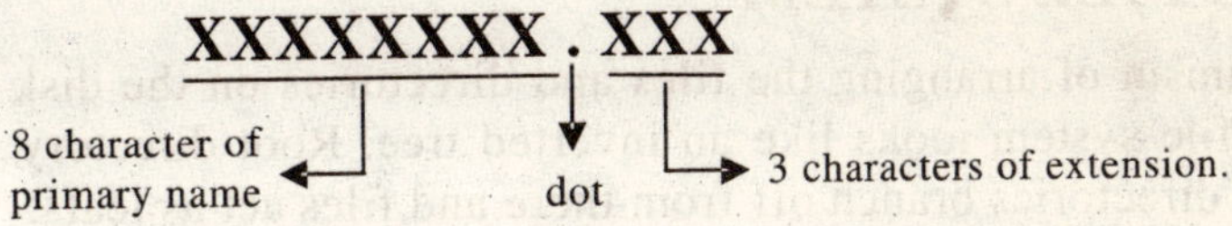


For example, DELL.COM, PLOT.DAT, MONTH12.BAK, YR.AS, RIGHT.DOC etc. are valid file names:

## Concept Of Directory

Directory can be conceptualized as special file, which could hold files and directories in it. Concept of directory is illustrated in figure 7.8. Refer this figure and note that D1 is a directory, which holds two files F1 and F2 in it. It also contains a directory D2 in it. D2 holds files F3 and F4 in it.

From this figure, it is quire clear that directories can be utilized for classified storage of files on the disk. For example, directory named LETTER could be utilized for storing all letters in it. Similarly the directory CPROGRAM could be used for storing all the programs written in C language.

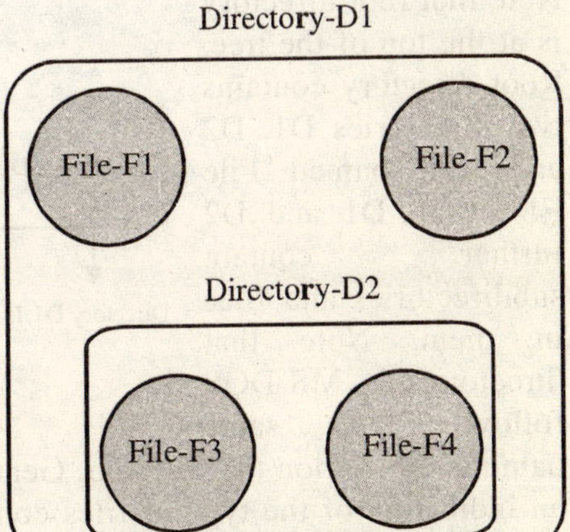


**Figure 7.8**

## Concept Of Root Directory

Root directory can be conceptualized as the top most directory, which holds other directories and files in it.

Whenever you use any disk for the first time, root directory gets automatically created on it. Other files and directories that are created thereafter are created within root directory. In MS-DOS, root directory is denoted by back slash (\) sign.

## Concept Of Parent Directory

Parent directory is the directory, in which you create another directory. For example, if directory D2 is created in directory D1 then D1 directory will be called parent directory of D2. Similarly if directory D3 is created in directory D2 then D2 will be the parent directory for D3. Parent directory is denoted by double dot (..) symbol.

## Current Directory

Once you have directories within directory you can always change your position from one directory to another directory. The directory within which you are currently placed is called current directory. For example, if you are currently placed in directory D1 then D1 will be called current directory. When you have moved from directory D1 to D3 then D3 will be the current directory.

# MS-DOS FILE SYSTEM

The mechanism of arranging the files and directories on the disk is called File System. In MS-DOS, file system looks like an inverted tree. Root directory appears at the top of the tree. Other directories branch off from there and files act as leafs.

Such a file arrangement is often referred to as hierarchical file system. A hierarchical file system consisting of few files and directories is illustrated in figure 7.9.

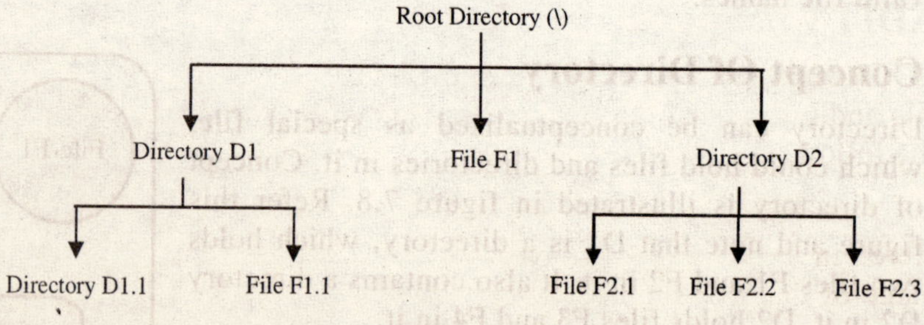


Figure 7.9

Note that root directory is at the top of the tree. Root directory contains two directories D1, D2 and a file, named "File F1" in it. D1 and D2 further contain subdirectories and files in them. Note that directories in MS-DOS follow the same naming convention that files do. Generally the directories are assigned a name, which gives an indication of the type of files contained in them. For example, the name DOCUMENT could be used for storing document files in it. Similarly SALARY directory could be used for storing pay related files.

## Concept Of Path

Path of a file or directory is the list of directory names in descending sequence, starting from root and each directory name separated by a back slash (\), following which you reach to the desired file or directory. For example, refer figure 5.8. The path for file F1.1 will be \D1\F1.1. Similarly the path name for F2.2 file will be \D2\F2.2. Note that first backslash in the path name denotes root directory, while other backslashes serve the purpose of separators.

## MS-DOS COMMANDS

Following are few commonly used MS-DOS commands:

## DIR Command

DIR command is used for displaying the names of all the files residing on media like floppy hard disk, CD etc. In its simplest form it can be executed in following format:

**C:\> DIR [Enter]**

This command will display the names of all the files present on C: disk, in the following format:

```
Volume in drive C is RAMESH
Volume Serial Number is 36F7-808
Directory of C:\
COMMAND     COM      54,645   06-11-99   7:21a
DATA        DAT         732   07-12-99   6:20p
REPORT      DOC       3,210   05-05-98   5:40p
DOIT        BAK         734   05-06-99   2:25a
TOTAL       DAT         351   10-10-99   5:25p
EXAM        LET         357   08-08-99   6:37p
       6 file(s)     60,029 bytes
                     18, 376, 520 bytes free
```

## EDIT Command

EDIT command is used to create a text file and type text in it. It is executed in the following format:

**EDIT [drive:] [file name] [Enter]**

Here, in this format, **"drive:"** is the name of that drive, on which the file is to be created and "file name" is the name, by which the file is to be created. For example, following command will create MYFILE.TXT file on default drive:

**EDIT MYFILE.TXT [Enter]**

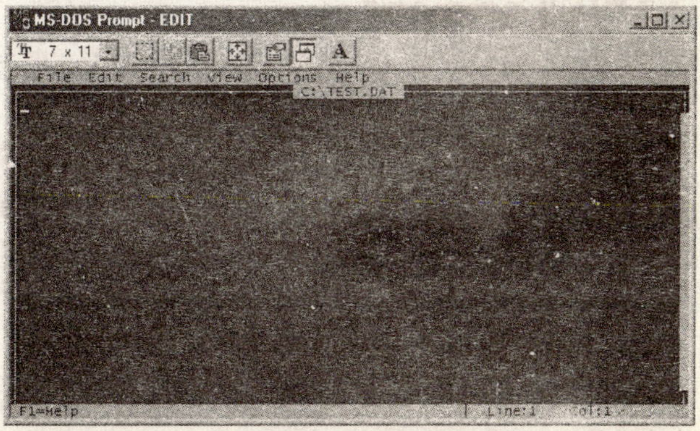


**Figure 7.10**

After execution of this command, a window, as illustrated in figure 7.10 will appear on the screen:

After getting the window on the screen, type the text as per instructions given in the following table:

| Sr.No | Task | Action |
|---|---|---|
| 1. | Type the text | Use character keys of the keyboard to type the text. |
| 2. | Take the cursor to the desired line and column. | Use arrow keys. |
| 3. | Delete a word | Take the cursor to the beginning of the word and press CTRL and T keys together. |
| 4. | Delete a line | Take the cursor to the line, which is to be deleted and then press CTRL and Y keys together. |
| 5. | Insert text | Take the cursor to the place where text is to be inserted and start typing the text. |
| 6. | Inserting a line between two lines | Take the cursor to the line, which is to be shifted below. After this, press Enter key. |
| 7. | Save the file | Press ALT and F keys together and then press S key. |
| 8. | Come out of EDIT | Press ALT and F keys together and then press X key. |

## TYPE Command

TYPE command is used for displaying the contents of a file on VDU. It is executed in the following format:

**TYPE <file name.> [Enter]**

Here, in this format, <file name> is the name of that file, contents of which are to be displayed on the screen. For example, if you wish to view the contents of MYFILE.TXT then following command is to be given:

**TYPE A:TEST [Enter]**

## COPY Command

COPY command is used to make a duplicate copy of a given file. It is executed in the following format:

**COPY <Source file name> <Destination file name> [Enter]**

Here in this format "Source file name" is the name of the file, which you wish to copy and "Destination file name" is the name of the file, in which the contents have to be copied. For example, if the contents of "MYFILE.TXT" file are to be copied in "MYFILE.BAK" then COPY command will have to be executed, as mentioned below.

**COPY MYFILE.DAT MYFILE.BAK**

## REN Command

REN command is used for renaming an existing file. General syntax for REN command is as follows:

**REN <Existing file name> <New file name> [Enter]**

To understand the usage of REN command, consider the following examples:

**REN DATA.DAT INFO.TXT [Enter]**

This command will change the name of DATA.DAT file to INFO.TXT.

## DEL Command

DEL command is used for removing a file from the disk. In its simplest form, it can be executed as follows:

**DEL <File name> [Enter]**

Here, in this format, "File name" is the name of the file along with drive reference, which is to be removed from the disk. To understand the usage of DEL command, consider the following examples:

**DEL A:DATA.DAT [Enter]**

This command will remove DATA.DAT file from A: drive.

## CD Or CHDIR Command

CD is short form of Change Directory. It is used for moving the current position from one directory to another directory. General format for CD command is as follows:

**CD <Directory name> [Enter]**

Here, in this format "Directory name" is the name of that directory, to which you wish to move. To understand the usage and functioning of CD command, consider the following example:

If current directory has a directory named SECOND in it and you wish to move to that directory then CD SECOND, command will have to be given.

If you wish to move to root directory then CD \ command will have to be given.

If you wish to move to the parent directory then CD .. command will have to be given.

If you execute CD command, without any file name, the name of current directory gets displayed.

## MD Or MKDIR Command

MD is short form of Make Directory. This command is used for creating a new directory. It is used in the following format:

**MD <Directory name> [Enter]**

Here, in this format, "Directory name" is the name that is to be assigned to the newly created directory. For example, if you wish to make a new directory with the name, "LETTERS", following command will have to be given:

**MD LETTERS [Enter]**

Note that when you make a directory, it is made in the current directory

## RD Or RMDIR Command

RD is short form of Remove Directory. This command is used for removing the directory from the disk, provided it is blank. It is used in the following format:

**RD <Directory name> [Enter]**

Here, in this format, "Directory name" is the name of that directory, which is to be removed. For example, if you wish to remove LETTERS directory, following command will have to be given:

**RD LETTERS [Enter]**

## SYS Command

SYS command is used to make a media, like floppy, hard disk etc. bootable. While making it bootable, it copies IO.SYS, MSDOS.SYS and COMMAND.COM files at specific location on the media. Since these files are copied at specific location hence the media becomes bootable. On the other hand, if you copy these files to the media, using COPY command, media will not become bootable. SYS command is used in the following format:

**SYS <Drive name> [Enter]**

Here, in this format, "Drive name" is the name of the drive that contains the media, which is to be made bootable. For example, if floppy in drive A: is to be made bootable, following command will have to be executed:

**SYS A: [Enter]**

## FORMAT Command

Format command makes tracks and sectors on the media, like floppy or hard disk. It is executed in the following format:

**FORMAT [Drive name]**

Here, in this format, "Drive name" is the name of that drive that holds the media, which is to be formatted. For example, if you execute "FORMAT A:" command then the floppy, present in A: drive will get formatted.

## MOVE Command

MOVE command, physically moves files and directories from one place to another. It is used in the following format:

**MOVE <Source file name> <Destination directory>**

Here, in this format, "Source file" is the name of the file, which is to be moved from its current location to another directory. "Destination directory" is the name of that directory, in which the file is to be transferred. For example, if you wish to move SUM file into GTOTAL directory of root then following command will have to be given:

**MOVE SUM GTOTAL [Enter]**

MOVE command can also be used for renaming the directories, as illustrated in the following example:

**MOVE JAN MONTH [Enter]**

Above-mentioned command will rename JAN directory to MONTH.

## DISKCOPY Command

DISKCOPY command is used for duplicating the contents of a given floppy. When you make use of this command, all files and directories of one floppy get copied to another floppy. In its simplest form, it can be executed as follows:

**DISKCOPY <Source drive> <Destination drive>**

Here, in this format, "Source drive" is the name of that drive, which contains the floppy, whose contents are to be duplicated. "Destination drive" is the name of that drive, which holds the floppy, on which the contents have to be copied. For Example, if you wish to copy the contents from drive A: to B: drive then following command will have to be given:

**DISKCOPY A: B: [Enter]**

## UNIX OPERATING SYSTEM

Unix is a multi user, multi tasking, time sharing operating system. It was designed and developed in between 1960 to 1970 by a group of AT & T employees at Bell Laboratories. Bell Laboratories in USA is a research and development wing of AT & T.

UNIX is quite versatile, powerful and standard operating system. It gained its popularity because of its portability and robustness. It is popular till date. Its features provide easy-to-operate working environment.

## ARCHITECTURE OF UNIX

UNIX is multi layered operating system, in which each layer performs specific type of function. A block diagram, showing multiple layers of UNIX operating system is illustrated in figure 7.11.

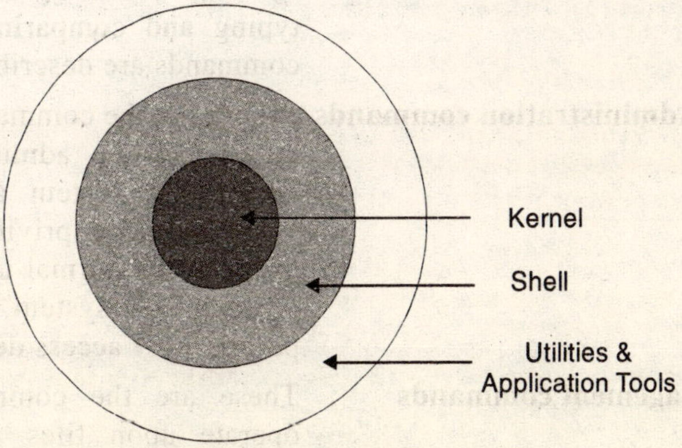


**Figure 7.11**

Refer this diagram and note that UNIX operating system consists of following three layers:

1. Kernel
2. Shell
3. Utilities & Application Tools

Role and functions of each layer are described below.

### Kernel

Kernel is basically group of programs, which directly interact with hardware and perform low level functions like reading the data from disk, transferring the data in the memory, executing programs, displaying the data, keeping track of all the activities that are being performed in the system, called System Accounting, providing security features etc.

### Shell

Shell is software that acts as an interface between user and the computer. When the user gives any command, from his terminal, shell interprets the command. If it finds the command correct then it executes the command, taking the services of the kernel. If it finds that the command is not correct then it displays a corresponding message on user's terminal. In other words, you can say that shell is basically a command line interpreter.

There are many shells available in UNIX. Bourn shell, C Shell, Visual shell etc. are few examples of commonly used shells. Although basic features of all the shells are same and all of them perform similar types of basic functions but they differ in terms of facilities, shell variables, shell programming language and commands. Each shell has its own advantages and disadvantages.

### Utilities and Tools

Unix supports more than 1000 utilities and tools for performing different types of functions. UNIX utilities and tools are classified into following groups:

**Commands for general use** : These are the most commonly used UNIX commands, using which you perform general functions like viewing the directory, copying, moving, deleting typing and comparing the files etc. Details of the commands are described later in this chapter.

**System Administration commands** : These are the commands, which can only be executed by the system administrator or the super user for performing system administration functions. These commands are privileged command and cannot be executed by normal users. It they try to execute these commands, system returns a message, saying permission / access denied.

**File Management commands** : These are the commands, which are designed to operate upon files and performing house keeping functions or extracting file related details.

| | | |
|---|---|---|
| **Text Processing commands** | : | UNIX provides number of commands, using which you can perform text processing functions. vi, ed, ex, awk, sed etc. are few commands, which are very commonly used for such types of tasks. |
| **Communication related commands** | : | Using these commands, either user of the same machine can communicate with each other or two different machines can communicate with each other. |
| **Graphics related commands** | : | Graphics related commands generally come in the form of library of graphics calls. By making uses of these calls in C language programs, you can perform drawing and sketching related tasks. |
| **Software development tools** | : | UNIX comes along with C compiler. On top of this, you may have different language compilers for developing different types of applications. |
| **Application tools** | : | Different types of third party softwares like Oracle, Ingres, WGS office etc. are available on UNIX. |
| **Games** | : | Different types of games come along with UNIX, which are generally available in games directory. |

## FEATURES OF UNIX

UNIX is a powerful operating system. It derives power from its features. Following are some important features of UNIX:

1. UINIX is multi-user operating system. It provides facility for connecting more than one terminal to single CPU. Concept of multi user feature is illustrated in. figure 7.3.

   Since UNIX provides the facility for connecting more than one terminal to single CPU hence multiple users can work on the system simultaneously. The number of terminals that can be attached to the system can vary from 2 to 256 or more.

2. UNIX is multitasking operating. It provides facility to invoke multiple tasks simultaneously, without waiting for the first one to finish. For example, you may do payroll processing and editing of the files together. At the same time, you could also start printing your cash book, without waiting for both of them to get finished. Remember that not only the single user of one terminal enjoys this facility but all the users of all the terminals can make use of this facility, simultaneously

3. UNIX is an operating system, which can run on variety of hardware platforms, without much change. This is because UNIX is written in C language, which is hardware independent language. Recall that prior to UNIX, all the operating systems were written in Assembly language, which is machine dependent language, so all the operating systems were machine dependent and could not be ported to another machines. For implementing the operating system on a new type of hardware, operating system had to be rewritten. UNIX is free from such type of troubles and enjoys portability to a large extent.

4. UNIX is well known for its richness of utilities. It has a collection of over 350 commands and utilities, which are integral part of the operating system. Practically it provides one or more utilities for performing a task.
5. UNIX provides multi level security features. Under first level of security, only authorized persons are allowed to login on to the computer.

   Under second level security, user's files are protected from other users in terms of reading, writing and their execution. When a user gives permission to other users to use his files then only others are able to access them else all the files remain inaccessible to others.
6. UNIX provides redirection features. Using this feature, input or output of a program can be redirected to any other desired destination. For example, ls command of UNIX displays the names of all the files present in current directory on VDU. But using redirection feature, this output can be redirected to the printer so that the names get printed on the printer. Similarly, say a program, PAY takes data from the keyboard. Its input can be redirected to a file so that it reads the data from specified file.
7. UNIX provides pipe feature. Using this feature, output of a program can be connected to the input of another program. This feature enables us to make powerful use of existing commands and build new commands from them.
8. UNIX provides hierarchical mountable file system. This file system provides easy expandability of the hardware and unrestricted growth of files and directories on the disk, under super user controlled working environment.
9. Concept of super user has been incorporated in UNIX to build secured and systematic working environment.
10. System accounting feature is part of UNIX operating system. This feature keeps track of how all the system resources were used by the users who logged on to the computer.
11. Spooling is a powerful UNIX feature. Using this feature printing tasks are performed efficiently in controlled environment. In the event of multiple printers being connected to the computer, any user can print his printouts on any desired printer. Spooler also allows super user to perform printer management tasks like adding a new printer to the system, removing an existing printer from the system, shifting printing queue of a printer to another printer, changing printing priority of a printing task etc.

## UNIX COMMANDS

UNIX has an ocean of commands and utilities. They can be executed by typing the command at the command prompt in the following format and pressing Enter key.

**Command Name [-switch] <enter>**

Here Command Name is the name of the command that you wish to execute.

Switch is a valid alphabet that produces some variation in the output. Note that switch is optional that means if you wish, you can include it in the command or you can omit it also.

For example ls is valid Unix command. It displays the name of all the files and directories present in current directory. Without any switch it can be executed on follows:

**ls <Enter>**

On execution it will display the output in following format:

| | | | |
|---|---|---|---|
| **data.dat** | **myfl.txt** | **text.dat** | **upper.dat** |
| **jan.dat** | **feb.txt** | **mar.dat** | **employee** |
| **dir1** | **yurl.txt** | **dir2** | **zen.txt** |

-l is a valid switch, which can be used along with ls command as mentioned below:

**ls -l <Enter>**

On execution, this command will generate the output as shown below:

```
[Ajit@localhost Ajit]$ ls -l
total 1200
-rw-rw-r--   1 Ajit   Ajit        23 Apr 21 15:09 abc
drwxrwxr-x   2 Ajit   Ajit      4096 Apr 21 15:34 Ajit
-rw-rw-r--   1 Ajit   Ajit     33280 Apr 21 15:36 ajit1.doc
-rw-rw-r--   1 Ajit   Ajit    158208 Apr 21 17:36 ajit2.doc
-rw-rw-r--   1 Ajit   Ajit     74240 Apr 20 17:02 ajit.doc
drwxrwxrwx   5 Ajit   Ajit      4096 Mar  4 14:23 CBOOK
-rw-rw-r--   1 Ajit   Ajit    173056 Mar 23 10:37 Cchap4.doc
```

Compare both the outputs from ls command and note the variation that is caused by -l switch. Often more than one switch are applicable for a command. Few commonly used commands of UNIX are described below.

| Command | Example | Description |
|---|---|---|
| ls | ls <Enter> | Displays the name of all the files present in current directory. |
| ls -l | ls -l <Enter> | Displays the names of all the files present in current directory, in long format. |
| ls- a | ls -a <Enter> | Displays names of all the files, including special files (that otherwise remain invisible) present in current directory. |
| mkdir | mkdir x <Enter> | Makes a new directory x. |
| cd | cd x <Enter> | Changes the current position to specified directory x. |

| Command | Example | Description |
|---|---|---|
| rmdir | rmdir x <Enter> | Removes specified directory x if it is empty. |
| mv | mv f d <Enter> | Moves file f from its current location to another directory d. |
| mv | mv d1 d2 | Renames directory d1 to d2 |
| passwd | | changes the current password. While changing the password. |
| exit | | Terminates current session. |
| lp -d | lp -d p <Enter> | Print file f on printer p. |
| sleep | sleep n <Enter> | Halts the process for n seconds. |
| chmod | chmod <who> <operator> <mode> <file> | Changes file access permissions. In this command, who can either be u, g, o or a i.e. user, group, other or all. The operator can be +, - or = i.e. adding, removing or assigning the permission. The mode can be r, w or x i.e. read, write or execute. For example, if for file abc read and write permission is to be given to others, then **chmod o + rw abc** command will have to be executed. |
| ps | ps <Enter> | Displays status of all the processes running on the computer that are related to you. |
| ps-e | ps -e <Enter> | Display status of all the processes of all the users that are running on the computer. |
| ps -f | ps -f <Enter> | Same as ps but more details are displayed when switch f is specified. |
| kill -9 | kill-9 p1 <Enter> | Kills process p1 immediately |
| date | date <Enter> | Displays current system date and time. |
| who | who <Enter> | Displays names of all the users, who are currently working on the system. |
| whoami | whoami <Enter> | Display login name of the user, who invokes the command. |
| pwd | pwd <Enter> | Print the name of current working directory. |
| vi | vi f <Enter> | Edits file f. |
| we | we f1 <Enter> | Displays the number of lines, number of words and number of characters that constitute file f1. |
| cat | cat f1 <Enter> | Display the contents of file f1 |

| **Command** | **Example** | **Description** |
|---|---|---|
| cat | cat f1 f2 > f3 <Enter> | Adds the contents of file f1 and f2 and puts them in file f3. |
| cmp | cmp f1 f2 <Enter> | Compares two files f1 and f2 and prints the difference. |
| cp | cp f1 f2 <Enter> | Copies the contents of file f1 into another file f2. |
| sort | sort f1 <Enter> | Sorts files f1 alphabetically. |
| split | split -n f <Enter> | Splits file f1 in n line pieces. |
| rm | rm f1 <Enter> | Removes file f1. |
| grep | grep 'pattern f1 <Enter><br>e.g.<br>grep '*' f1 <Enter> | Displays only those lines of file f1, which have specified pattern in them e.g. if you mention * in place of pattern then it will display those lines of file f1 which contain * in them. |
| chown | chown u1 f1 Enter> | Changes the ownership of file f1 to another user u1. |

# WINDOWS OPERATING SYSTEM

Windows is the most popular operating system of modern times. It has been designed and developed by Microsoft Corporation of United States of America for Personal Computers.

Windows is single user multitasking GUI based operating system. It not only activates computer and makes its devices functional, but it also provides numerous applications for performing wide variety of tasks.

## File Naming Convention In Windows

As you know, file is a medium to store data and information on computer. File is always identified by its name and the same name is used while operating upon it.

Recall that MS-DOS allows maximum 8 characters in primary name and three characters for extension name. Primary name and extension name are always separated by dot (.). The pattern of file names in Windows is slightly different from MS-DOS. Following points apply on file names in Windows:

1. File name can be up to 255 characters long.
2. Blank and dot (.) can also be used in file names. For example "My File" is a valid file name.
3. Characters like + ; [ = ] etc. can also be used in file names
4. Characters like \ / : * ? " are still prohibited in file names.

## Concept Of Folder

Folder is a special file, which is capable of holding other files and folders into it. In MS-DOS operating system they were called directory. In Windows, they are called folders.

## Features Of Windows Operating System

Following are the features of Windows operating system:

### GUI based Operating System

Windows is Graphical User Interface based operating system. It provides very simple, yet powerful and interesting mechanism for interacting with computer. It is a user-friendly operating system.

### Multitasking Operating System

Windows is multitasking operating system. Thus in Windows, you can perform multiple tasks simultaneously. For example, while typing the letter through keyboard, you can view a movie on VDU and print the report on printer too.

### Drag And Drop

Drag and Drop is a powerful feature of Windows. Using this feature, you can drag objects like files, folders etc. from their original location and drop them at some other place, with the help of mouse so that they could either be copied or moved there. For example, dragging file A, present in folder F1 and dropping it on folder F2 will move the file A from F1 to F2.

### CUT COPY And PASTE

Cut, Copy and Paste is powerful feature of Windows. Using this feature, selected items like text, graphics, icons etc. can either be copied from their original place to a temporary memory location called Clipboard or from Clipboard to the desired location. Thus you can either move these items from one place to other or make duplicate copies of them. For example, when you select an item and perform CUT operation on it, it physically moves from its original place to the Clipboard. When you move to some other location and perform PASTE operation there, the item from Clipboard gets copied to the current location. Thus at the end of CUT-PASTE operation, selected item changes its place. Similarly when you perform COPY operation on a selected item, it gets copied into Clipboard (doesn't get removed from its original place, as it happens in the case of CUT operation). After this, when you change your current location and perform PASTE operation there, contents of Clipboard get copied to the new location. Thus a duplicate copy of the selected description gets made.

### Object Embedding

This is one of the most powerful features of Windows operating system. It enables two or more programs to exchange objects like text, graphics etc. among them. For example, a drawing made in Paint program could be brought into the document, created through WordPad program so that figure could be embedded in the text. For example, you may type the description of Kutubminar in WordPad program and may draw the figure in Paint program. Using object embedding feature of Windows, you could place the figure of Kutubminar in Kutubminar text, so that it becomes complete text in itself.

### Object Linking

Object linking is yet another powerful feature of Windows operating system. Using this feature, one object can be linked to the other object. Thus whenever linked object is modified or changed modifications/changes get visible in other object also. For example, you may type the description of Personal Computer in WordPad program and draw its figure in Paint program. Using object linking feature, you may link the computer figure to computer description so that it appears in the document (as it did in case of object embedding). Now whenever you will make modifications / changes in computer figure they will automatically get reflected in the description. You may note that overall effect of object embedding and object linking is same but automatic reflection of changes / modifications doesn't take place in case of object embedding.

### Plug And Play

It is an important feature of Windows operating system. It makes the operating system, slightly intelligent. Whenever you attach a new device to the computer, operating system automatically senses its attachment and loads its device driver. Thus due to plug and play feature, new devices become automatically operational in Windows operating system.

### Command Prompt

Using this feature, Windows provides MS-DOS working environment, within Windows environment. Thus all MS-DOS commands can also be executed in Windows.

## Useful Programs Of Windows

Windows is a powerful operating system. It not only makes the computer functional, it also provides many useful programs. Using these programs, you can perform different types of operations. Following are commonly used programs of windows:

**Notepad** This program is used for creating small text files.

**WordPad** This program is used for creating large text files. It also provides many formatting options, using which the document can be formatted and made attractive.

**Paint** It is basically a drawing and coloring program. Using this program, you can create shapes and drawings and color them with attractive colors.

**Calculator** Using this program, you can use your computer as calculator and perform arithmetic and scientific calculations.

**Media Player** Using this program, you can play audio and video files on you computer.

**Movie Maker** Using this program, you can put graphical pictures in sequence and prepare multimedia presentations.

**Windows Explorer** Windows Explorer is a program, using which file and folder maintenance activities are performed on computer. Creating new folders, deleting old folders, moving or copying files from one folder to another are the activities, which are easily performed with the help of Windows Explorer.

**Internet Explorer** Internet Explorer is basically a browser, using which activities like net surfing, net search, e-mail etc. are performed.

Different versions of Windows are available for use. They all are look-alike GUI based operating systems. They support all above mentioned features and utilities, but differ from each other in terms of their capabilities and additional facilities.

## EXERCISES CHAPTER 7

### Short Type Questions

**A. Select most appropriate answers for following questions:**

1. Which software activates the computer?
   (a) Commands (b) User Interface (c) Operating System (d) Utilities
2. What is booting?
   (a) Carrying out a self test (b) Loading operating system in computer's memory
   (c) Checking computer's memory (d) Executing a command
3. Which of the following management function is not performed by operating system?
   (a) Information management (b) Software management
   (c) Device management (d) Memory management
4. Which of the following is single user multitasking operating system?
   (a) MS-DOS (b) Windows-XP (c) Windows-2000 (d) Unix
5. Which of the following is not a GUI component?
   (a) Radio button (b) Menu (c) Command (d) Text box
6. Which type of user interface does MS-DOS provide?
   (a) GUI (b) CUI (c) ABI (d) MGI
7. Which type of user interface is provided in Windows-XP?
   (a) GUI (b) CUI (c) ABI (d) MGI
8. What that feature is called, which permits a user to execute multiple programs simultaneously?
   (a) Multi user (b) Multi tasking
   (c) Multi programming (d) Multi process

9. Maximum how many options can be selected from a radio button list?

   (a) As many as required (b) all the option present in list

   (c) 1 (d) 0

10. Maximum how many options can be selected from a check box list?

    (a) As many as required (b) all the option present in list

    (c) 1 (d) 0

## B. Fill in the blanks:

11. .................................... is the software, which enables the computer to perform read / write functions.
12. That user interface, in which user interacts with computer, using commands is called .................... ................... ......................
13. That user interface, in which user interacts with computer, using graphical items is called ................. .................. ......................
14. ...................................... is an example of network operating system.
15. .................................. is an example of multi user operating system.
16. GUI is acronym for ................... ............... ....................
17. ........................... command of MS-DOS will be used to view the names of all the files present in current directory.
18. Directory in MS-DOS serves same purpose, which ................. does in Windows.
19. In Linux ................................. command is used for creating a new directory.
20. In MS-DOS ...................... command can be used for viewing the contents of a file.

## C. State, true or false for following statements:

21. UNIX operating system was developed by Microsoft corporation of USA.
22. MS-DOS is single user, multi tasking operating system.
23. Windows is single user multi tasking operating system.
24. UNIX provides CUI interface to interact with computer.
25. Process management function is one of the responsibilities of operating system.
26. Multi-user operating system and Network operating system, both, provide the facility that multiple users could use the same system.
27. UNIX is an example of multi user operating system.
28. In GUI, user has to type command for interacting with computer.
29. Writing data and reading it are memory management activities that operating system performs.

## D. Answer the following questions in short:

30. In which type of user interface, user has to type the command to interact with computer?
31. In which type of user interface, user has to interact with computer through graphics and pictures?

32. Name at least four types of management functions that operating system performs.
33. Name any two operating systems, which are multi-user operating systems.
34. Name any two operating systems that are network operating systems.
35. Name that program of Windows, using which small text files are created.
36. What do you call that software, which activates computer?
37. What do you call that process, which loads the operating system into computer's memory?
38. Name any five GUI components that are generally provided in any GUI based operating system.
39. Which command of UNIX can add the contents of two files and put them in a new file.?
40. What is Windows-NT?

## Detailed Answer Type Questions

### E. Answer the following questions in detail:

41. What is operating system and what is its purpose? Name any four popular operating systems?
42. What do you understand by following? Explain them in brief, giving a suitable example:

    (a) Process management function. (b) Information management function
43. What is booting? Why it is required?
44. What is the function of following MS-DOD commands? What are their equivalent commands in UNIX?

    (a) DIR (b) CD (c) TYPE (d) COPY
45. How MS-DOS is different from Window? Give at least two arguments.
46. Describe multitasking feature? What is plug and play feature of Windows operating system?
47. What do you understand by multi user feature of operating system? Name any two operating systems that support this feature.
48. What is CUI interface? Explain it by giving a suitable example.
49. What is GUI interface? Explain it by giving a suitable example.
50. Describe at least three features of multi-user operating systems. How multi-user operating systems are different from network operating systems?
51. Describe object embedding features of Windows.
52. Describe object linking feature of Windows.
53. Differentiate between object embedding and object linking features of Windows.
54. Describe cut, copy and past features of Windows.
55. Explain at least 5 new features of Windows operating system.

****

# COMPUTERS

## CHAPTER 8

# *Types of Computer Language*

For every communication or interaction that takes place between two identities, there has to be a language. Language of communication can either be a natural language like Hindi, English etc. or a symbolic language, like the one that is used for deaf and dumb people, in which gestures are made by hands and lips.

Be it any language, the rule for communication is that both the persons should understand the language. For example, if you have to communicate with a person in English, he should know English. Unless he knows English, you will not be able to communicate with him.

Now let's take the case of our communication with computer. In this communication, following conditions prevail:

1. Computer understands only the language of 0s and 1s i.e. machine language.
2. We understand few natural languages like Hindi, English, Tamil etc.

Till the time such conditions will prevail no communication between computer and us will take place. Neither the commands given by us, in English, will be carried out by computer nor the results displayed by it, in 0 and 1, will be understood by us. Thus a communication gap will prevail. In order to communicate with computer, somehow this gap will have to be filled.

**Apparent solution for communication gap**

Now the question is how proper communication can take place between computer and us? Probably the two solutions that strike the mind, at first instance, are as follows:

1. We communicate with computer in its language of 0s and 1s. That means, we should learn the language of 0s and 1s and give all the instructions to the computer in this language only.
2. Make the computer capable of understanding:

   (a) all the words of the dictionary

   (b) all grammatical rules

   (c) meaning of the sentences.

The first solution mentioned above is not practical from our (user's) point of view. It is a difficult language. What to say of big and complicated tasks, even for the smallest task, long series of 0s and 1s will have to be typed on computer and executed. A single mistake here and there would ruin the task. So for all practical purposes, this is not a feasible solution.

Second solution mentioned above is also not practical, because it will make the computer very large and unaffordable.

**Practical solution for communication gap**

Now the question is, if both the methods are infeasible then what is the practical solution for the problem? Well, to solve the problem, mid way path is taken. Not all the words from the dictionary are chosen, but only few selected words are taken (about 30 to 40 words). For this limited vocabulary few new rules of grammar are framed. For all practical purposes, communication with computer is kept limited to these words and rules only.

These words and rules are so chosen that commands for almost all the operations that have to be performed through computer can get framed using them. In this solution the only thing, which comes in way, is that for single task we may have to give multiple instructions. Because due to limited vocabulary, there may not be a single instruction available in newly framed language to complete the task. To get a feel of the method, recall the vocabulary of a child who has just learned speaking. Although he has limited words at his disposal, but somehow he manages to communicate almost every thing. Similarly with the help of this limited vocabulary, we can write any program for performing any task through computer. This limited set of words along with specific rules of grammar is called computer language.

When the words and the rules of the grammar change, a new language is formed. COBOL, BASIC, FORTRON, Pascal, C, C++, Java etc. are some examples of computer languages. Note that each language is designed for a specific type of work and it delivers best results if that type of work is done, using that language. For example, COBOL language was designed for commercial applications, which involved lot of data and formatted report printing. So the programs for such types of applications can be best written in COBOL language. Similarly FORTRAN was specifically designed for scientific applications, which involve less data volumes but lot of calculations. For example, matrix inversion, regression analysis etc. are the applications that require lot of calculations. Thus programs for them can be very well written in FORTRAN language. Each language has its own advantages and disadvantages. This is the reason why we have so many computer languages.

## NEED FOR INTERPRETER

Note that in above mentioned process, we have only restricted ourselves to the use of few words, instead of full dictionary. We have not filled the communication gap, because these words are still in English and computer understands the language of 0s and 1s only.

Now the question is how do we fill this gap? Well, the gap is filled by using the same process, which is adopted, when two persons of different language-land communicate with each other. Recall that they take services of Interpreter. Interpreter listens to one person and translates his sentences for the other person. Thus both of them communicate through interpreter.

In computer also, an interpreter or compiler is designed for each language. It performs following activities:

1. Takes English like commands from us.
2. Converts them in 0 and 1 equivalent form, so that computer could understand the command and execute it.

## COMPUTER PROGRAM

When few sentences are framed in computer language, for the purpose of instructing the computers to perform a specific task then all those sentences put together are called computer program.

For example, 10 sentences written in FORTRON language for finding LCM of given numbers will be a program. Similarly other 25 sentences written in C language for adding two arrays will be another program.

## INTERPRETER

The software that converts a computer program into machine language is called interpreter. It follows following strategy:

1. It reads first line of the program.
2. Converts it into machine language.
3. After conversion it executes that line.
4. After execution, it reads the next line and repeats above mentioned processes.
5. Reading conversion and execution of lines continues one by one, till the last line of the program is executed.

In short, it can be said that interpreter follows step-by-step approach for execution of the program.

## COMPILER

Compiler performs the same task that interpreter performs. It converts computer program into equivalent program written in 0s and 1s. But it performs this task differently. It adopts following strategy:

1. It reads all the lines of the program in one shot.
2. Converts the whole program in 0 and 1 equivalent form.
3. Executes the whole program line by line.

In short, it can be said that compiler follows one shot approach for execution of the program.

## CLASSIFICATION OF COMPUTER LANGUAGES

Computer languages are broadly classified into following two categories:

1. Low level languages.
2. High level languages.

The language of 0 and 1 is called machine language. **Those computer languages, which are quite near to the language of 0 and 1, are called low level languages.** Machine language and Assembly language are the examples of low level language.

**High level languages remain quite near to natural languages** and they use English like words. It is quite easy to write programs in these languages. It takes lesser time and fewer words to write the program in these languages. C, C++, BASIC, COBOL, FORTRAN, Pascal, Visual Basic, Java etc. are few examples of high level language.

## CONVENTIONAL COMPUTER LANGUAGES

Computer languages like BASIC, COBOL, FORTRAN, Pascal etc. are called conventional compiler language.

They follow conventional method of creating data structures and focusing the attention on flow of data and manipulating the contents of data structures, as per requirement. For example, if contents of two arrays are to be added then conventional languages would require following activities to be done within the program:

1. Create the arrays
2. Store data into arrays
3. Perform all logical steps that will add the contents of the array.
4. Display the result.

On the other hand, APL (acronym of **A** **P**rogramming **L**anguage) is a non conventional programming language.

It is array programming language, base on notations invented in 1957. This language is used for developing vector related applications. Such applications find use in advanced field of mathematic, science, engineering, robotics, data visualization etc. Unlike conventional programming (mentioned above), APL programs are written as chains of dynamic functions and operators.

## ADVANTAGES OF CONVENTIONAL LANGUAGES

Conventional languages are quite popular in computer world. They are commonly used for developing applications. Following are the advantages of conventional computer languages:

1. They are easy to learn.
2. They follow simple approach for declaring and processing the data.
3. Simple applications can be developed very fast, using conventional languages.
4. Programs written in conventional languages can be written, debugged and maintained easily.

5. They do not require special hardware. Conventional computers are sufficient for their working.
6. Their cost is generally low.

## LIMITATIONS OF CONVENTIONAL LANGUAGES

Following are the limitations of conventional languages:

1. They deal with only straight forward problems.
2. Only conventional type of data, like name, age, salary etc. can be processed by them.
3. Data that has multi dimensional aspects associated with it cannot be handled by them. For example, vector processing applications cannot be developed using conventional computer languages.
4. Knowledge based applications like artificial intelligence, expert systems, which require special hardware and processing can not be developed using conventional programming languages.

## INTRODUCTION TO C LANGUAGE AND C PROGRAMMING

Till quite some time back, low level languages like machine language, assembly language etc. were the only languages using which instructions could be provided to the computer for performing its internal functions. High level languages were exclusively used for automating human tasks. Truly speaking, there was no single computer language available; using which both types of tasks could be performed.

In the year 1970, two computer scientists Dennis Ritchie and Brain Kernighan, at Bell Laboratories of USA, developed C language. It was quite powerful computer language. Its scope of operations was very wide. It not only provided means and commands for writing programs for computer's internal functioning but it also provided instructions for automating manual tasks. In other words, you can say that it had the penetrating power of low level language and had the features of high level language too. Within short span of time, C language became quite popular all across the world. Although it came into existence, long time back, but it is still being used.

### Components of C Language

C language is basically a very small instruction set, coupled with few standard rules and predefined formats. To write a C program, one needs to use appropriate instructions and follow the standard rules.

To make the C programming easy, many programs for primitive functions (that need to be performed in most of the cases) have already been written and compiled in the form of libraries (also called header files). These libraries remain available for public use. Over a period of time these libraries have become industry standard. Programs of these libraries are so commonly used that they are almost considered as part of C language. stdio.h, string.h, stdlib.h etc. are the names of few standard C libraries. Likewise there are many other libraries, which are considered as part of C language. Complete shape of C language is illustrated in figure 8.1.

To make use of any program available in library, the name of the library need to be included in the main program. Following is the format, using which function libraries are included in any program:

```
#include <library-name>
```

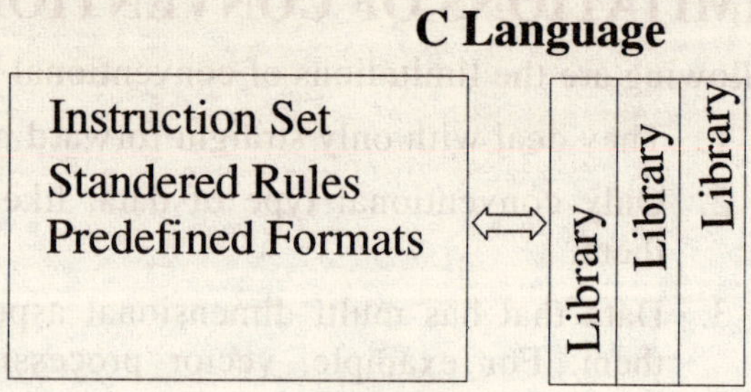


**Figure 8.1**

Here #include is keyword, which is to be written as such and library-name is the name of that library, which is to be included in the program. For example, to include stdio.h library in the program, following statement will be written, at the beginning of the program.

```
#include<stdio.h>
```

## Structure of C Program

All C programs are written in the form of function. Structure of C program is illustrated below:

```
#include<library-name>
main()
{
Statement-1;
Statement-2;
:
:
Statement-n;
}
```

Refer above illustrated structure of C program and note that all the libraries that need to be used within the program are include at the beginning of the program, using #include statement. The function, from where the execution will start is named as main(). The scope of main() function extends from open curly bracket ({) to closing curly(}) bracket.

## C Programming Rules

Following are the rules of C programming. They should be strictly followed. In case of any violation, compilation error occurs and the program doesn't execute.

1. A C program may consist of one or more functions. One of them must be named as main(). Execution of the program always starts from main() function.

2. All the statements of main() function should be written, within the opening and closing brackets, which mark the beginning and end of the function, respectively.
3. Practically there is no limit to the number of statements that could be included in the main() function. However to make the program manageable, use of multiple functions should be promoted instead of too many lines in one function.
4. Each statement should be terminated with a semicolon sign (;).
5. C is case sensitive language. A differentiation is always made between upper case and lower case letters. For example, Andhra, ANDHRA and andhra will be treated as three different identities in C language.
6. Comments can always be included in the program by placing the matter in between /* and */ delimiters. Following is the general format for placing comments in the program:

   **/* Comments */**

   **e.g. /* This program calculates simple interest */**

Now you are familiar with the rules and structure of C program, so let's write a simple C program, which multiplies speed by time and prints the traveled distance.

```
#include<stdio.h>
main()
{
/* This program calculates distance
   for given speed and time. It also demonstrates
   the general structure of C program */
int speed , time, distance ;
speed = 75 ;
time = 4 ;
distance = speed * time ; /* Calculating distance */
printf("\nDistance is %d ", distance) ; /* Displaying distance */
}
```

Read above mentioned program carefully and note the following points:

1. The speed, time and distance are the variables, which are utilized in the program for holding the values.
2. As a rule of C programming, all the variables that are used in the program should be declared before they are used. Note that this is done through the statement, "**int speed, time, distance**". This statement states that these three variables will hold integer (int) values.

3. printf is standard output function, which displays the value of said variable on the VDU. It remains available in stdio.h library, which has been included at the beginning of the main() function.

## Data Types In C language

Following four basic data types are permitted in C language:

1. Integer denoted as int.
2. Character denoted as char.
3. Real denoted as float.
4. Real with double precision denoted as double.

Other user defined data types can be generated from these basic data types. Few qualifiers, such as short, long, signed and unsigned etc. can augment int data type. For example, integer quantities can be defined as:

```
short int
long int
unsigned int
signed int
```

Short int will require less memory space than long int because long int requires double the amount of memory.

If the need be, two qualifiers could be used together so as to make a logically correct statement:

```
unsigned short int
unsigned long int
```

## Declaring Variables

Following format is used for declaring variables in C program:

```
data type variable name ;
```

Following are few examples of valid variable declarations:

```
Int roll
int  rate, quantity,  amount ;
int age = 45 ;
float area=10.5,  radius=2.45, perimeter=200.00 ;
char     religion_code1, religion_code2, religion_code3, religion_code4 ;
```

## Operators

C is operator rich language. It provides wide variety of operators. C operators are described below:

### Arithmetic Operators

To operate upon numerical values, C provides, following operators:

| Operator | Symbol | Purpose |
|---|---|---|
| Addition operator | + | Adds given values. |
| Subtraction operator | _ | Subtracts one value from other value. |
| Multiplication operator | * | Multiplies given values. |
| Division operator | / | Divides one value by other. |
| Modulus operator | % | Provides remainder value that results from integer division |

### Unary Operators

Those operators, which operate upon single operand, are called unary operators. Following unary operators are available in C language:

### Unary Minus (-)

Unary minus operator is denoted by minus sign (-).

Although arithmetic subtraction operator and unary minus appear to be same but truly speaking, they are different. Subtraction operator requires two operands, while unary minus requires only one operand. Unary minus converts positive value to negative value. For example, -150, -(-700) etc .are the examples of unary minus.

### Increment Operator (++)

Increment operator operates upon numerical values or numerical expressions. It increases their values by one. For example, if i=100 then i++ will give 101.

Note that ++ operator can either be used as ++i (pre increment) or i++ (post increment). In both the forms, same result is obtained but the time of execution differs.

In pre increment case (i.e. ++i), the value of the operand (ie i) gets incremented before the execution of the statement in which it is used. For example, if i=100 then print ++i will print 101.

In post increment case (ie i++) the values of the operand (ie i) gets incremented after the execution of the statement, in which it is used. For example, if i=100 then print i++ will print 100 and then increment the value of i to make it 101.

### Decrement Operator (--)

Decrement operator works in the same way as increment operator does, except that it decreases the value of its operand. --i and i-- are the examples of pre decrement and post decrement operators.

### Relational Operators

Relational operators operate upon an expression and evaluate it to find whether it is true or false. In the event of expression being true, they return 1 else if it is found false they return 0.

Following relational operators are available in C language. Their usage with examples and results is illustrated in following table.

Assume a=100, b=50, c=150, d=50

| Operator | Symbol | Usage | Result | Value Returned |
|---|---|---|---|---|
| < | Less than | a<b | False | 0 |
| <= | Less than or equal to | b<=c | True | 1 |
| > | Greater than | a>b | True | 1 |
| >= | Greater than or equal to | b>=c | False | 0 |
| = = | Equal to | a= =b | False | 0 |
| | | b= =d | True | 1 |
| != | Not equal to | a!=b | True | 1 |
| | | b!=d | False | 0 |

## Logical Operators

Following are logical operators provided in C language:

<table>
<tr><th>Operator</th><th>Symbol</th><th>Purpose</th></tr>
<tr><td>AND</td><td>&&</td><td>Connects two conditions and gives true only if both the conditions are true.</td></tr>
<tr><td>OR</td><td>||</td><td>Connects two conditions and gives false only if both conditions are false.</td></tr>
<tr><td>NOT</td><td>!</td><td>Negates the given condition.</td></tr>
</table>

## Simple And Compound Assignment Operators

Assignment operators, operate upon operands and assign the resultant value to a variable. Following are the assignment operators, available in C language. Their usage, with example is summarized in the following table:

Assume that x=100, y=200

| Operator | Expression | Equivalent Expression | Result |
|---|---|---|---|
| = | z=x | | z=100 |
| += | x+=50 | x=x+50 | x=150 |
| -= | y-=10 | y=y-10 | y=190 |
| *= | x*=3 | x=x*3 | x=300 |
| /= | y/=2 | y=y/2 | y=100 |
| %= | a%=3 | a=a%3 | a=1 |

### The Conditional Operator (?:)

Conditional operator, denoted by ? sign, is ternary operator. It requires three expressions and is used in the following format:

```
expression1 ? expression 2 : expression3 ;
```

While evaluating the conditional expression, expression1 is evaluated first. If expression1 is found to be true then expression2 is evaluated and its value becomes the value of the whole expression. If expression1 is found false then expression3 is evaluated and its value becomes the value of the whole expression. To understand the usage of conditional operator, consider the following code:

```
total = (x==y) ? 700 : 900 ;
```

When x will be equal to y then the expression will evaluate to 700. Thus the value of total will become 700 (i.e. total=700). On the other hand, if x is not equal to y then the expression will evaluate to 900. Thus the value to total will become 900.

### Bitwise Operators

As mentioned earlier, C language is a powerful and versatile language, it provides facilities for low-level programming and high-level programming too. These facilities are extended to the user through bitwise operators. Following types of bitwise operators are available in C language:

1. One's complement operators
2. Bitwise logical operators
3. Bitwise shift operators

## Statements and Compound Statements

All the statements, in C language have to be terminated with a semicolon (;) sign. C statements can be broadly classified into following two categories:

1. Simple statements.
2. Compound statements.

Simple statement comprises of single statement, followed by a semicolon. For example, "int speed ;" is a simple statement. Similarly, "distance=speed*time;" and "printf("%d",distance) ;" are two other statements.

Compound statement comprises of more than one statement enclosed within { and }. For example, following is a compound statement:

```
{
    distance=speed*time ;
    printf("%d",distance) ;
}
```

## Reading Data

C language provides many functions for reading the data. The functions remain available in stdio.h library. To make use of these functions, you need to include this library at the beginning of your program, using following command:

```
# include <stdio.h>
```

Following are the functions, using which data is read in C program:

### getchar function

The getchar function accepts single character from the keyboard. It is used in the following format:

```
character variable = getchar()
e.g.  a=getchar() ;
```

This command will take one character from the keyboard and assign it to the variable a.

**scanf function**

Function scanf accepts all types of data in the program. General format for scanf function is as follows:

```
scanf (control string, variable1, variable2, variable3---------)
```

Here control string is a string of characters, which contains information of the format, in which data will be accepted from the keyboard. The variable1, variable2, etc. are the variables, which will hold the values that will be given from the keyboard, when the program runs.

Control string may consist of multiple % signs and conversion characters. One % sign and one conversion character put together is called character group. Note that one character group will hold formatting information for one data item only (i.e. one variable only). For example, if four data items are to be accepted from the keyboard then scanf statement will have four character groups and four variables in it. Each character group will consist of one % sign and one conversion character. Thus control string of such a statement will have the following format:

```
%conversion character %conversion character %conversion character%conversion character
```

First character group (i.e. % conversion character) will be associated with first variable, second character group will be associated with second variable and so on.

Now the question is, what are conversion characters? Conversion characters are the characters that indicate the type of data that the corresponding variable has to hold. In other words you can say that it indicates the type of data that is supposed to be entered from the keyboard, for this variable. For example, first conversion character will indicate the type of data that first variable has to hold. The second conversion character will indicate the type of data associated with second variable and so on. Note that different characters are used for different types of data. Following table illustrates a list of conversion characters that are used for accepting different types of data.

| Conversion Character | Meaning |
|---|---|
| c | Data item is a single character. |
| d | Data item is a decimal number. |
| e | Data item is a floating point value in scientific notation. |
| f | Data item is a floating point value in normal notation. |
| i | Data item is an integer. |
| h | Data item is a short integer. |
| s | Data item is a string followed by white space characters. |
| u | Data item is an unsigned decimal integer. |

So if the first value to be entered is an integer, second is floating point, third value is again a floating point and the fourth value is a character then the control string will have to be written as follows:

**"%i %f %f %c"**

Note that control string is always written within double quotes. Another important point to be noted for scanf function is that, while writing the variables within the bracket, in scanf statement, each variable name is preceded by an & sign. For example, variable names for above mentioned control string will be written, as mentioned in the following scanf statement:

**scanf ("%i %f %f %c", &pay &da &itax, &religion);**

Here pay, da, itax and religion are the variables of int, float, float and char type respectively, which will hold the respective values.

## DISPLAYING DATA

C language provides many functions for displaying the data. These functions remain available in stdio.h library. To make use of these functions, you need to include this library at the beginning of your program, using following command:

**# include <stdio.h>**

Following are the functions, using which data is displayed in C program:

### putchar function

The putchar function displays single character on the monitor screen. It is used in the following format:

**putchar(variable)**

**e.g. putchar(a) ;**

This command will display the value of variable a on monitor screen.

## printf function

The printf function displays all types of data on the screen. General format for printf function is as follows:

**printf (control string, variable1, variable2-----------)**

Here control string contains formatting information and variable1, variable2 etc. are the variables, values of which have to be displayed. Control string of printf function is formed in the same way as that of scanf function (explained in the last section). The characters used for forming the control string of printf statement are listed in the following table:

| Conversion Character | Meaning |
|---|---|
| c | Data item is displayed as a single character. |
| d | Data item is displayed as decimal number. |
| e | Data item is displayed as floating point value with an exponent. |
| g | Data item is displayed as a floating point value, using either e-type or f-type conversion, depending upon the value. Trailing zeros and trailing decimal points will not be displayed. |
| f | Data item is displayed as floating point value without an exponent. |
| i | Data item is displayed as integer. |
| s | Data item is displayed as string. |
| u | Data item is displayed as an unsigned decimal integer. |

For example, for printing four variables i.e. pay, da, itax and religion, which are of integer, float, float and character data types respectively, following printf statement will be written:

**printf ("%i %f %f %c", pay da itax, religion);**

Following program illustrates the usage of printf and scanf functions. It takes roll number, physics, chemistry and maths marks from the keyboard and then calculates and prints percentage.

***Program***

```
#include <stdio.h>
main()
{
    int roll, phy, chem, maths, total ;
    float per ;
```

```
        printf ("\n Give Roll number");
        scanf ("%i", &roll);
        printf("\nGive Physics, Chemistry and Maths marks ");
        scanf ("%i %i %i", &phy, &chem, &maths);
        total = (phy + chem + maths) ;
        per = total * 100.00 / 300.00 ;
        printf ("\nTotal  marks are %i + %i + %i  %i ", phy, chem, maths ,total) ;
        printf ("\n Percentage = %f ", per) ;
        printf ("\n Percentage = %g ", per) ;
}
```

# DECISION MAKING

The **if** command is used for making decisions in C programs. It is used in the following format:

**if (expression) statement**

Here statement can either be simple statement or compound statement, as mentioned in following examples:

```
if (marks>75) printif ("Distinction")
if (Sale>=10000)
    {
        comm=1000;
        incentive=200;
        increment=30;
    }
```

## if else Statement

The **if** statement can also include an **else** statement, which specifies the statements that have to be executed when the given condition is found false. It is used in the following format:

**if (expression) Statement 1 else statement2**

Statement 1 and Statement 2 can either be simple statements or compound statements.

## Nested if Statements

In nested if statements, another if statement is placed either in if branch or in else branch or in both the branches. Thus there are various versions of nested if statement. Their formats and examples are given below:

***Format 1***

```
if (expression-1) if (expression-2)
        Statement-1 ;
        else
        statement-2;

e.g. if (marks>=90) if (phy_marks>=95)
            award_amount=10000 ;
            else
            award_amount=5000 ;
```

***Format-2***

```
    if (expression1) statement1
    else if (expression2) statement2

e.g.    if (marks>=90)
            award_amount=1000;
        else if (category = ='r')
            award_amount=500;
```

***Format-3***

```
if (expression1) if (expression2) statement1
else statement2;
else statement3;
e.g.    if (marks > 50) if (maths_marks > 75)
        maths_award_amount=1000;
        else
        maths_award-amount=0;
        else
        printf("Student failed the test");
```

***Format-4***

```
if (expression1) statement1
else if(expression2) statement2;
    else statement3;
```

```
e.g.        if (marks>50) printf ("Pass")
            else if (marks>45) printf ("Can Get Grace Marks")
            else
            printf("Failed Badly");
```

## SWITCH STATEMENT

Switch statement provides means for checking a given expression and selecting one course of action among various actions available. It is used in the following format:

```
switch (expression)
{
    case(value-1) Statement;
    case(value-2) Statement;
    |
    |
    case (value-n) Statement;
}
```

***EXAMPLE***

Let's write a program, which accepts month in number and prints it name in words. For example prints January for 1 or June for 6.

```
#include<stdio.h>
main()
{
    int mth ;
    char code, junk ;
    printf("\nGive Month in number");
    scanf("%i",&mth);
    switch(mth)
    {
        case 1 :
            printf("\nJanuary");
            break ;
        case 2 :
            printf("\nFebruary");
            break ;
```

```
            case 3 :
                printf("\nMarch");
                break ;
            case 4 :
                printf("\nApril ");
                break ;
            case 5 :
                printf("\nMay");
                break ;
            case 6:
                printf("\nJune");
                break ;
            case 7 :
                printf("\nJuly ");
                break ;
            case 8 :
                printf("\nAugust ");
                break ;
        case 9 :
            printf("\nSeptember ");
            break ;
        case 10 :
            printf("\nOctober ");
            break ;
        case 11 :
            printf("\nNovember ");
            break ;
        case 12 :
            printf("\nDecember ");
            break ;
            default:
            printf("Wrong month ");
        }
    }
```

## GOTO STATEMENT

The goto statement makes the program control jump from its current location to some other, named location.

General format of goto statement is as follows:

**goto label ;**

Here label is the named line, to which the program control has to jump. For example, to transfer the program control to a line named merit, goto command will have to be used as follows:

**goto merit ;**

The labels can be marked in the program in following manner:

**label name:**

**e.g. merit:**

## WHILE LOOP

While loop always has an expression and one or more than one commands associated with it, till the time, expression evaluates to true, commands associated with it get executed repeatedly. The moment expression becomes false, while loop terminates. General format of while loop is given below:

```
while(expression)                 while (expression)
or                                {
command;                          command-1;
                                  command-2;
                                  command-3;
                                  }
```

***Example***

Following program illustrated the use of while loop:

```
#include<stdio.h>
main()
{    int start=2, limit = 20, sq=0 ;
     while (start <= limit)
     {    printf("\nNunber = %i",start) ;
          sq = start * start ;
          printf("  Its square  = %i",sq) ;
          start = start + 2 ;
     }
}
```

## DO WHILE LOOP

The do while loop is variant of while loop. In this loop, expression is placed at the end of the loop rather than at the beginning. General format of do while loop is given below:

```
do
{
    Statement1;
    Statement2;
    etc.
}
while (expression) ;
```

```
do
{
    value = pow(start,start);
    tot = tot + value;
    printf("\n%i",value);
    ++start ;
}
while(start <= target) ;
```

## FOR LOOP

The for command of C language is another means of forming a loop within the program. General format of for loop is as follows:

**for(expression1; expression2; expression3)**

**Statement or Group of statements**

Here in this loop, all the there expressions involve a control variable of integer type, say x.

Expression1, initializes the value of the control variable and is generally of the type x=value e.g. x=100.

Expression3 increments or decrements the value of control variable, say x in each iteration and is generally of the type x=x+value or x=x-value. If the value of the control variable is to be incremented by 1 in each iteration then it is of the type, x++. On the other hand, if the value is to be decremented each time then it could be of the type x--.

Second expression is used for framing the target value for the control variable. That means the loop will continue till the time expression is true. The moment this expression becomes false the loop terminates. To understand the functioning of for loop, consider the following example:

```
for (x=1; x<25; x++)
{   sq=x*x ;
    printf("%i",sq) ;
}
```

In this example, for loop will start with x=1. In the body of the loop, squre of x(i.e. 1 at present) will be calculated and assigned to the variable, sq. The printf statement will display the square (of 1).

After printing, the value, x will be incremented by 1 and the second expression (x=25) will be evaluated to check if value of x has reached 25 or not. Since current value of x will be 2 hence this expression will be true. Thus the loop statements will be executed again. Again the value of x will be incremented and process will continue, till the time x becomes 25.

The moment x becomes 25, loop will break and the statement next to for statement will get executed.

***Example***

Following program illustrates the usage of for loop:

```
#include<stdio.h>
#include<math.h>
main()
{   int i, value, total=0;
    printf("\nSeries Total\n");
    for(i=1 ; i<=10 ; i++)
    {   value = (i+(i+1)) ;
        (printf("(%i+%i)+",i,(i+1));
        total = total + value ;
    }
    printf(" =%i",total) ;
}
```

On execution this program will give the following output:

(1+2) + (2+3) +(3+4)+ (4+5) + (5+6) +(6+7)+ (7+8) + (8+9) +(9+10)+(10+11)+ = 120

## ADDITIONAL FACILITIES PROVIDED IN C LANGUAGE

Apart from above mentioned features and facilities, C language provides following additional facilities:

1. Defining arrays.
2. Predefined functions.
3. User defined procedures.
4. Defining structures (records).
5. Pointers and their manipulation.
6. File reading and writing mechanism.

Using all these features data is processed as per requirement and output is generated.

## INTRODUCTION TO PASCAL AND PASCAL PROGRAMMING

Pascal is conventional programming language. It was originally developed by Niklaus Wirth. He developed Pascal to provide features that were lacking in other languages of that time. His prime objective was to come out with a language that could be used for teaching important concepts of computer programming. Pascal language was named after the mathematician Blaise Pascal.

## Features Of Pascal

Following are few powerful features of Pascal language:

1. **Built in Data Types -** Pascal supports Integer, Real, Character, and Boolean data types.
2. **User defined Data Types** – Pascal provides mechanism for defining your own data types, derived from built in data types.
3. **Provides a defined set of Data Structures** - Data structures like arrays, records, files and sets can be defined in Pascal.
4. **Has a strong data typing element** - Pascal compliers can diagnose all incompatible assignments in which one type of variable is assigned a value of another data type. For example, assigning alphabetic value to a numeric variable can be easily traced.
5. **Supports Structured Programming -** This is accomplished through the use of subprograms called procedures and functions.
6. **Simplicity and Expressivities -** Because of its simplicity and expressiveness it can be very well used for teaching the concepts of computer programming and demonstrating programming techniques.

## General Structure Of Pascal Programs

Every Pascal program is written in following format:

```
program TITLE ( input, output );
begin
    program statements;
    program statement
end.
```

Program is the first word of all Pascal programs. It is a keyword. TITLE is the name that the programmer assigns to the program. It is an identifier. Identifiers begin with a letter. Then they are followed by any digit, letter or the underscore character ( _ ). (input, output) states what the program will do. The word begin defines the starting point of the program, and provides means of grouping statements together (i.e. all statements between begin and end are considered part of the same group or block). Program statements are commands or instructions to the computer, which perform various tasks. They are basically Pascal language verbs. The word end marks the end of the program. It must always be the last statement of a Pascal program.

Note that all program statements and lines are terminated with a semi-colon, except begin and end keywords. Program statements preceding an end statement do not require a semi-colon.

## Your First Program In Pascal

Now you are familiar with general format of Pascal program, so let's write a program to display 'I Love India' on VDU.

```
program MYFIRST (output);
begin
    writeln('I Love India')
end.
```

The keyword *writeln* writes text to the VDU. The text to be displayed is written within single quotes. After printing the text the cursor is positioned in the beginning of the next line.

## Adding Comments In The Program

Comments are inserted into Pascal programs by enclosing the comment within { and } braces. Comments are ignored by the computer, but they improve readability of the program. A Pascal program with comments is given below:

```
program MYPROG (output);
begin
        write('I am waiting for a call');
        {the write statement does not set the cursor
         to the beginning of the next line. }
        writeln('I wish he gives me a call')
        { This will be printed in the same line, because of time writeln command }
end.
```

## Data Types In Pascal Language

Following four basic data types are permitted in C language:

1. Integer
2. Character denoted as char.
3. Boolean
4. Real

## Pascal Variables

All variables must be declared before they are used. They are declared in the following format:

```
var name : type;
```

*Here var is the keyword and name* is the name of the variable being declared, and *type* is one of the four data types mentioned above. Following are few examples of variable declaration:

```
var  age: integer;
     salary: real;
     name: char;
```

## Assigning Values To The Variables

Special operator := is used for assigning values to Pascal variables. An example of assignment is given below:

```
age := 43;  { it sets age variable to 43 }
letter  := 'W'; {char values are given within single quotes}
```

## Operators

+, -, * and / operators are used for addition, subtraction, multiplication and division respectively.

## Reading data from the keyboard

Read command is used for accepting data from the keyboard. It is used in following format:

```
read( variable );
```

Following piece of code illustrates the usage of read command:

```
writeln('Please enter age and pay separated by a space');
          read( age );
          read( pay );
```

On execution of the program, if you give 25 8000, the value 25 will get assigned to age variable and value of pay variable will become 8000.

## Displaying the Data

The *write* or *writeln* statements display the value of variables on monitor screen. To print the given text, it is enclosed within single quotes. Usage of these statements is illustrated in the following program:

***Program***

```
program DISPLAYVARIABLES (output);
var    number1 : integer;
       letter  : char;
       money   : real;
begin
```

```
        number1 := 23;
        letter  := 'W';
        money   := 23.73;
        writeln('number1 = ', number1 );
        writeln('letter  = ', letter  );
        writeln('money   = ', money  )
    end.
```

***Result***

On execution this program will display following output on monitor screen:

```
number1 = 23
letter  = W
money   = 2.3730000000E+01
```

## DECISION MAKING

The **if then** command is used for making decisions in Pascal programs. It is used in the following format:

```
if (condition) then   statement;
```

If the given condition is found true the statement gets executed else it is bypassed. Following is a valid if statement:

```
if number >= 50 then  writeln('You are pass');
```

## IF THEN ELSE

**if** statement can also include an **else** statement. It specifies the statement that is to be executed when the given condition is found false. It is used in the following format:

```
if (condition) then statement1
    else
        statement2 ;
```

If the given condition is found true, statement1 gets executed else statement2 gets executed. Following is valid if then else statement:

```
if number >= 50  then
    writeln('You are pass')
else
    writeln('Sorry, you are fail')
```

## Loops

Loop is basically a mechanism of executing few lines of the program given number of times. Most common loop in Pascal is the **for** loop. Statements written within **for** block are executed number of times, depending on the control condition. General format for the **for** command is as follows:

**FOR var_name := initial_value TO final_value DO program_statement;**

Following is a valid For loop:

```
FOR i := 1 TO 10 DO
    writeln( i );
```

Here loop will start with i=1 and go on till i=10. Thus loop will get executed 10 times. In each execution, value of i will get incremented by 1. Statement writeln( i ) written within the loop will thus get executed 10 times. Each time it will print the value of i. So as a result, numbers from 1 to 10 will get printed through this loop.

Following program illustrates the usage of for loop. It converts 20 temperature values, given in Celsius into equivalent Fahrenheit values:

***Program***

```
program CELCIUS_TABLE ( output );
    var    celcius : integer; farenhiet : real;
    begin
        writeln('Degree''s Celcius   Degree''s Farenhiet');
        for celcius := 1 to 20 do
        begin
            farenhiet := ( 9 / 5 ) * celcius + 32;
            writeln( celcius:8, '      ',farenhiet:16:2 )
        end
    end.
```

## Nesting of Loops

Placing a **for** loop within a **for** loop is called nesting of **for** loops. When nested, inner loop gets executed for each iteration of outer loop. Rules for nesting are as follows:

1. Each loop must use a separate control variable.
2. The inner loop must begin and end entirely within the outer loop.

***Program***

Following is a program that demonstrates the use of nested **for** loop:

```
program NESTED_FOR_LOOPS (output);
    var   line, column : integer;
    begin
        writeln('LINE');
        for line := 1 to 6 do
        begin
            write( line:2 );
            for column := 1 to 4 do
            begin
                write('COLUMN':10);  write(column:2)
            end;
            writeln
        end
    end.
```

***Result***

When executed, this program will generate following output:

```
LINE
 1   COLUMN 1   COLUMN 2   COLUMN 3   COLUMN 4
 2   COLUMN 1   COLUMN 2   COLUMN 3   COLUMN 4
 3   COLUMN 1   COLUMN 2   COLUMN 3   COLUMN 4
 4   COLUMN 1   COLUMN 2   COLUMN 3   COLUMN 4
 5   COLUMN 1   COLUMN 2   COLUMN 3   COLUMN 4
 6   COLUMN 1   COLUMN 2   COLUMN 3   COLUMN 4
```

## WHILE LOOP

The **while** loop is another type of loop that is provided in Pascal. It involves a condition and a set of commands with it. Loop continues till the time condition remains true and commands placed within the loop get executed in each iteration. When condition becomes false, loop breaks and the statement written immediately after the loop gets executed. The structure of the WHILE statement is as follows:

```
while  (condition) do
begin
    program statement;
    program statement
end;
```

## REPEAT

The **repeat** statement is similar to the **while** loop, in which the condition is evaluated at the beginning of the loop. However, in **repeat** statement, the conditional test occurs at the end of the loop. Thus program statements, which constitute the body of the loop, are executed at least once. The format of **repeat** command is as follows:

```
repeat
    program statement;
until condition_is_true;
```

## CASE command

The **case** command is a substitute for multiple **if else** commands. It involves an operator and matching statements for different values of operator. When case statement gets executed, operator's value is checked and then depending upon the value of operator, matching statement is executed. . Consider the following code portion written using **if else** statements:

```
if operator = '*' then result := number1 * number2
  else if operator = '/' then result := number1 / number2
    else if operator = '+' then result := number1 + number2
      else if operator = '-' then result := number1 - number2
        else invalid_operator = 1;
```

***Example***

Above mentioned complex statement can be replaced by simple case statement, given below:

```
case operator of
    '*' : result:= number1 * number2;
    '/' : result:= number1 / number2;
    '+' : result:= number1 + number2;
    '-' : result:= number1 - number2;
otherwise   invalid_operator := 1
end;
```

The value of *operator* is compared against each of the values specified. If a match occurs then the program statement(s) associated with it are executed.

If *operator* doesn't match, it is compared against the next value. The purpose of the *otherwise* clause ensures that appropriate action is taken when *operator* does not match against any of the specified cases.

## Other Facilities In Pascal

Apart from above mentioned features and facilities, Pascal provides following additional facilities:

1. Defining arrays.
2. Predefined functions.
3. User defined procedures.
4. Defining record structures
5. File reading and writing mechanism.

Using all these features data is processed as per requirement and output is generated.

# INTRODUCTION TO FORTRAN AND FORTRAN PROGRAMMING

FORTRAN is conventional programming language. It was specially designed and developed by IBM in 1950 for scientific and engineering applications. In 1966 the language was standardized and FORTRAN IV was born. Revision of the language led to FORTRAN 77, the language we use today. Although FORTRAN 90 is now available but it is not widely used.

FORTRAN remained quite popular for next fifty years or so after its invention. Climate modeling, computational fluid dynamics, computational physics, computational chemistry, statistical analysis etc. are few areas, in which it enjoyed its popularity and was extensively used for various applications.

FORTRAN is an acronym for FORmula TRAnslation. Original FORTRAN was number oriented programming language but later versions added support for processing of character-based data, array programming, module-based programming and object-based programming.

## FORTRAN Program Layout

FORTRAN program basically comprises of following four segments:

1. **Heading** : Heading section marks the beginning of the program and gives it a name.
2. **Comments** : This segment serves the purpose of opening documentation and gives a brief idea of program's input, output and purpose.
3. **Specifications** : This segment contains the names and types of the constants and variables that are used in the program.
4. **Main Logic** : This segment contains the executable statements, using which the problem is to be solved.

Following is a sample FORTRAN program:

```
PROGRAM MYPGOGRAM
****************************************************************
* This program calculates payable amount for purchased items  *
* if the rate of the item is given. Variables used are:* * QTY : Quantity
*
* RATE : Rate *
* AMT : Payable amount * * Input : none * * Output: AMT
*
****************************************************************
REAL QTY, RATE, AMT

RATE = 90
QTY = 55
AMT = RATE * QTY
PRINT 'AMOUNT TO BE PAID =', AMT
END
```

## FORTRAN Program Format

All FORTRAN programs are entered in the following format:

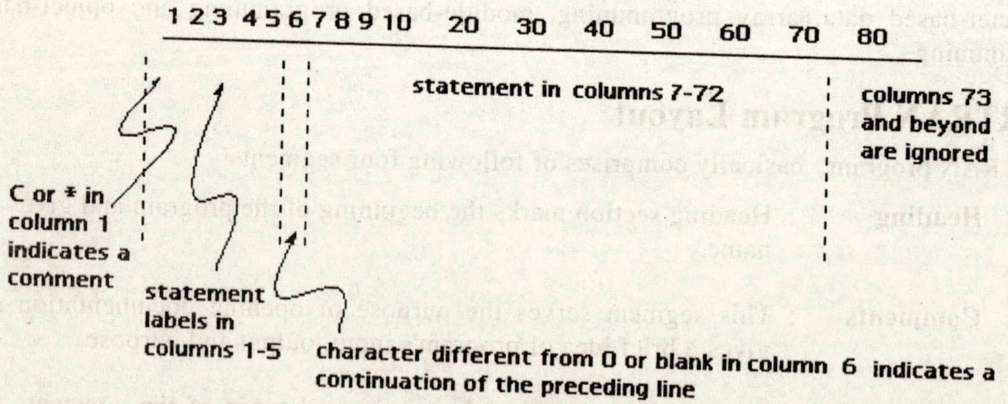


Refer above mentioned format and note that for the sake of understanding, columns are numbered as 1, 2, 3, . . ., 80. As strict rule, all FORTRAN statements are positioned in columns 7 through 72. Characters that appear in columns 73 and beyond are ignored. If a statement has to be labeled (say line has to be assigned a number) then label must appear in columns 1 through 5. Statement labels must be integers in the range 1 through 99999.

Often a FORTRAN statement may continue in multiple lines. In such cases, the statement may be continued on another line or lines (up to a maximum of 19), provided that a continuation indicator is placed in column 6 of the line(s), on which the statement is continued. This continuation indicator may be any alphabetic or numeric character other than a zero or a space. A zero or a space in column 6 indicates the first line of a statement.

Lines that contain only blanks or that have the letter C or an asterisk (*) in column 1 represent comment lines. In standard FORTRAN, comments themselves may not be continued from one line to another by using a continuation indicator in sixth column; instead, all comments must begin with a C or * in column 1.

## FORTRAN Variables

Variables in FORTRAN are of different types. You specify the type and name of each variable that you intend to use, at the top of your program, after the PROGRAM statement and before any other executable line.

If the variable declarations are omitted, the compiler assumes that variables, whose names start with I, J, K, L, M, N are integers. For example, variable INAME will be taken as integer.

## Data Types In FORTRAN

Following data types are allowed in FORTRAN:

1. Integer
2. Real
3. Character
4. Double

Variables of these type can be declared as follows:

**Data Type variable1, variable2 ........**

Note that when a variable is declared of specific type then basic assumption that variables starting with I, J, K, L, M, N are taken as integers, remains no more applicable. For example, variable INAME can be declared as character type.

## Operators

Following operators are used in FORTRAN language:

1. + operator for addition and unary plus
2. - operator for subtraction and unary minus
3. * for multiplication
4. / for division
5. ** for exponentiation

## Reading Data From The Keyboard

Read statement is used for reading the data from the keyboard into the variables. It is used in the following format:

```
READ *, input-list
```

Here input-list is a single variable or multiple variables separated by commas. Following is a valid Read statement:

```
READ *, RATE, QTY
```

## Displaying Data

PRINT statement is used for displaying the data on VDU. It is used in the following format:

```
PRINT *, output-list
```

Here output-list is a single expression or a list of expressions, separated by commas. Each of these expressions can be a constant, a variable, or a formula. Following is valid PRINT statement:

```
PRINT *, 'The values are ', RATE, QTY, AMT
```

PRINT statement can also be used in following form:

```
PRINT *
```

This statement will print a blank line.

## Decision Making

**IF THEN** command is used for making decisions in FORTRAN programs. It is used in the following format:

```
IF (logical expression) THEN
    block
END IF
```

If the given condition is found true the whole block of statements is executed else the block is bypassed. Following is a valid if statement:

```
IF (ALOK .GT. BABLI) PRINT*, 'BABLI is older than ALOK'
```

## IF THEN ELSE

The **IF** statement can also include an **ELSE** statement, which specifies the block of the statements that is to be executed when the given condition is false. It is used in the following format:

```
IF (logical expression) THEN
block 1
```

```
ELSE
    block 2
END IF
```

In this case, if the given condition is found true, block 1 of statements is executed. On the other hand, if the given condition is found false, block 2 of statements is executed.

## IF ... THEN ... ELSE IF ... THEN ...

This structure is used when more than one block of code is to be executed if the initial logical expression is false or if you need to make additional decisions. If none of the decisions are true, then none of the blocks are executed. General format of this statement is given below:

```
IF (decision 1) THEN
    Block 1
ELSE IF (decision 2) THEN
    Block 2
ELSE IF (decision 3) THEN
    Block 3
END IF
```

## IF ... THEN ELSE IF ... THEN ELSE

This is another decision structure, which can be used to make multiple decisions. If none of those decisions are true then Block 4 is executed. Its general structure is given below:

```
IF (decision1) THEN
        Block 1
    ELSE IF (decision2) THEN
        Block 2
    ELSE IF (decision 3) THEN
        Block 3
    ELSE
        Block 4
    END IF
```

## DO Loops

DO loop allows repetition of a section of program for specified number of times. DO statement marks the beginning of the loop and CONTINUE statement marks the end of the loop.

All the statements written within DO and CONTINUE statement get executed as many times as the loop iterates. How many times the loop will iterate is governed by the values of START, STOP and STEP that are included in DO statement. General format of DO loop is given below:

```
DO n X = START, STOP, STEP
.
.
Statements
.
.
n
CONTINUE
```

*Here* **n** is a statement line number defining the region for the repetitive loop. It can be a positive integer of up to five digits. **X** will be incremented in each iteration by the value of STEP. When the STEP value is omitted from the expression, its default value is set to 1. Its initial value is set to the value of START.

Looping will break when X will exceed the value of STOP. After that the statement, written just after CONTINUE statement will get executed.

**X** is called the loop index and must be an integer variable. Following is an example of DO loop:

```
    DO 20 I = 10,0,-1
    Print *, I
20  CONTINUE
```

On execution, this program will print all the numbers starting from 10 to 0 in reverse order.

## Array Declarations

Arrays must be declared at the beginning of the program. The easiest way to do this is to state what type of values the array will contain (real, integer, character) followed by the name of the array, followed by the number of elements in the array enclosed in brackets. Arrays are declared in the following format:

```
INTEGER NAME(n)
REAL FNAME(n)
```

Here NAME and FNAME are array names and n specifies number of elements in the array. Following is a valid array declaration:

```
INTEGER MARK(100)
REAL AMT(25), SALARY(50)
```

In the above example, MARK array will hold 100 integer values with subscript beginning at 1. Thus MARKS(1) will refer to the first element of the array. Similarly MARKS(7) will refer to seventh element of this array.

AMT array will hold 25 real values with AMT(1) as its first element. SALARY will hold 50 real values in it. SALARY(1) will be its first element and SALARY(50) will be its last element.

## Reading Elements Of Array

Values of array elements can be read from the keyboard, using READ command. Following example clearly shows the method of reading the array elements:

```
      NUM = 25
      DO 20 J = 1, NUM
      READ (1,*) ROLL(j), MARKS(J)
20    CONTINUE
```

On execution, this loop will read roll numbers and marks of 25 students.

## Printing Of Arrays

All the elements of an array will be printed if the array name is used without a subscript. However individual elements can be printed in the same way as other variables are printed. The method is described below:

```
   DO 20 I = 1,50

      Print *, ROLL(I)
   20 CONTINUE
```

## Other Facilities In FORTRAN

Apart from above mentioned features and facilities, FORTRAN provides following additional facilities:

1. Predefined functions
2. User defined procedures

Using all these features data is processed as per requirement and output is generated.

## EXERCISES CHAPTER 8

### Short Type Questions

**A. Select most appropriate answers for following questions:**

1. Which of the following is machine language?
   (a) FORTRAN (b) BASIC (c) COBOL (d) Language of 0 and 1
2. Which of the following language was developed exclusively for scientific calculations?
   (a) FORTRAN (b) BASIC (c) COBOL (d) Language of 0 and 1
3. Which of the following language provides facility for high level programming and lower level programming too?
   (a) C (b) FORTRAN (c) BASIC (d) Pascal
4. Which of the following software converts high level programs into machine language?
   (a) Compiler (b) Interpreter (c) Both of the above (d) None of the above
5. In which of the following language programs are written as functions?
   (a) C (b) FORTRAN (c) Pascal (d) BASIC

**B. Fill in the blanks:**

6. That software, which reads program, converts it to machine language and executes line by line is called..................................
7. That software, which reads program, converts it to machine language and executes in one shot is called..................................
8. Those languages, which are very near to machine language, are called.................languages.
9. Those languages, which are quite near to natural language, are called.......................languages.
10. The statement # include<stdio,h> belong to ..........................language.
11. Conditional operator (? ) is part of ....................language.
12. The being and end clauses are used in ....................language to mark the beginning and end of group of statements.
13. FOR I: =1 to 100 is loop command of ........................language.
14. While writing the program, all commands and statements are written from column 7 to 80 only in ............................language.
15. READ *RATE is data reading command in ...................language.

### C. State, true or false for following statements:

16. All computer programs that are written in high level language need to be converted into language of 0 and 1, for execution.
17. FORTRAN, PASCAL etc. are examples of conventional computer languages.
18. In C language, printf function is used for taking the data from the keyboard.
19. ++ and == are two valid operators of FORTRAN language.
20. The if command is provided by all three languages i.e. FORTRAN, C and PASCAL.

### D. Answer the followings questions in short:

21. What name is given to the language of 0s and 1s?
22. Name any two conventional languages.
23. Name any two high level languages.
24. What is the full form of FORTRAN?
25. Which functions are used for reading and displaying data in C language?
26. Which commands are used for reading and displaying data in FORTRAN language?
27. Which commands are used for reading and displaying data in PASCAL language?
28. Name the two classes, into which programming languages have been classified.
29. Which type of variable names represents integer quantities in FORTRAN?
30. Which language provides bit wise operators?

## Detailed Answer Type Questions

### E. Answer the following questions in detail:

31. What is computer language? Why it is required? Name any two computer languages.
32. Differentiate between high level language and low level language. Name one high level language and one low level language.
33. What is the role of computer? Why it is required?
34. What is interpreter? Why it is required?
35. Differentiate between compiler and interpreter.
36. What is C? Write at least two characteristics of C language. Name any three operators of C, which are not found in other languages.
37. Name the data types that are supported in C language. Write C statement to declare three variables that are of different data types.
38. What is FORTRAN? Describe the general format of FORTRAN program. Write the significance of different column i.e. starting from 0 to 80 in FORTRAN programming.
39. Who invented Pascal language? Describe the general format of a Pascal program. Describe any three commands of Pascal.

****

# COMPUTERS

## CHAPTER 9

# *Introduction To Computer Networks*

When two or more identities of similar type, work in coordination with each other, performing similar types of tasks, they form a network. For example, when sales executives of an organization work together to sell the product of the company, they form company's sales network. Similarly when two or more computers are connected together to work in coordination with each other they form a computer network.

To form a computer network and establish coordination among the computers of the network, certain hardware and software elements need to be present in the network. For proper communication among them, certain rules of communication, called protocol, need to be established.

Resource sharing and instant data communication are the immediate results of a computer networking. Advantages of these two simple-looking factors are so many that in short span of time numerous computer networks have come into existence and millions of applications have been developed around them.

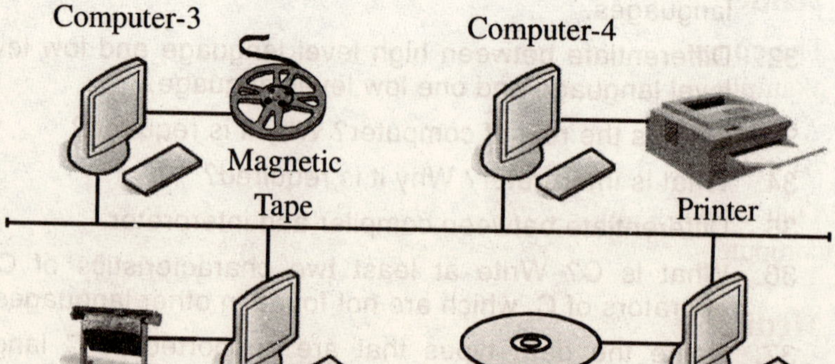

**Figure 9.1**

What computer network is, in what form its computers are arranged in the network, what protocols are used to establish communication among them, and what type of services are available on computer networks, are the issues, which are described in this chapter.

## WHAT IS COMPUTER NETWORK

When two or more computers are connected together with the objective of communicating with each other and sharing their resources, they form a computer network. A simple computer network is illustrated in figure 9.1.

Refer figure 9.1 and note that 4 computers are connected together to form a computer network.

In this network, any computer can communicate with any other computer. For example, computer-1 can send message to computer-4 and computer-4 can reply back. Similarly computer-2 and computer-4 can exchange data among them.

In computer network, computers share their resources among them. For example, computer-3 can print its documents on the printer attached to computer-4. Similarly computer-1, computer-3 and computer-4 can store their data on computer-2's disk.

## NETWORKING BASICS

Basic model of network communication is illustrated in figure 9.2. It shows that network comprises of following essential components:

1. Sender
2. Receiver
3. Medium
4. Message
5. Protocol

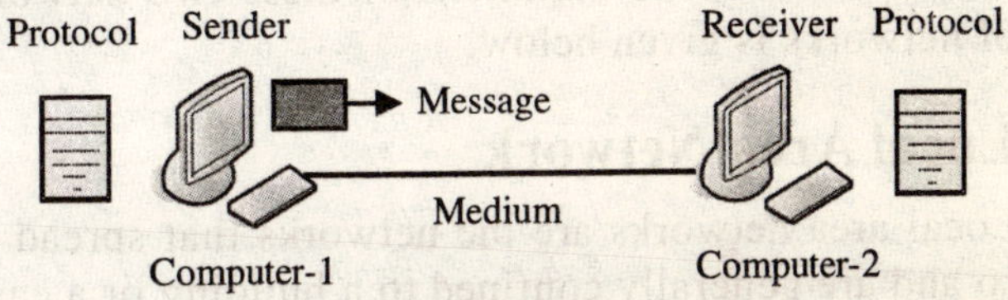


**Figure 9.2**

A brief description of each component is given below.

### Sender

The computer, which sends the message (or starts the communication) is called sender. For example, if computer-1 sends message to computer-2 then computer-1 will be termed as sender.

### Receiver

That computer of network, which receives the message is called receiver. For example, if computer-2 receives the message sent by computer-1 then computer-2 will be termed as receiver.

### Medium

Medium is basically that substance through which message propagates and reaches the destination (receiver). In some of the networks, it could be wire. In other networks it could be air.

### Message

What travels from sender to receiver is called message. As an essential rule of communication, message should reach the destination in its original form. No deformation or manipulation should take place in the middle.

### Protocols

A protocol is basically a set of standards that are decided between the sender and receiver for error free communication between them. A message, though received in its original form, is of no use, if receiver cannot understand the message.

For example, if the sender sends an encrypted message to the receiver and the receiver doesn't know that the message is in encrypted form, then in this case all the manipulations that are done at receiver's end, may lead to errors. Thus before communication starts, rules must be set that all the communication will be done in encrypted form and decryption has to be performed before the message is manipulated. It is quite obvious that if the rules of communication are set properly there will not be any problem in interpreting the message.

## COMMON TYPES OF NETWORKS

Computer networks are broadly classified into two categories:

1. Local Area Networks (LAN).
2. Wide Area Networks (WAN).

Other networks are variations of these two networks. A detailed, description of both types of networks is given below.

### Local Area Network

Local area networks are the networks that spread within a short distance of a kilometer or so and are generally confined to a building or a campus. For example, computers of various departments of an organization that confine within the factory building itself, if connected together, will form a Local Area Network.

The data transfer speed of Local Area Network is generally high i.e. large amount of data can be transferred from one computer to the other in a very short span of time. Roughly the data transfer rate in LANs is generally of the order of Mega bits per second (MBPS). In Local Area Networks you connect the computers using dedicated wires of your own which are only used for transferring the data/information and are not used for any other purpose. Since these communication lines run, in private, undisturbed area hence LANs are relatively error free. (For example, 1 bit error in $10^9$ bits transmitted). A rough sketch of Local Area Network is illustrated in figure 9.3.

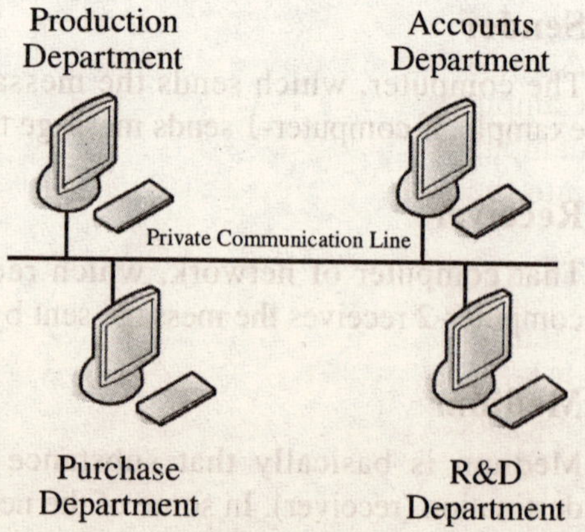


**Figure 9.3**

### Wide Area Networks

Wide Area Networks or WAN are the networks, in which computers are spread over long distances (more than a kilometer). Computer of Wide Area Network could either be in different countries, cities or localities. Since they spread over wide geographical area hence you cannot connect them through your own lines. You have to take help of the connections, which already exist in one form or the other. For example, telephone connection is one type of link that is already available in ready-made form. It can be used to connect the computers, which are far apart. Similarly microwave link and satellite link are other types of links that can be used to connect the computers that are in remote locations or are situated across the countries.

Note that these ready-made links were originally designed for some other purpose not for connecting computers, so their transfer speeds are not as high as that of dedicated lines (LAN lines). Roughly their speed is of the order of few Kilo bits per second say 32 Kbps or 64 Kbps.

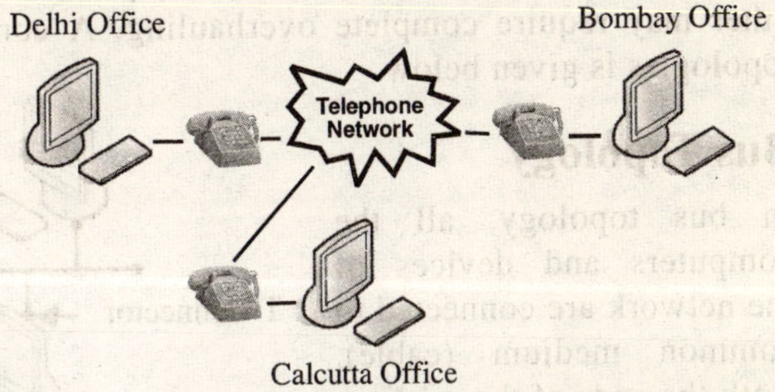

**Figure 9.4**

Since these lines run into noisy environment where humidity, heat and electrical signals cannot be controlled hence Wide Area Networks are more error prone than Local Area Networks (For example a bit error rate of 1 in 100,000 bits transmitted). A rough sketch of Wide Area Network is illustrated figure 9.4.

## VARIATIONS OF LAN AND WAN

Various variations of Local Area Network and Wide Area Network exist with different names. Some of them are described below.

### Campus Area Network (CAN)

When the computer of various departments of an organization, such as factory, college, university etc. are connected together. They form Campus Area Network or CAN in short. Since computers of Campus Area Network remain confined within the campus boundary (which is short) hence they fall into the category of LAN.

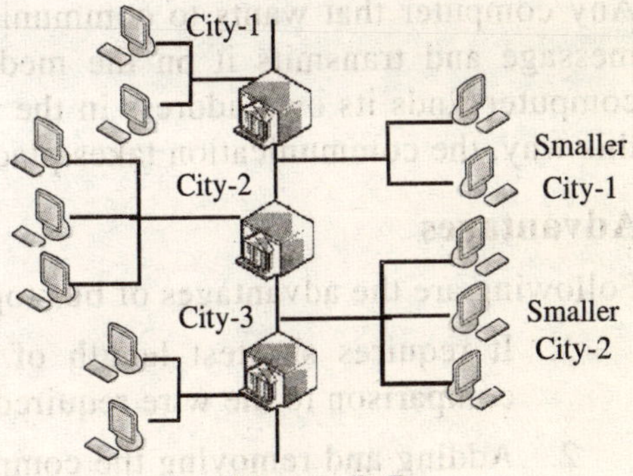

**Figure 9.5**

### Metropolitan Area Networks (MAN)

Metropolitan Area Networks are basically Wide Area Networks, which connect computers of larger cities. They are generally high-speed networks and work as backbone for connecting other local area networks (say smaller cities which are located between two big cities of MAN). A block diagram of Metropolitan Area Network is illustrated in figure 9.5.

### Home Area Network (HAN)

When two or more computers located within a house are networked together, they are given the name Home Area Network or HAN in short. Obviously HAN falls into the category of Local Area Network.

# NETWORK TOPOLOGIES

Topology is basically the physical arrangement of computers in the network. There are various topologies, suggested for computer networks. Each topology has its own advantages and disadvantages. One topology suits well for one set of given conditions while the other doesn't. Some of the topologies undergo network expansion easily while other may require complete overhauling. A complete overview of most commonly used topologies is given below.

## Bus Topology

In bus topology, all the computers and devices of the network are connected to common medium (cable). Both the ends of the medium are terminated by a special device called terminater. Computers are connected to the cable, using other device called T connector. Block diagram of bus topology is illustrated in figure 9.6.

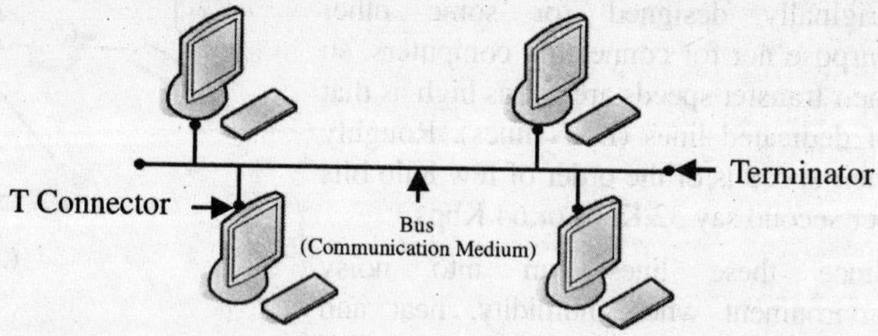


**Figure 9.6**

### Functioning

Any computer that wants to communicate with other computer includes its address in the message and transmits it on the medium. Every computer receives the message. If any computer finds its own address in the message, it picks up the message else it rejects it. In this way, the communication takes place between the computers of the network.

### Advantages

Following are the advantages of bus topology:

1. It requires shortest length of wire to connect the computers of the network in comparison to the wire required in other topologies.
2. Adding and removing the computers in the network is easy. It does not affect other computers of the network.
3. Even if one computer of the network becomes down, it doesn't affect the functioning of other computers. The network keeps on functioning.

### Disadvantages

Following are the disadvantages of bus topology:

1. Terminators often malfunction, due to loose connections.
2. A failure in the cable makes the entire network down.

## Ring Topology

In ring topology computers of the network are connected to each other in circular fashion.

The first computer is connected to the second computer, second is connected to the third and so on and the last computer is connected to the first computer, so as to complete the circle. Computers of the network are connected to the cable, using a special device called repeater. A computer network, in ring topology is illustrated in figure 9.7.

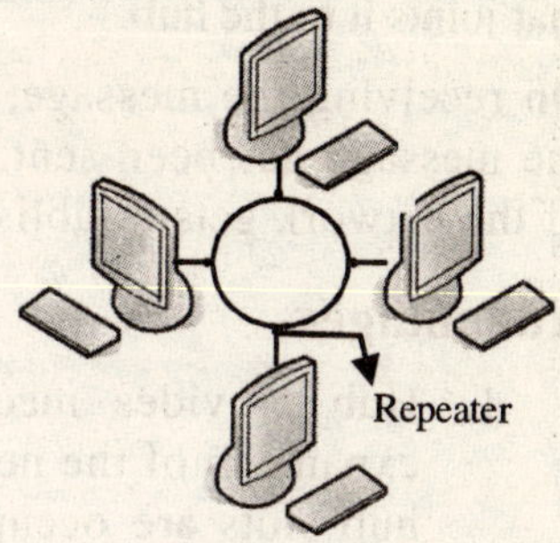


**Figure 9.7**

### Functioning

In this type of topology, a bit pattern, called token, keeps on moving from one computer to another in circular fashion. Only that computer, which has the token, is allowed to transmit the message.

Transmitting computer introduces destination computer's address in the message and pushes it on the medium. The message moves forward in one direction, on the medium. When a computer receives the message it checks the destination address. If the destination address matches with its own address it picks up the message else it rejects it.

After transmitting the message on the medium, computer passes the token to next computer to communicate.

### Advantages

Practically there are no highlights of this topology but there are lots of disadvantages.

### Disadvantages

Following are the disadvantages of ring topology:

1. Malfunctioning in any of the repeater or cable makes the entire network down.
2. Addition and removal of computers in/from the network is a difficult task.
3. Long cables are required to connect the computers.
4. This type of network cannot be expanded beyond a limit. Not many computers can be added in this arrangement.

## Star Topology

In star topology, computers of the network are connected together through a special device called hub. Each computer is connected to the hub. Thus the link of one computer is established with other computers via hub. A computer network in star topology is illustrated in figure 9.8.

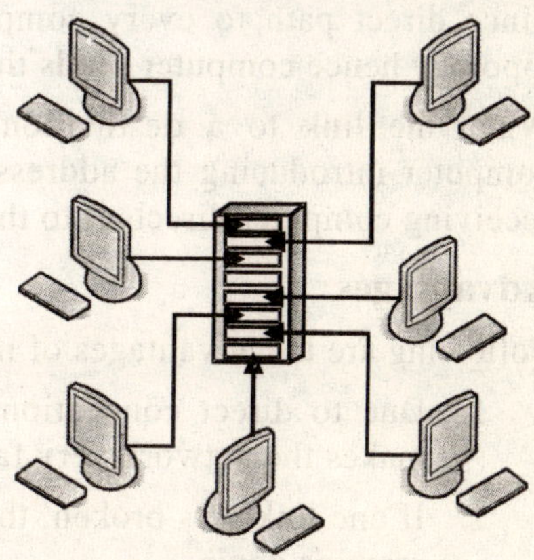

**Figure 9.8**

### Functioning

Hub in this topology works as intelligent switch. When a computer wish to transmit a message it introduces destination computer's address in it and pushes the message, on the link, that joints it to the hub.

On receiving the message, hub checks the address and routes it to the computer, to which the message has been sent. In this way, proper communication between the two computers of the network gets established.

### Advantages

1. Hub provides mechanism for connecting many computers to the network. So expansion of the network becomes very easy, if a slot is vacant in the hub. If all the hub slots are occupied then another hub can be added into the network, so as to make provision for many more new slots. Thus network growth doesn't have any limitation in this type of topology.
2. If the link between the hub and a computer is down then it doesn't effect the functioning of other computers of the network. Rest part of the network works fine.
3. All the hubs generally have a status light for each link. Glowing light of a link indicates that the link is working fine. Thus identification of malfunctioning link, in this topology, becomes very easy.

## Mesh Topology

In mesh topology each computer of the network remains connected with other computers through dedicated medium (wire). A computer network, in mesh topology, is illustrated in figure 9.9.

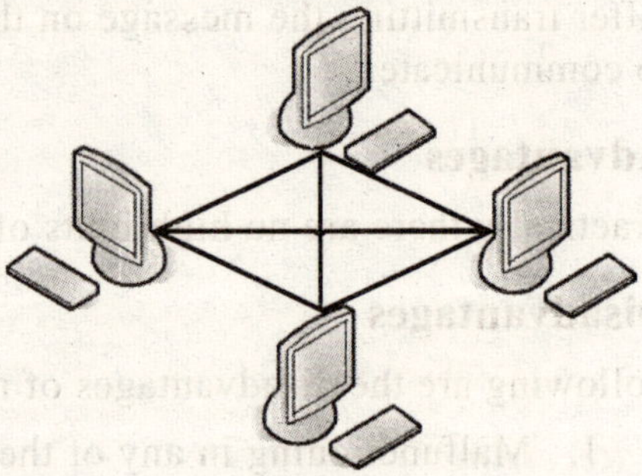

**Figure 9.9**

### Functioning

Since direct path to every computer exists in this topology hence computer sends the message directly to the required computer.

When the link to a destination gets broken, computer sends the message to another computer introducing the address of the destination computer. On receiving the message, receiving computer directs it to the computer, to which the message belongs.

### Advantages

Following are the advantages of mesh topology:

1. Due to direct connection between the computers, no switching is required. This makes the network very fast.
2. If one link gets broken, the data is routed through other links. Thus network seldom remains down.

### Disadvantages

Following are the disadvantages of Mesh topology:

1. It doesn't suit well for large number of computers.

2. Very large length of cable is required to connect the computers, so the cost turns out to be very high.
3. This topology is almost obsolete now.

## PROTOCOLS

Protocol is basically a set of rules that are formed for the computers of network for establishing proper communication among them. To understand the need and concept of protocol, consider a the smallest form of computer network, in which two computers say Com-1 and Com-2 are connected together, with the objective of communicating with each other. Say their addresses are 100.105.110.115 and 200.220.240.250 respectively. Say the computer Com-1 wants to send the message "MY ACCOUNT. NUMBER is 12376" to Com-2. For all practical purposes and security reasons they may decide the following rules for sending and receiving the messages.

***Sending The Message***

1. Complete message will be broken into two equal parts.
2. Contents of the messages will be reversed.
3. Total number of digits in the message will be added at the end.
4. Address of the computer, sending the message, will be added at the beginning of the message.
5. Address of the computer, to which the message is being sent, will be added at the end of the message.

***Receiving The Message***

1. Takeout the address from the end of the message and compare it with own address. If the addresses match, read the message else reject it.
2. Remove the addresses from the end and beginning of the message. Also remove total number of characters from the message to get the reversed message.
3. Count the number of characters in the message and compare it with the number received along with the message. If both the numbers match reverse the message to get the original message else ask the computer (whose address was attached at the beginning of the message) to sent the message again.

Following these rules Com-1 will break its message into two packets, as follows and send it to the other Com-2.

Packet-1 : 100.105.110.115 UN TNUOCCA YM13 200.220.240.250.

Packet-2 : 100.105.110.115 67321 SI REBM 13 200.220.240.250.

When Comp-2 will receive this message, it will perform all those steps, which have been agreed upon for receiving the message.

Note that this simple agreement between two computers (protocol) ensures following points:

1. Small messages are transmitted over the communications medium so that it doesn't get heavily loaded and other computers also get time to send their messages.
2. Message reaches the right destination.
3. If the message doesn't reach destination in original form, it is traceable and recovery is made by requesting the source computer to send the message again.

This example clearly illustrates that protocols are necessary in computer networks. They not only establish communication rules but they also promote efficient use of communication components. Numerous protocols are available for computer communication network. Some of them are described later in this chapter.

## ARCHITECTURE OF SEVEN LAYERS OF COMMUNICATION

Computer network doesn't mean that, all the computers will be of same type, they will use same operating system and communicate for same application. The truth is that a computer network may comprise of heterogeneous hardware, different operating systems and diversified applications. Such a situation demands that multi layered protocol should be defined to take care of the issues like, how the computers should be physically connected, how should data be transported from one computer to another, how two different applications should exchange data among them etc. To resolve this issue, International Organization for Standards (ISO) defined a seven layers model, called Open System Interconnection reference model or OSI model in short.

Each layer of OSI model is basically a specific issue. Related to each issue, it defines certain rules and thus sets parameter for network connectivity. OSI reference model is illustrated in figure 9.10.

Refer figure 9.10 and note that OSI reference model is a seven-layer mode. Each layer uses the functions of the layer below and exports the functionality to the layer, above. A brief introduction of each layer is given in figure 9.10.

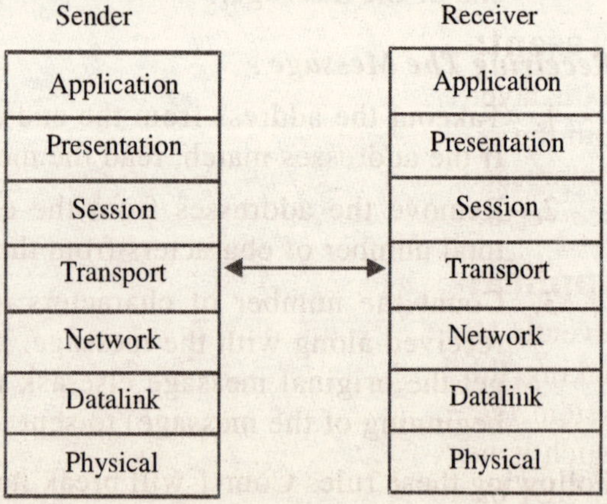


**Figure 9.10**

### Physical Layer

First layer of OSI model is called physical layer. It defines all the electrical and physical specifications for the devices that are part of computer network. This includes pin specifications, voltage details, cable specifications etc. Specifications related to hubs, repeaters, network adapters etc are also specified in physical layer.

### Data Link Layer

Second layer of OSI reference model is called data link layer. This layer deals with the issues related to data transfers. It takes care of speed mismatches that may occur while data gets transmitted in the physical layer. It detects and possibly corrects errors that may occur in the physical layer.

## Network Layer

Third layer of OSI reference model is called network layer. It mainly deals with routing functions and congestion control.

To understand the concepts consider figure 9.11 that shows that there are various paths, which a message can follow, if it has to travel from A to D. It can either follow A->D or A->B->D or A->C->D path. Network layer decides, in which circumstance, which path should be followed. It takes care that no path should be heavily loaded.

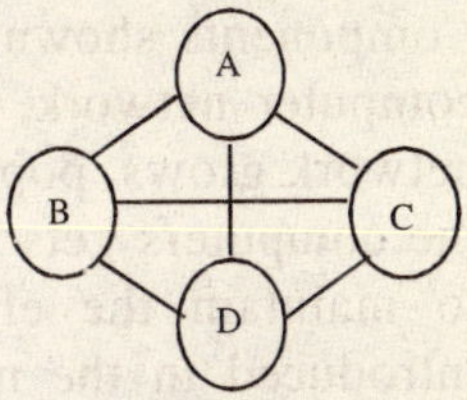


**Figure 9.11**

## Transport Layer

Fourth layer of OSI model is called transport layer. It mainly deal with flow control of data and provides error detection and correction mechanism.

To understand the concept, consider figure 9.11. D may receive data from A, B and C together for three applications, say Ap1, Ap2 and Ap3. Transport layer keeps track of each data stream and ensures that they maintain their individuality and do not get mixed up with each other.

## Session Layer

Fifth layer of OSI model is called session layer. It deals with the individual sessions that are established between the computers for communication. To understand the concept, refer figure 9.11. Say A wants to be in dialog with D then it will open a session for communication and when communication finishes it will gracefully terminate the session. Session layer will perform these functions.

## Presentation Layer

Sixth layer of OSI reference model is called presentation layer. This layer mainly deals with the presentation aspects of the data. For example, the data may have to be presented in compressed form or coded form or EBCDIC data may have to be converted into ASCII or vice versa. All such type of data presentation activities are performed by presentation layer.

## Applications Layer

Seventh layer of OSI reference model is called applications layer. It mainly deals with making the information available to the user. For example, files residing on CUI based system (say a computer with Unix operating system) have to be viewed on GUI system, which is networked with it. It is quite obvious that these files have to appear as icons here on GUI based system. Now the question is, how this becomes possible? Well, application layer takes care of such issues. Telnet, FTP (File Transfer Protocol), Simple Mail Transfer Protocol (SMTP) etc. are few protocols that are related to applications layer.

# NETWORK PROTOCOLS

You are familiar with the word protocol and are aware of its significance in relation to computer networks. To establish a proper communication network, many protocols are used. Few commonly used protocols are TCP/IP, IPX/SPX, Net BUIOS/Net BEUI, FTP, Telnet.

# NETWORK COMPONENTS

Computer communications network, shown in figure 9.2 is the simplest form, of network. Components shown therein are essential components of a network. Be it a small or large computer network, those components will always be present in the network. But as the network grows, population of computers in the network increases and the distance between the computers gets wider. Thus to make the communications possible in all situations and to maintain the efficiency of communication, additional devices and equipments are introduced in the network. These articles are often referred to as network components. Details of few commonly used network components are described below.

## Computers

Computers are the most essential part of computer network. They send and receive the message. In addition to this they also take part in routing the message.

## Communications Medium

As mentioned earlier, communication medium is that channel, through which message passes from one computer to another. Communication medium can be broadly classified into two categories:

1. Wired medium.
2. Non-wired medium.

Depending upon various factors, such as distance, type of message, price etc. communication medium may vary from network to network. A brief introduction of both type of medium is given below.

### Wired Medium

When a cable is used for physically connecting the computers of the network the medium is said to be wired medium. Following types of cables are generally used for connecting the computers of the network.

#### Twisted Pair Cable

Two independently insulated wires twisted around one another form twisted pair cable. Often both the wires are encased within plastic case. Wires pick up electromagnetic signals from the atmosphere and thus noise gets created in the signal, passing through the wire. When two wires are wound over each other, noise in one wire destroys the noise in other. Thus twisting of wires helps in reducing cross talks. It is the cheapest cable. It can support up to 1Gbps data transmission speed. Twisted pair cable is illustrated in figure 9.14(a).

#### Coaxial Cable

Coaxial cable is basically a conducting wire surrounded by insulation material, like hard plastic. Conducting sheath surrounds this assembly and the whole assembly is then encased within insulating plastic. It is used as high frequency transmission line to carry high frequency or broadband signals. Noise is reduced to bare minimum in such type of cables. The cable that comes to our home from local cable TV network is an example of coaxial cable. It supports data transmission up to 10 Mbps. A coaxial cable is illustrated in figure 9.14(b).

### Optic Fiber Cable

Optic fiber consists of a very fine (as fine as hair) transparent fiber, usually made of glass or plastic. Light can pass through the fiber even it if is not straight.

Optic fiber cable consists of optic fiber enclosed within a rubber shield. Above it is placed braided copper or foil layer and the assembly is enclosed within jacket. An optic fiber cable is illustrated in figure 9.14(c).

In optic fiber cable data transmission takes place, using light. Light doesn't gather any type of noise from the atmosphere. Thus quality of the signal / message remains unchanged throughout the distance. Optic fiber cable support up to 100 Gbps data transfer rates.

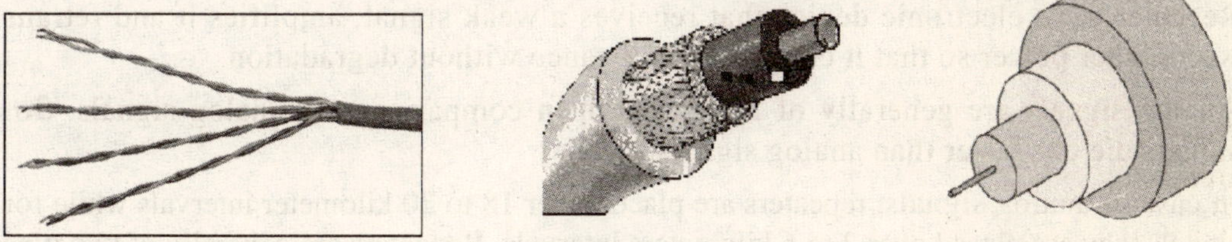

(a) (b) (c)

**Figure 9.14**

### Non Wired Medium

The medium other than cable, which is used to transmit the message from one computer to another in a network, is called non-wired medium. Usually it is nothing other than air. Message from one computer, in the form of infrared waves or radio waves or high frequency waves passes through air and reaches other computer of the network. Thus a connection gets established among them.

This medium of communication is becoming popular day by day in computer networking. Wireless networking, including wireless Internet connectivity are very popular these days. LANs and WANs both are making use of this medium for computer connectivity.

## Modem

Modem is a device that enables computer to transmit data over telephone or cable lines. Note that computer internally stores the data in digital form, whereas data transmitted over telephone lines, is transmitted in the form of analog signals. A modem at sender's end converts digital form into analog form and the modem at receiver's end performs opposite function i.e. converts analog form into digital form. The functioning of modem is illustrated in figure 9.15.

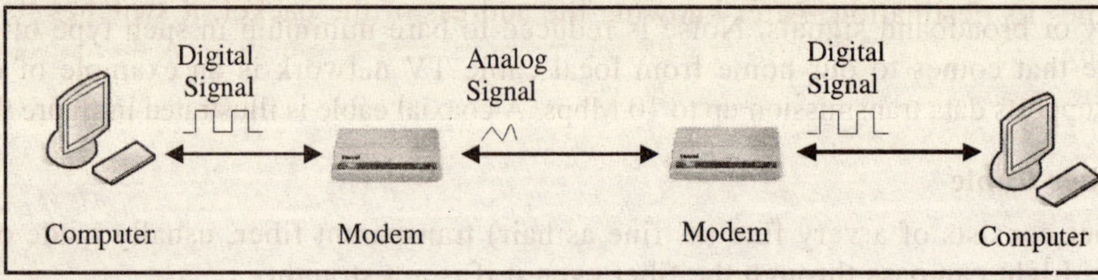


**Figure 9.15**

## Repeater

An electronic signal on passing some distance gets gradually weaker and finally dies out. Thus if the distance between the sender and receiver computer is large in a network and no arrangement has been made to keep the signal alive, receiver doesn't receive signal in its original form and the communication becomes faulty.

Repeater is an electronic device that receives a weak signal, amplifies it and retransmits it with higher power so that it covers large distance without degradation.

Analog signals are generally of low strength in comparisons to analog signals. So digital signals die out faster than analog signals.

In case of analog signals, repeaters are placed over 18 to 20 kilometer intervals while for digital signals they are placed over 2 to 6 kilometers intervals. Repeaters are generally of two types:

1. Digital repeaters.
2. Analog repeaters.

Analog repeater is an analog device that amplifies an input signal regardless of its nature i.e. analog or digital.

Digital repeater is a digital device that amplifies and reshapes the deformed signal and retransmits it.

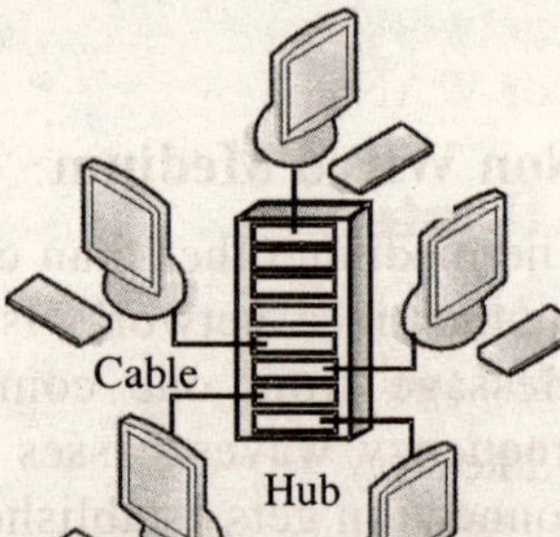


**Figure 9.16**

## Hubs

Hub is basically a connecting device. Multiple computers of the network are connected to the hub, as illustrated in figure 9.16.

When the hub receives a packet from a computer on one of its ports (to be sent to another computer of the network), it copies the packet to all the ports so that all the computers can see all the packets. Computer whose address is stamped on the packet takes the packet.

## Switch

A network switch is basically a small hardware device, to which all the computers, of the local area network are connected.

Network switch in its physical appearance resembles a hub but it contains more intelligence. On receiving a packet from the computer, it inspects the packet and determines its destination. After knowing the address of the packet, it switches its internal arrangements to establish a direct path for the required computer. Thus the packet reaches its destination.

To understand the concept, refer figure 9.17, which is a block diagram of a switch. All the eight computers of a LAN are connected to the switch.

Initially no computer has a direct link to another computer. When a computer, say A sends a packets for another computer say H, it first goes to the switch. One receiving the packet switch determines its address and internally sets a link between A and H. Thus the packet travels from A to H.

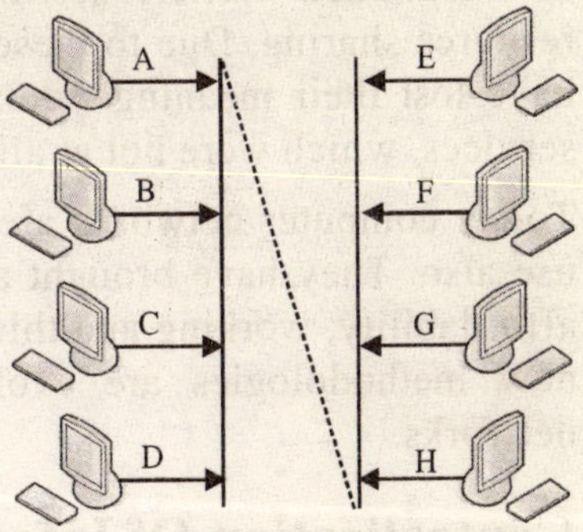


**Figure 9.17**

## Bridges

A network bridge is a device that connects multiple Local Area Networks or segments of networks. It is also called network switch. Network traffic from one LAN is diverted to another LAN through it. Bridges operate at data link layer (i.e. second layer) of OSI model. They monitor incoming traffic and decide whether to divert it or discard it.

## Gateways

Gateway is a computer that acts as main gate from where the network traffic gets diverted into multiple paths connecting to LANs or WANs, as illustrated in figure 9.18.

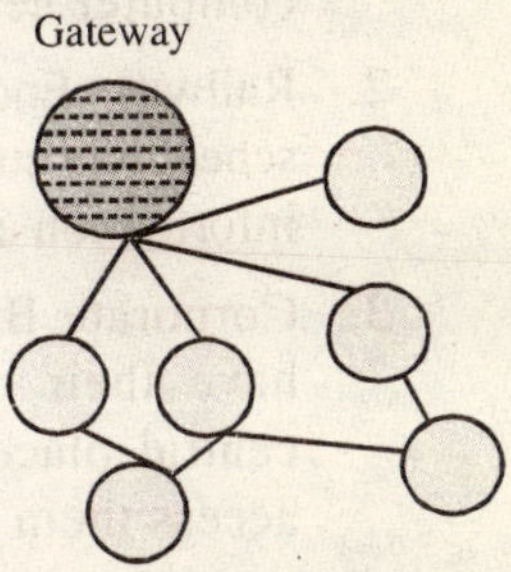


**Figure 9.18**

## Routers

Routers are the devices placed at gateways that divert the packet on the path, traveling on which, it will reach its destination.

They try to select the best possible path available at any point of given time, as there may be multiple paths available, for the packet to reach its destinations. The reason for not selecting the nearest path at any point of time could be that the nearest path may be busy, sending some other packet or the link may be broken etc.

## Network Interface Card

Network interface card is used for connecting the computer to the network. It generates proper signals and transmits them in the medium. It also receives signals from other computers of the network and accepts them within the computer. It is placed in one of the expansion slots available in computers. When placed in it, it provides socket for connecting transmission medium to the computer. Network Interface Card, called NIC in short is illustrated in figure 9.19.

**Figure 9.19**

# USES OF NETWORK

As mentioned earlier, networking results into instant data communication and on-line resource sharing. Due to these two factors, geographical boundaries have crushed, distance have lost their meanings, communication barriers have broken and such applications and services, which were not at all possible few years back, have now come into existence.

Today computer networks are being used for not only business purposes, but for personal use also. They have brought a revolution in the society. Due to their easy availability and affordability, working and thinking patterns of human being are continuously changing and new methodologies are evolving. Below given are few important uses of computer networks.

## Centralization Of Information

Networks are very commonly used for storing the data / information, centrally and making it available for others. Centralization of information doesn't sound so great but applications developed around this concept are really great.

1. Railway Reservation System is one of the most useful application that is serving masses. Reservation in any train from any city has been made possible, because of computer networks.
2. Railway Enquiry System: Seat availability in trains, train fare information, train schedules, current train position etc. are made available to general public by placing the information at central place in the network and allowing its accessibility to the masses.
3. Corporate Business System: These days companies, organizations, institutions etc. have their own computer networks. Company's information is maintained at a central place by head office. Their regional offices, business clients, vendors etc. access them for their uses.
4. Education Service System: Making admission forms, examination notices, results of examination etc, available to the masses are the common applications, which educational institutions make possible through computer networks. For this they make the information centrally available on network server. Virtual universities have now come into existence, in which services related to registration, teaching, problem handling, examination etc. are provided to the students over computer network.

## Resource Sharing

One of the main uses of computer networks is resource sharing. Costly devices are made centrally available for use to the network users. For example, costly laser printers are put into network so that other users could also use it. Similarly large disk storage is also made available on the network for common use. Network users can either put their applications on them or take data backups on such disks. Resource sharing ultimately leads to cost reduction and easy network maintenance.

## Fast Communication

Instant data transfer over network leads to fast communication among network users. It also ensures data delivery. Transfer losses are never there. Following are few commonly used communication applications, which are quite popular in computer network environments:

1. **Video Conferencing :** Video conferencing is an applications, in which not only voice communication takes place among network users but they also exchange their live movements through web camera.
2. **Voice Chat :** There are many network applications, using which network users talk to each other on network, like they do on normal phone. For this, they make use of microphone (speaking) and speakers (listing) connected to their computer.
3. **Text Chat :** Using text chat applications, network users exchange their thoughts and feelings by typing the text on their computers. What one user types on his computer, appears on other user's computer and what other user types on his computer appears on first person's computer.
4. **e-mail :** e-mail is the most commonly used network application. Using this application, network users exchange letters over network. Letters sent through e-mail reach their destination instantly and do not get lost in the transit.

## Efficient Services

Organizations and institutions make use of computer networks for providing efficient services to either internal people or external persons. Following are few services, which are common over computer networks:

1. On-line advice.
2. On-line help.
3. On-line enquiry.
4. FAQ (i.e. answer to frequently asked questions made available in the form of list).

## EXERCISES CHAPTER 9

### Short Type Questions

**A. Select most appropriate answers for following questions:**

1. Which of the following is not an essential element of communication system?
   (a) Sender (b) Receiver
   (c) Wire (d) Message
2. What is protocol?
   (a) A message (b) Rules that have been set.
   (c) A code (d) A message
3. Which of the following network remains confined to a building?
   (a) LAN (b) WAN
   (c) MAN (d) SAN
4. In which type of network, computers of five different cities are connected together to form a computer network?
   (a) LAN (b) WAN
   (c) HAN (d) SAN
5. Which of the following topology makes use of a special device called hub?
   (a) Bus (b) Star
   (c) Ring (d) Mesh
6. What is TCP/IP?
   (a) A compression techniques (b) A protocol
   (c) A type of network (d) A topology
7. How many layers of protocol are there in OSI model of network?
   (a) 7 (b) 6
   (c) 5 (d) 10
8. Which layer of OSI model is related to physical connection of computers, within the network?
   (a) First (b) Second
   (c) Fifth (d) Seventh
9. Which of the following is related to file transfers across the network?
   (a) FTP (b) TCP/IP
   (c) CSMA/CD (d) Telnet

10. Which of the following will have fastest rate of data transmission in it?
    (a) Twisted pair of wire (b) Coaxial cable
    (c) Optic fiber (d) Starlight wires

## B. Fill in the blanks:

11. ...................................... is a device, which is used for transmitting computer signals, over normal telephone line.
12. ..................... is a device that is used to amplify the signal that gets weak, while traveling from one computer to another, in a network.
13. ................. is device, which is used for connecting two branches of LAN.
14. ........................ is a device that decides, which data packet will follow which root in the network.
15. LAN is acronym of ................. .......................... ..........................
16. WAN is acronym of ................... ............... ....................
17. MAN is acronym of ................... ............... ....................
18. That topology of network, in which computers are connected to each other in circular fashion, is called ................. topology.
19. In ................ topology, each computer of the network remains physically connected to other computers with a separate medium.
20. OSI is acronym of ................... ..................... ....................

## C. State, true or false for following statements:

21. Network layer of OSI model deals with routing functions and network congestion.
22. TCP/IP protocol provides the facility to remote login on another computer of the network.
23. NetBIOS/NetBEUI is a network routing device.
24. IPX/SPX is a protocol for networking the computers.
25. Coaxial cables offer better data transfer rates than twisted pair of wires.
26. In coaxial cable data is transmitted using laser light.
27. Hub is a device, using which multiple computers are connected together over single point.
28. Switch in a computer network, physically connects two computers together for data transfer.
29. NIC is acronym of Network Inter Connection.
30. In peer to per network model, one computer acts as server and other acts as client.

## D. Answer the following questions in short:

31. What are the two main objectives of networking?
32. Name five essential components of networking.
33. Name any three mediums, which are generally used for connecting the computers of the network.

34. Name any two variants of LAN.
35. Which type of network serves as backbone for connecting the computers of smaller cities?
36. Name any four topologies, in which computers of the network could be connected.
37. Draw a block diagram of Ring topology.
38. Draw a block diagram of Star topology.
39. Name any two protocols of networking.
40. Name a medium, in which data is transmitted, using light.

## Detailed Answer Type Questions

### E. Answer the following questions in detail:

41. What is computer network? Draw a block diagram showing basic elements that are required for networking.
42. What do you understand by protocol? What is TCP/IP? Explain its role in computer network.
43. What is LAN? Explain its characteristics, giving suitable example.
44. What do you understand by the term topology? Draw a block diagram of bus topology. State advantages and disadvantages of this topology.
45. What is MAN? How it is different from WAN?
46. Draw a block diagram OSI model for networking. Explain the role of its first three layers.
47. Explain the objective of following:

    (a) FTP (b) TELNET
48. Explain the role of following devices in network. When are they used?

    (a) Modem (b) Bridges (c) Router (c) Hub
49. What do you understand by client server mode of computing?
50. Describe at least five applications of networking. State their advantages also.

****

# COMPUTERS

## CHAPTER 10

# *Algorithms & Flowcharts*

You know that computer requires program for everything that it does. A program is basically a group of instructions written in computer language for performing a specific task. For example, instructions, written in computer language to swap the two given numbers will be a program. Similarly instructions written in computer language to find out the roots of a quadratic equation will be another program. Likewise you can have program for other problems.

Now the question is how do you write program? Do you simply translate spoken English sentences in computer languages to build a program? Will, the sentence, "Swap the two numbers" translated into computer language, form number swapping program? Well, the answer is no. This is not programming. Programming is a process, in which the solution of the problem is framed, in light of all those activities, which computer can perform. For example, computer cannot perform the command "print largest number out of given two numbers". But it can certainly read numbers, compare them and print the given number. Thus to print the largest number, out of given two numbers, following steps will have to be performed:

1. Read first number.
2. Read second number.
3. Compare the two numbers.
4. If first number is greater, print it else print second number.

Now the question is, are these four steps a program for printing the largest number out of given two numbers? No, these steps are not program. They are basically framework of program. This framework is often called algorithm. When each step of algorithm is translated into a computer language, it is called program. The process of writing the instructions in computer language, to solve a problem, is often called coding. The person, who does coding, is called computer programmer.

# ALGORITHM

Algorithm is basically a step-by-step solution for solving a given problem. When solution is sought for solving the problem through computer, all the steps are framed in the light of basic operations that computer can perform. What computer can do is described below.

# WHAT COMPUTER CAN DO

Basic operations that computers can perform are very limited. These operations are:

1. Read data
2. Write data
3. Make comparison
4. Create variables and assign value to them
5. Do processing
6. Loop formation

An introduction of all these activities is given below.

## Read Data

Computer can read the data from three sources i.e. either from program itself or from the keyboard or from data file. It totally remains on the discretion of the programmers to select the source from where the data will be fetched in the program. All the programming languages provide commands for reading the data from all the three sources.

## Write Data

Computer can display the data and print it too. Data can either be printed on printer or in a file. All the programming languages provide commands for all these activities.

## Make Comparison

Computer can compare only two given values at a time. For example, if A and B are two given values then it can make comparisons like "IS A>B", "IS A<B", "IS A=B" etc.

The result of comparison is always either true or false. If the given condition holds good, the result of comparison is true. On the other hand, if the condition doesn't hold good, result is false. For building the conditions, following types of operators are used:

**Conditional Operators**

| Sign | Meaning | Sign | Meaning |
|---|---|---|---|
| = | Equal to | >= | Greater than or equal to |
| > | Greater than | <= | Less than or equal to |
| < | Less than | <> | Not equal to |

**Relational Operators**

| | |
|---|---|
| **OR** | Joins two conditions and gives true if either of the condition is true or both of them are true. In all other cases it gives false. |
| **AND** | Joins two conditions and gives false if either of the condition is false or both of them are false. In all other cases it gives true. |
| **NOT** | If the condition is true, it gives false. If it is false, it gives true. |

## Create Variables And Assign Values

Variables can be defined as memory locations, which are assigned a name for reference. For example, Number1=0 means Number1 is a variable whose value is 0. Similarly Pay=10000 means Pay is a variable, which holds 10000 in it.

Note that in computers, statements like Pay=Pay+1000 are used. Statements of such type are algebraically wrong but in reference to computer they are correct. This statement simply increases current values of Pay variables by 1000. For example, if its value was 10000 then after execution of this statement, it will become 11000.

## Do Processing

Computers can do calculations. Calculations that they do are called processing. In support of processing, computer languages provide following operators:

| Operator | Meaning | Operator | Meaning |
|---|---|---|---|
| + | Addition | / | Division |
| - | Subtraction | * | Multiplication |
| ^ | Raised to the power of (Exponentiation) | () | Brackets |

## Loop

Computer can process a single or more steps repeatedly. When it does so, it is said that a loop has been formed. Following is an example of loop:

1. A=0
2. B=A+10
3. Display B
4. Repeat Step 2 and 3, 4 more times.

Here step 4 forms a loop and makes a provision that the value of B gets increased and displayed five times.

## CHARACTERISTICS OF AN ALGORITHM

Following are the characteristics of algorithm:

1. Algorithms always have a definite starting point and an end point. These points are generally marked with the words like Start, Begin, End, Stop etc.
2. They consist of finite number of steps.
3. They always relate to a specific problem or you can say that they are written for a given problem.
4. They serve as foundation stone for programming.
5. They are written in easy language. Their sentences are always unambiguous and are easy to understand.

## WRITING ALGORITHMS

You are now familiar with algorithm and their characteristics, so let's write few algorithms for following problems:

***Problem-1***

Let's write algorithm for swapping contents of two variables.

***Algorithm-1***

1. Start
2. Read value of A
3. Read value of B
4. C=A
5. A=B
6. B=C
7. Print A
8. Print B
9. End

This algorithm makes use of intermediate variable C. It holds value of A in it, temporarily and later assigns it to the variable, B. Thus values of variable A and B get interchanged.

There is another way of interchanging the values i.e. without making use of intermediate variable. Algorithm for it is given below.

1. Start
2. Read value of A and B
3. A=A+B
4. B=A-B
5. A=A-B
6. Print A, B
7. End

***Problem-2***

**Let's write an algorithm to find the largest of given three numbers.**

***Algorithm-2***

1. Start
2. Read three numbers A, B and C
3. Let Big=0
4. IF A>B Then Big=A Else Big=B
5. IF C>Big Then Big=C
6. Print Big
7. End

***Problem-3***

**Let's write an algorithm for calculating and printing factorial of a given number.**

***Algorithm-3***

1. Start
2. Read a number, A
3. Let I=0
4. Let Sum=0
5. Sum=Sum*I
6. I=I+1
7. If I<A perform step 5, 6 and 7
8. Print Sum
9. End

***Problem-4***

**Let's write an algorithm for solving a given quadratic equation, $ax^2+bx+c=0$. Note that roots are determined by following formula:**

$$x=\frac{-b\pm\sqrt{b^2-4ac}}{2a}$$

***Algorithm-4***

1. Start
2. Read value of a, b and c
3. If a=0 Stop
4. Calculate values of discriminant $D=b^2-4ac$
5. If D=0 then there is one root p=-b/2a

6. If D>0 then there are two real roots

$$p = \frac{b+\sqrt{D}}{2a} \quad \text{and} \quad q = \frac{-b-\sqrt{D}}{2a}$$

7. If D<0 then there are two complex roots

$$p = \frac{-b+i\sqrt{-D}}{2a} \quad \text{and} \quad q = \frac{-b-i\sqrt{-D}}{2a}$$

8. Print p and q
9. Stop

## PROBLEMS WITH ALGORITHMS

As mentioned earlier, algorithms are written in English like sentences. Sentences are always subject to misinterpretation.

If the sentence is complex, different readers may interpret it differently. This definitely leads to problems at the time of coding.

To overcome this problem, often the solution of the problem is provided in pictorial form. Pictures carry more meaning and don't lead to ambiguity. Such type of step-by-step solution, provided in pictorial form is called flowchart. Details of flowchart are given below.

## FLOWCHARTS

When a step-by-step solution of a given problem is illustrated in the form of graphical chart that chart is called flowchart.

## FLOWCHARTS SYMBOLS

Flowchart is a universal tool, using which solution providers communicate the solution of a given problem to the programmers so that they could translate the steps (of flowchart) into computer language and transform it into a computer program. To make the process universal and unambiguous, symbols of flowcharts have been standardized. These symbols and their meaning are explained below.

## Read

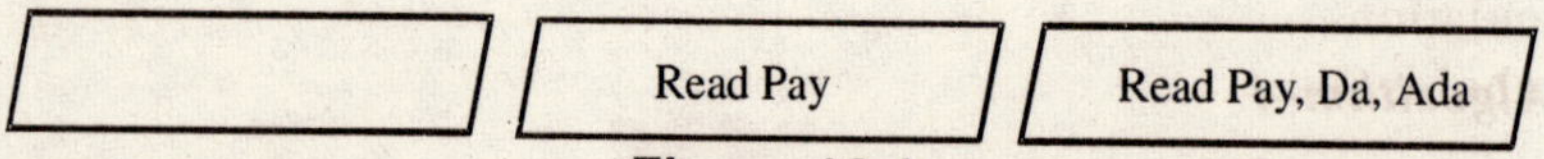


**Figure 10.1**

A parallelogram is used to depict reading process in the flowchart. To make the process more illustrative, the word "Read" is written along with variable names, within the parallelogram. For example, the word "Read Pay", written within the parallelogram states that value of Pay has to be read and stored in Pay variable. Similarly "Read Number" states that a number has to be read and the value has to be stored in a variable named, Number. Use of read symbol is illustrated in figure 10.1.

Often multiple values have to be read one after the other. In such cases, instead of making use of multiple read symbols and making the flowchart lengthy, one read symbol is used and name of all the related variables are written in it. This is illustrated in last symbol of figure 10.1.

## Write

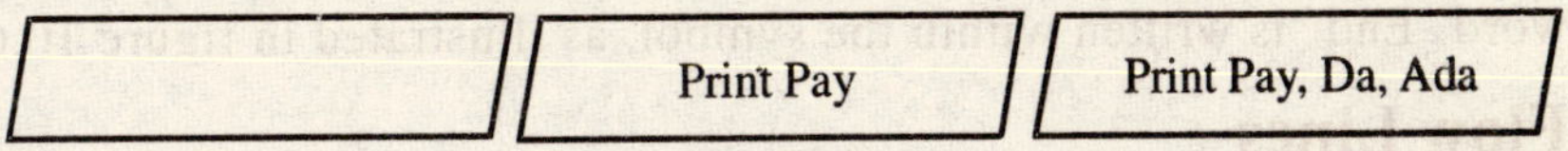


**Figure 10.2**

In flowcharts, parallelogram is also used to depict writing / displaying process. To make the writing process more illustrative, the word Write, along with names of the variables (whose values have to be printed / displayed) are written within the parallelogram. Use of write symbol is illustrated in figure 10.2.

Values of multiple variables can also be displayed / printed, using single symbol, as illustrated in figure 10.2.

## Condition Checking / Decision

Symbol shown in figure 10.3 is used for depicting comparison of two values or condition checking or decision-making etc. The condition that is to be checked is written within the figure.

It is quite obvious that whenever a condition will be checked there will be two outcomes i.e. either the condition will be true (condition holds good) or it will be false (condition doesn't hold good).

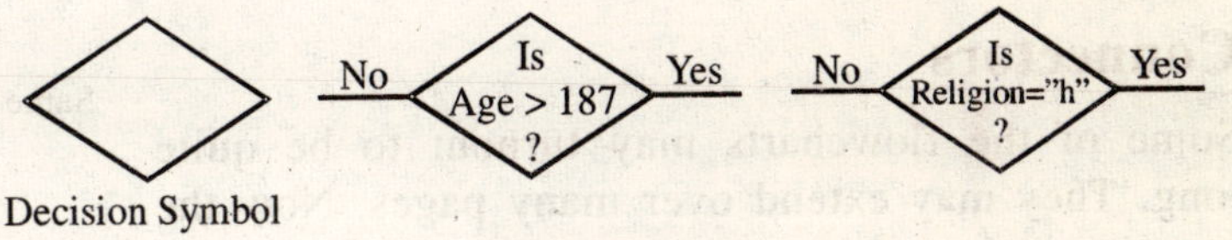


**Figure 10.3**

These two outcomes are shown as two branches coming out of the symbol. To clearly illustrate, which branch relates to true and which branch relates to false, the words Yes (or Y or True or T) and No (or N or False or F) are written near the branches. This process is illustrated in figure 10.3.

## Processing

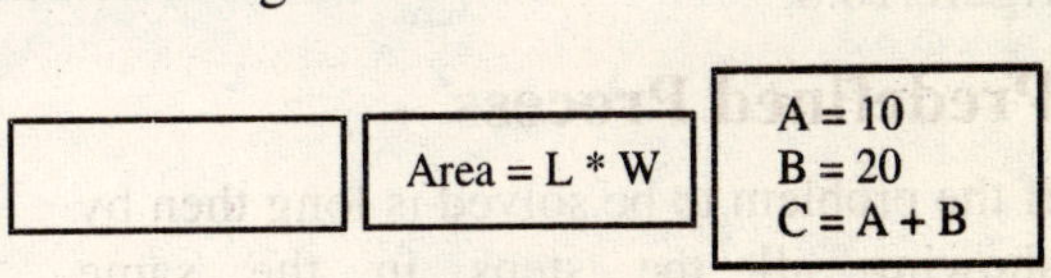


**Figure 10.4**

In flowcharts, processing or calculation activities are depicted, using a rectangle. To make the process more illustrative, calculations that are being done are written within the rectangle. The use of processing symbol is illustrated in figure 10.4.

## Start

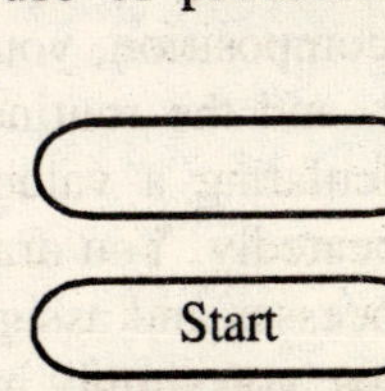


**Figure 10.5**

Like algorithm, flowchart must have a definite starting point. Starting point of the flowchart is depicted through a flat oval shape symbol, as illustrated in figure 10.5. To make the symbol more illustrative, the word "Start" is written within the symbol, as illustrated in figure 10.5.

## End

Like algorithm, flowchart must have a definite end point. End point of the flowchart is depicted through a flat oval shape symbol, as illustrated in figure 10.6. To make the symbol more illustrative, the word "End" is written within the symbol, as illustrated in figure 10.6.

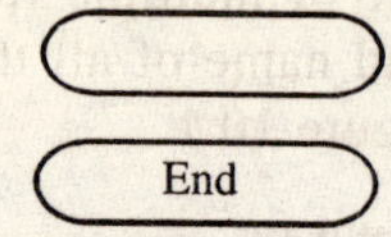


**Figure 10.6**

## Flow Lines

Joining many symbols together makes flowchart. As you know, each symbol of the chart represents an activity. Which activity will be conducted after which activity, is depicted with help of flow lines. A flow line is a simple line with an arrow at its end. The head of the arrow depicts the direction of flow. Flow lines are illustrated in figure 10.7.

For example, if the third step is connected to seventh symbol, using a flow line, it will indicate that after executing third step seventh step is to be executed. Step four, five and six are not to be executed in this case.

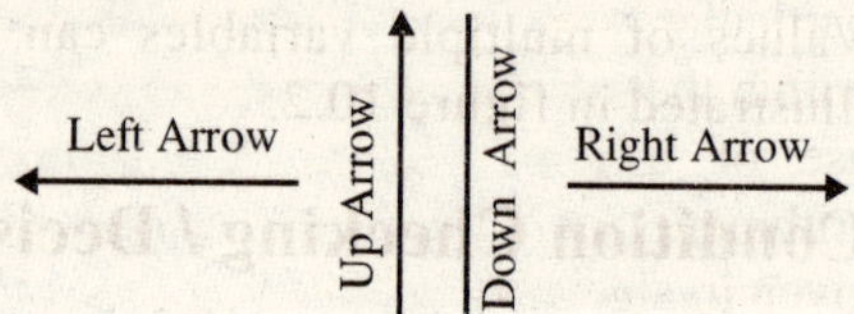


**Figure 10.7**

An up arrow is used for depicting a loop in the flowcharts. For example, if fifth step is connected to the third step, using an up arrow flow line, it will mean that after executing fifth step, third step is to be executed again. Note that this is nothing but a loop.

## Connectors

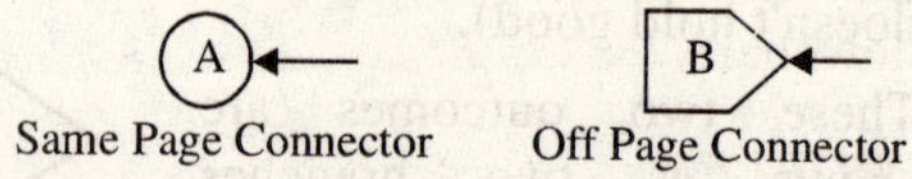


**Figure 10.8**

Some of the flowcharts may turnout to be quite long. They may extend over many pages. Now the question is, how do you connect the process, which are either far apart or are off the pages? For such type of requirements, two types of connectors are used. One is called same page connector and another is called "Off Page Connector". Both the connectors are shown in figure 10.8.

## Predefined Process

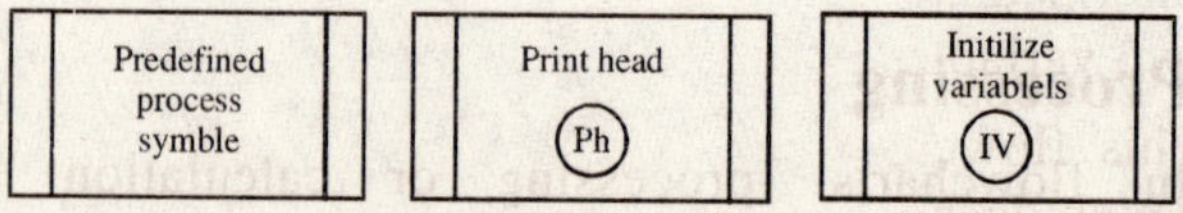


**Figure 10.9**

If the problem to be solved is long then by showing all the steps in the same flowchart may make the chart complex. To avoid the complexity and keep the flowchart simple, you have to decompose the problem. In decomposition, you identify the group of mundane routines, such as initialization of variables etc. and the routines, which are repeatedly required. For example, printing the header lines, calculating a value based on certain formula etc. are the tasks that have to be performed repeatedly. You draw a flowchart for these processes separately and declare them as predefined processes and assign them name for reference. In the main flowchart, instead of writing all the steps, you simply include the symbol of predefined process along with the name. Whenever one wants to go into the details of predefined process, he refers corresponding chart separately. The symbol and use of predefined process is illustrated in figure 10.9.

Note that each process has a unique name. With the help of name you identify the process and refer it in the flowchart. In figure 10.9, we have shown two predefined processes with the name "PH" and "IV".

From the description of flowchart, given above, it is quite clear that flowchart is nothing but a graphical representation of the solution to a given problem, in which you make use of standard graphical symbols and within the symbol, you write the details of the operation. Note that for communicating the details of the activities of a particular step, you have to include some text within the symbol. For this, you have to often make use of different types of operators. A list of arithmetic, relational and conditional operators is given in algorithm section.

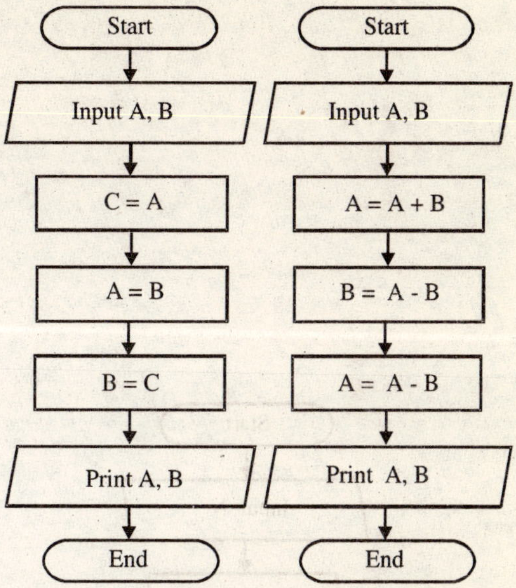


(a) (b)

**Figure 10.10**

## DRAWING THE FLOWCHART

Now you are familiar with flowchart definition and flowchart symbols, so let's draw few flowcharts for given problems.

### *PROBLEM-1*

Let's draw a flowchart for swapping the contents of two variables.

Flowchart for this problem is illustrated in figure 10.10(a). Note that this flowchart makes use of intermediate variable, C to interchange the values of two variables A and B.

This flowchart can be redrawn to depict the method, which doesn't make use of intermediate variable. Such a flowchart is illustrated in figure 10.10(b).

### *PROBLEM-2*

Let's draw a flowchart to find the larger of the two given numbers. Flowchart for this problem is illustrated in figure 10.11.

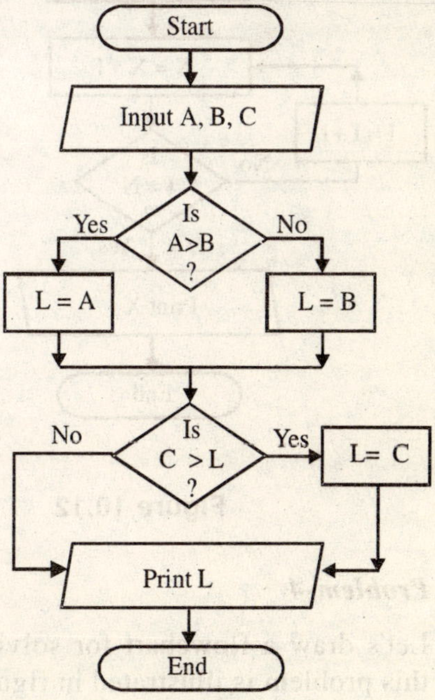


**Figure 10.11**

### *PROBLEM-3*

Let's draw a flowchart for calculating and printing factorial of a given number. Flowchart for this problem is illustrated in figure 10.12.

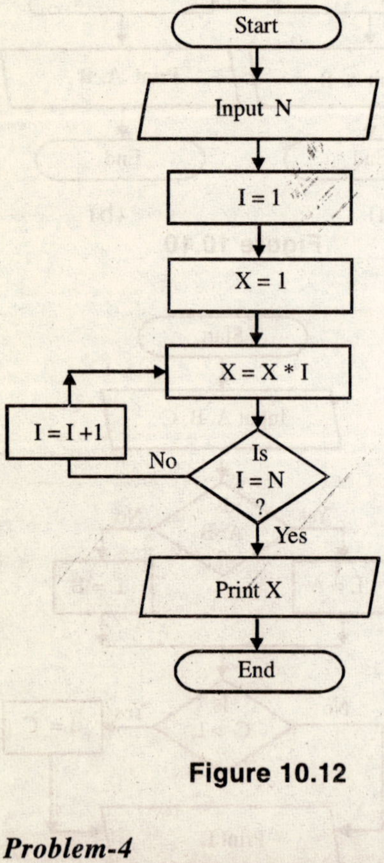


**Figure 10.12**

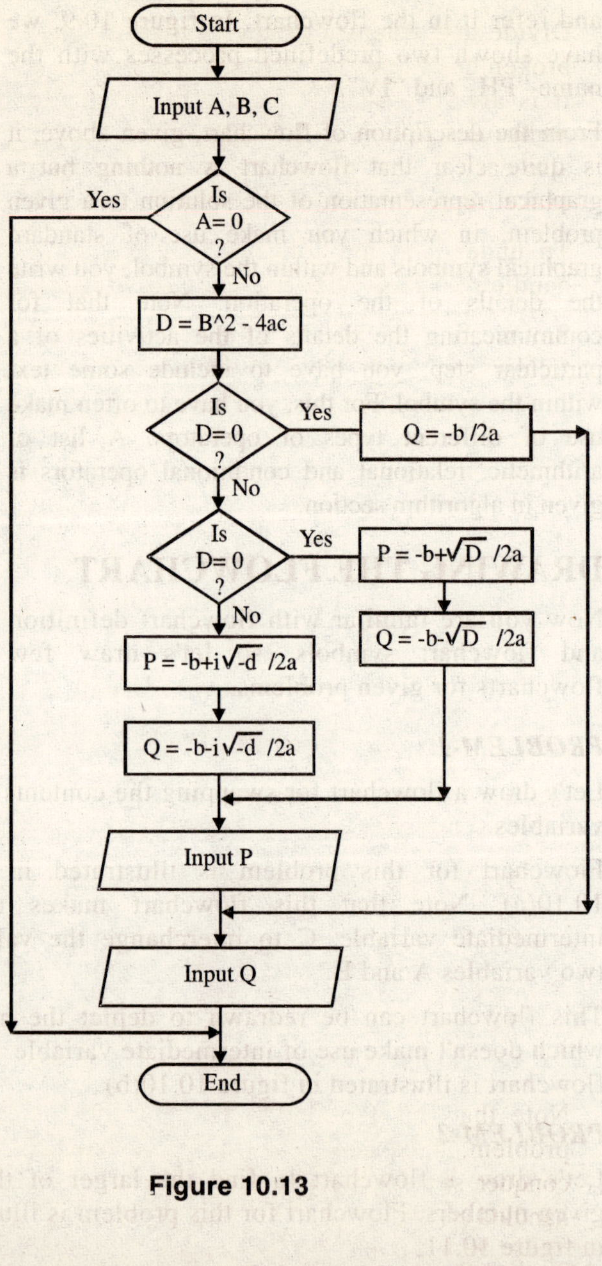


**Figure 10.13**

***Problem-4***

Let's draw a flowchart for solving a given quadratic equation $ax^2+bx+c=0$. Flowchart for this problem is illustrated in figure 10.13.

## DIVIDE AND CONQUER STRATEGY

Divide and conquer strategy suggests that a large or complex problem, should always be broken down into multiple smaller problems and each problem should be dealt individually. It is always easier to handle many smaller problems rather than one large problem. In addition to this, if suggested solution has some errors then finding errors and fixing them is also easy in case of smaller problems. In short, the thumb rule is, "Small Is Manageable, so Make Many Smalls and Manage Them".

To understand the concept, let's take an example of a large and complex problem and handle it, using divide and conquer strategy.

Say we have to calculate the net pay of the employee. Rules for calculating the pay are as follows:

1. Employee is given some amount of basic salary.
2. In addition to his basic salary, he also gets DA. DA is calculated @150% of his basic salary.
3. Employee also gets few other benefits, as follows:
    (a) CCA (City Compensatory Allowance) @ 20% of basic pay.
    (b) LTC (Leave Travel Concession) @ 15% of basic pay.
    (c) CA (Conveyance Allowance) @ 7% of basic pay.
    (d) HMA (House Maintenance Allowance) @ 5% of basic pay.
4. Apart from above mentioned earnings, some deductions are also made from employee's salary. Deductions are made as per following rules:
5. WA (Welfare Amount) is deducted @2% of basic pay.
6. IT (Income Tax) is deducted @5% of the total earning i.e. 5% of (basic pay + DA + Sum of all benefits – WA)
7. Net salary is calculated by adding all the earnings and deducting all deductions from it (i.e. Basic salary + DA +benefits - WA- IT).

Note that this problem is lengthy and bit complex also. If it is treated as single large problem, there may be difficulty in handling the problem. On the other hand, if divide and conquer strategy is applied and the whole problem is fragmented into small units, it can be handled easily.

To handle this problem in divide and conquer manner, we will have three modules. First module will be the main module, which will take care of primary activities. Other two modules will handle benefit calculation and income tax calculations. This strategy is applied in the given flowchart, show in figure 10.14.

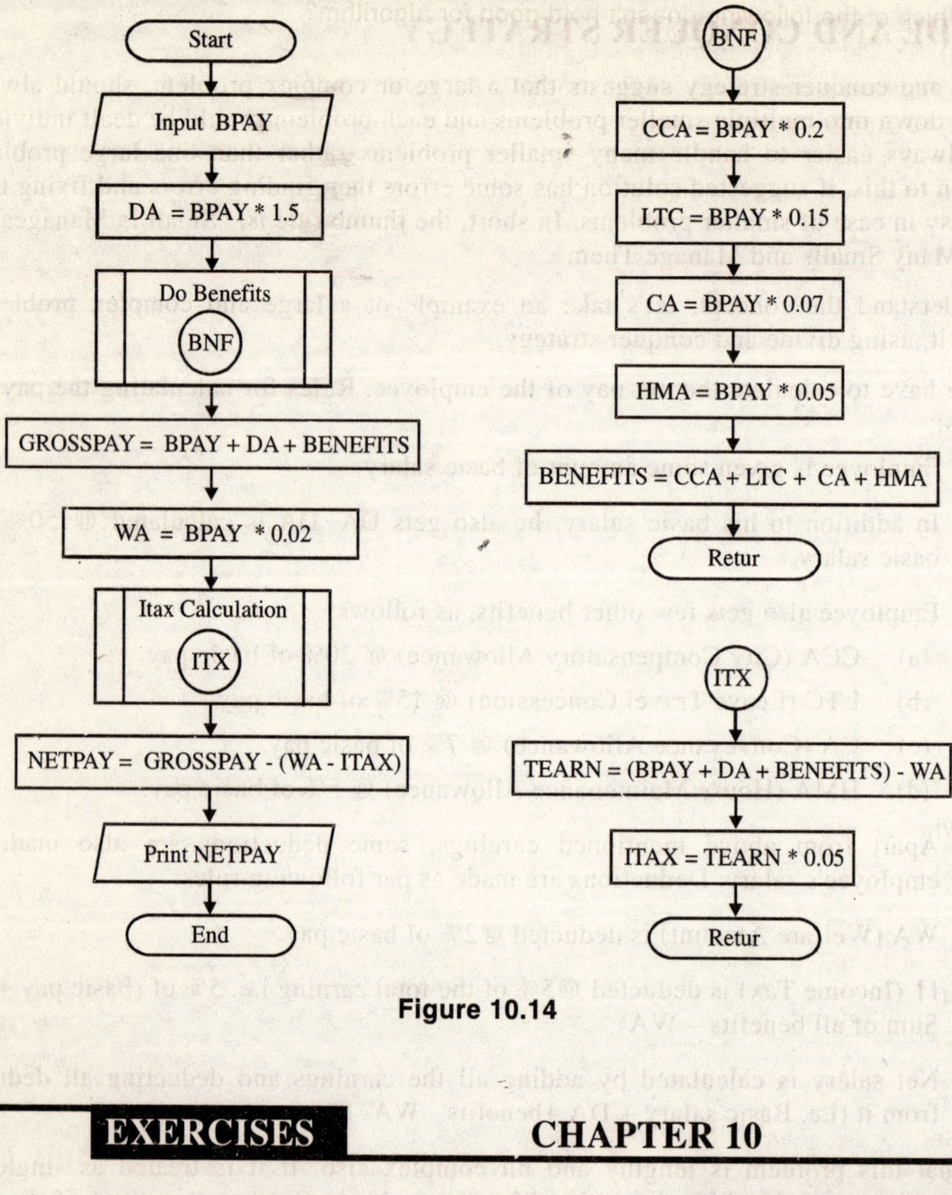


**Figure 10.14**

## EXERCISES — CHAPTER 10

### Short Type Questions

**A. Select most appropriate answers for following questions:**

1. Which of the following describes algorithm in best possible way?

   (a) Solution of a problem  (b) Best solution of a problem

   (c) Solution of a problem in simple sentences

   (d) Solution of a problem in simple and unambiguous sentences

2. Which of the following doesn't hold good for algorithm?
   (a) Should have a definite starts point
   (b) Should always present the most efficient solution
   (c) Should be written in simple sentence
   (d) Should have a definite end point
3. Which of the following cannot be included in an algorithm?
   (a) Reading data (b) Selecting one choice among may available
   (c) Comparison of two numbers (d) Processing
4. Which of the following is an example of variable creation?
   (a) a=100 (b) a<100 (c) a=<100 (d) a>=100
5. If A is true and B is false then what will be the result of A and B?
   (a) True (b) False
   (c) Depends upon other things (d) None of the above
6. If A is true and B is false then what will be the result of A OR B?
   (a) True (b) False
   (c) Depends upon other things (d) None of the above
7. What is graphical representation of the solution of a given problem, called?
   (a) Algorithm (b) Flowchart (c) Icon (d) Wizard
8. What does a parallelogram in flowchart represent?
   (a) Read (b) Process (c) Start (d) Condition evaluation
9. What does an oval sign represent in a flowchart?
   (a) Start (b) End
   (c) Both of the above (d) None of the above
10. To show a=100 *No + RT in flowchart, which symbol will be used?
    (a) Rectangle (b) Parallelogram
    (c) Diamond shape (d) Oval shape

**B. Fill in the blanks:**

11. Getting the value of age will be depicted as ...................... activity in algorithm.
12. Is a>b will be shown as ............... in flowchart.
13. First line of an algorithm always starts with .....................
14. Last line of algorithm is always, .........................
15. If A=10 then A=A+10 will make the value of A, .......................

**C. State, true or false for following statements:**

16. Before writing an algorithm, the problem has to be studied in detail.
17. Algorithm is always related to a problem.
18. Algorithm is always written in terms of basic activities that computer can perform.

19. Three outcomes of a given condition can be shown in a flowchart.
20. OR operator gives false only if all the associated conditions are false.
21. Flowchart is a pictorial representation of algorithm.
22. Read and write activities in flowchart are depicted using same symbol.
23. Processing activity in flowchart is depicted, using a square shape.
24. Loop in a flowchart is depicted using up arrow line.
25. Divide and conquer strategy fragments the complete flowcharts into many predefined processes.

## D. Answer the following questions in short:

26. Name the basic activities of computer, using which algorithms are written.
27. How will the condition, "Is A>B and B<C" be depicted in flowchart? Draw the symbol.
28. What will be written in algorithm, if the three steps 6, 7 and 8 are to be performed 10 times?
29. What will be the outcome of NOT A if A is false?
30. In which condition OR operator gives false?
31. In which condition AND operator gives false?
32. Draw the symbol, which will illustrated a=b+c statement.
33. Draw the symbols that are used for on page and off page connectors, in flowcharts.
34. Draw the symbol that is used for illustrating predefined process in flowcharts.
35. How loops are depicted in flowcharts?

**Detailed Answer Type Questions**

## E. Answer the following questions in detail:

36. What is algorithm? Write at least three characteristics of algorithm.
37. Write an algorithm for checking if the given number is an even number or odd number.
38. What are the similarities between algorithm and flowcharts?
39. Write an algorithm for converting feet into centimeters.
40. Write an algorithm to generate the sum of all even numbers starting from 1 to 100.
41. What is flowchart? How it is different from algorithm?
42. What is the use of predefined process symbol that is used in flowcharting? Explain with example.
43. Draw a flowchart to accept a number and print "Even" if it is an even number and "Odd" if it is an odd number.
44. Draw a flowchart to calculate and print the sum of following series:

    $1^2+2^2+4^2+16^2$
45. What is divide and conquer strategy? Explain using a suitable example.

****

# COMPUTERS

## CHAPTER 11

# *Structured System Analysis and Design*

Structured System Analysis and Design (SSAD) is a complete methodology that suggests practical guidelines for preparing a paper design of the application that is required to be developed for computerizing a given activity.

When we say application, we take few assumptions into considerations, i.e. the activity is at least of moderate size, and requires multiple programs to be development and integrated together in the form of a package. These assumptions have been specially elaborated here to highlight a fact that SSAD is for projects not for signal program or two. Note that design aspects of single program can be very well handled by flow charting methods. But designing of larger applications cannot be handled by flow charting. The issues involved in designing large systems, are more complex, they need more careful attention, more areas to be addressed and they require different type of treatment and this is the concerned work area for SSAD.

## WHAT SSAD DOES?

SSAD suggests proper guidelines for application development. Using these guidelines, you can:

- ❑ Generate a feed back mechanism that ensures that there is no communication gap between client's requirements and the deliverables provided by the developer.
- ❑ Analyze client's requirements properly and design the system accordingly.
- ❑ Create a design that suits current requirements and is flexible enough to take care of future modifications and expansions of the system.
- ❑ Ensure proper progress of the system, as per client's requirements.

## PHASES OF SYSTEM DEVELOPMENT LIFE CYCLE

SSAD mainly deals with application development. Application development processes is not just writing programs for the activity to be computerized. It comprises of many phases. These phases are listed below:

1. Conception phase
2. Initiation phase

3. Analysis phase
4. Design phase
5. Construction phase

All these steps put together are called System Development Life Cycle (SDLC). Each phase of SLDC has its own objective and comprises of well defined activities.

If these activities are carried out as per specifications of SSAD then at the end of each phase, a well defined document gets generated, which helps in carrying out the activities of the next phase.

These documents serve as communication media between the client, system analyst and the programmer. Using these documents, one can ensure that software development proceeds in the right direction and the programs are developed as per client's requirements. A detailed description of all the phases is given below.

## Conception Phase

Often client faces some kind of inefficiencies in day-to-day functioning of his organization. When he finds that these inefficiencies cannot be resolved internally at human level, he appoints a consultant.

System Analyst, from consultant's side visits customer's organization and meets the client. He studies the problem in detail and tries to define the problem in short and clear cut sentences, so that all his efforts, suggestions and relations get oriented towards the problem only.

He also tries to quantify the solution in terms of measurable parameters. Quantifying the solution in measurable parameters means that he tries to list down, how the effect of the proposed solution, would come to notice. For example, the identified problem is that, "Dispatch of electricity bills get delayed every month". Solution could be "Computerization of bill generation process". But the question is how do we quantify the solution? Solution can be quantified by saying that, "By computerizing electric bill generation process, you will be able to produce all the 50,000 bills by 20$^{th}$ of every month". Note that this figure of 50,000 and 20$^{th}$ is not a random figure for him. By observing the job volume, taking various factors into consideration and doing few standard calculations, he arrives at this figure.

After this, he prepares a document called Feasibility Report and submits it to the client for review and approval.

Note that all these activities carried out up till now, fall under conception phase of SDLC.

After this, client goes through feasibility report and if he finds it convincing then he asks the System Analyst to study the functioning of all related areas in detail. From here onwards the initiation phase starts.

## Initiation Phase

System Analyst studies the complete functioning of the manual system and proposes a complete solution. He also prepares the list of items, on which the investments will be required and makes a comparison chart of benefits verses investments (called cost benefit analysis). He also prepares schedule of time and cost estimates for the completion of the projects.

Based on the details of this study, he prepares a detailed document called **system proposal** and submits it to the client for approval.

Note that at this stage, Initiation phase ends and Analysis phase starts.

## Analysis Phase

After receiving the system proposal from consultant's side, client goes through system proposal and if he finds it worth, he places an order for software development. From here onwards analysis phase of SDLC starts.

After receiving the order, consultant again sends System Analyst to client's site for finalizing the following:

- ❑ Outputs and their formats.
- ❑ Input and their formats.
- ❑ Nature of data and their retention period in the system.
- ❑ Procedures that will transform the input into output.
- ❑ Acceptance criteria for the system (on what basis the system will be accepted by client).

After studying these details and finalizing the procedures, System Analyst prepares a document called **functional specifications** and submits it to the client for approval. Acceptance from the client's side is a commitment that this is the final system that has to be developed as per the procedures listed in functional specifications document. This is a written assurance that there is no information gap between client and consultant.

At this stage, analysis phase finishes and design phase starts.

## Design Phase

In design phase, System Analyst designs the complete architecture (paper design) of the package. He decides file structures, codification structures, program specifications, menu designs etc. All these design related details are written down on paper and are illustrated through standard symbols in the form of a diagram. This design is often called **system design.**

SSAD provides detailed guidelines, procedures and symbols for finalizing the system design.

The finalization of System design completes design phase.

After this, a team of programmers is constituted. Generally the System Analyst is the group leader of the team. Team members are assigned specific tasks with specific time schedules. They are also provided with test data, using which they have to test their programs.

## Construction Phase

Programmers write the programs, compile them and execute. After satisfactory completion of the program, they test the program, using test data given to them. For any error found, the program is checked and mistakes are rectified. Side by side programmers keep on doing documentation also.

Periodic meetings are held among the team members and the group leader. Group leader is apprised of the progress. On the basis of the feed back given by the team members, if required, he takes corrective actions and tries to maintain the time schedules. He also provides guidance to all the members and keeps them motivated and in high spirit.

In this phase **product manual and user manuals** are also written for the package.

After completion of all the programs the complete system is tested on the basis of test data. Any mistake found at this level is corrected by modifying the corresponding program. After successful run of the system, programs are taken to the client's site and the package is loaded on client's computer.

Package is demonstrated running. System acceptance procedures are carried out (which were listed in functional specification) and the system is accepted by the client.

User's training is also carried out so that they get familiar with the new working environment.

The progress of system development life cycle is illustrated in the following figure:

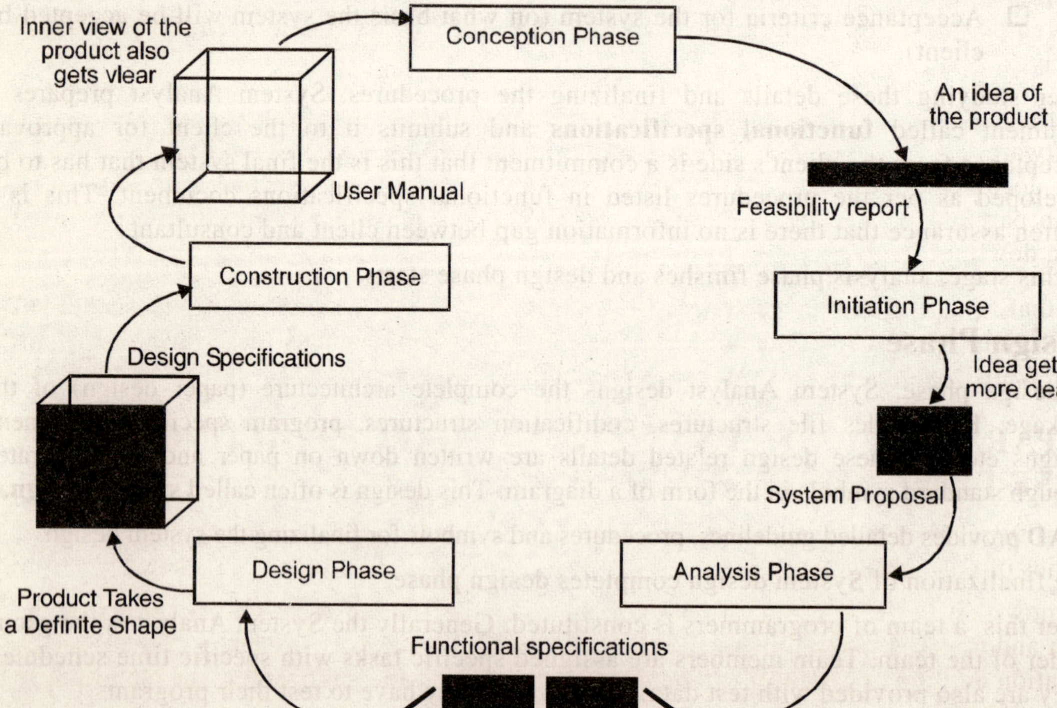


**Figure 11.1**

## NEED FOR PROPER DOCUMENTATION

Documentation means writing the things in detail and in understandable format, so that in future, if any clarification is to be sought then the related documents can be referred. It also serves as means of communication between two persons. Since software development is a large activity and requires communication between different persons hence documentation plays a major role in the success of the project. Better is the documentation; higher are the chances of success of the project.

You may note that in practical situations, it may often so happens that the person who interacts with the client initially and studies the system may not be the same person who designs the system. He may be a different person. In majority of the projects, the person who designs the system is not the person, who writes the program. One, who writes the program, is not the user of the system. So the persons keep on changing from one stage to another. But at the same time software development activity demands that there should not be any communication gap at any stage. The moment, communication gap occurs, it is likely that software will deviate from its specifications. For example, if the system analyst does not understand client's need properly then a system with unwanted features will get designed. Similarly, if System Analyst does not convey the requirements properly to the programmers, programmers will not develop correct programs. Similarly, if the users are not given proper training and proper manuals, they will not be able to operate the package properly.

Now the question is how to ensure that a proper communication takes place at each stage. In Software Development Life Cycle a proper communication is ensured by the means of proper documentation. In all the stages of Software Development, documentation has to be done in parallel. All the facts, findings, procedures and outcomes of meetings have to be written down properly and filed for future reference.

Note that the documents such as project request form, system proposal, functional specifications, design specifications, run manual and system manuals (described above) are part of Well-Documentation strategy, suggested by SSAD methods.

## PREPARATION OF FEASIBILITY REPORT (CONCEPTION PHASE DOCUMENT)

Feasibility report is conception phase document. It is generated after carrying out all the activities of conception phase.

You know that in the conception phase, System Analyst makes a preliminary study of the system and prepares a feasibility report. In this report he makes cost benefit analysis of the application and the expenditures to be done in developing the application. He submits it to the client for his approval. Feasibility report generally has the following contents in it:

| **Feasibility Report** | |
|---|---|
| For : <Client's name><br>Project title : | From <Consultant's name><br>Date : |
| Description of the problem | 1-----<br>2-----<br>3----- |
| Solution | 1--<br>2--<br>3-- |
| List of benefits | 1.<br>2.<br>3. |
| Expected Cost<br><br><br><br>Total | Break up of all expenditures -<br>-<br>-<br>———<br><br>——— |
| Cost Benefit Analysis | -<br>-<br>- |
| Scope of the Project | Name of the identities getting affected<br>1<br>2<br>3 |
| Signature of System Analyst | 1<br>2 |

## PREPARING SYSTEM PROPOSAL REPORT (INITIATION PHASE DOCUMENT)

System proposal report is initiation phase document. It is generated after carrying out all the activities of initiation phase.

After doing the system study and finishing other tasks like, finding the alternate solutions, cost benefit analysis and time schedules etc. System Analyst prepares a report, called system proposal report. Contents of system proposal report are shown in following report format:

| **System Proposal** | |
|---|---|
| Project Title ___________ | Date ________ |
| Study Methodology | |
| Recommended Solution | |
| Alternative Solutions considered | |
| Cost benefit Analysis | |
| Time Schedules | |

## SYSTEM SPECIFICATION REPORT (ANALYSIS PHASE DOCUMENT)

System specification report is analysis phase document. It is generated after carrying out all the activities of analysis phase.

After finalizing all these issues they are documented in the form of a report, called system specification report. Contents of system specification report are shown in following report format:

| **System Specification Report** | |
|---|---|
| Project Title ___________________ | Date _______________ |
| Outputs to be produced | |
| Inputs to the system | |
| Procedures | |
| Data to be retained | |
| Audit & Control requirements | |
| System Acceptance criteria | |

This report is prepared in duplicate. One copy is submitted to the client and one copy is retained in consultant's office for future reference. Client goes through the report, thoroughly and makes sure that all the outputs are as per his requirements, inputs are in order and the procedures mentioned are exactly same as those, which are being followed in the organization. He also reviews, system acceptance criterion and audit controls. If he finds all the contents in order, he gives his consent (in writing) to proceed further else he asks System Analyst to make modifications at all the required places.

## Data Flow Diagrams (DFD: Analysis Phase Document)

You know that at the end of analysis phase, System Analyst has to be sure of the way, data flows in the physical system and the way it gets transformed into outputs like reports, statements etc. In order to be very clear about all these issues, SSAD suggests that Data Flow Diagrams, called DFDs in short, should be prepared.

A data flow diagram is basically a chart, which is made using standard graphical symbols to illustrate the flow of data in a system. When the chart is made for illustrating the flow of data in physical system, it is called physical DFD. When it is made to illustrate the data flow in computer system, it is called logical DFD. What symbols are used for making DFDs, how they are made and what rules are followed for making the DFDs are the issues that are explained below.

## DFD SYMBOLS

Following are the symbols that are used for making DFDs. These symbols are called Gane-Sarson symbols for DFD.

Explanation of each symbol is given in figure 11.2.

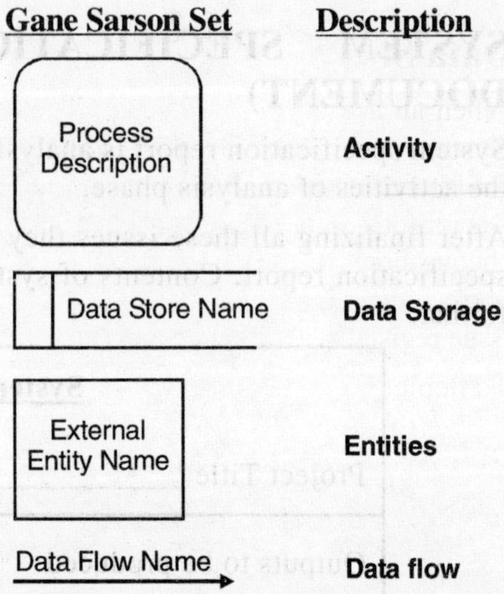


**Figure 11.2**

## Process

Any operation that takes place in the system is represented as process. Each process may require one or more data input and may produce one or more data outputs. You can say that process transforms the input data into output data. To understand the concept, consider an example, say bank manager accepts a cheque book issue request from the client and issues him a cheque book. This activity of accepting the request and issuing the check book is a process. In data flow diagramming, process has to be represented by the symbol, as illustrated in figure 11.3.

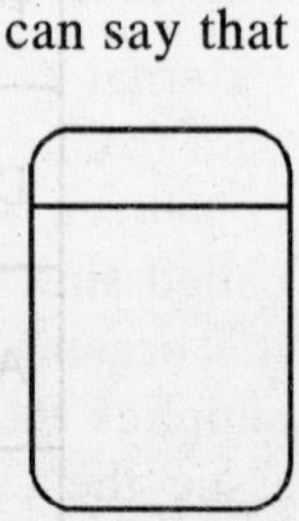

**Figure 11.3**

The above illustrated symbol states that some process is taking place in the system, but it doesn't indicate which type of process is taking place. So to make the process indicative of its functions, the process has to be given a name. Name, which is indicative of process's function, is written in the lower part of the symbol, as shown in figure 11.4.

If the activity is large, many such processes will be there in the system. Obviously each one of them will have its own name and indicate, the type of function being performed. For example, after getting the cheque book, person gets the facility of withdrawing the money from the account, where he fills the cheque and gives it to the cashier; cashier checks the balance amount in his account and gives him money. So to depict this activity, you would need another process, say "Pay money by cheque". So, now the question is how we depict, which process is first process and which process is second process. For this, a process number is included in upper portion of the process box, as illustrated in figure 11.5.

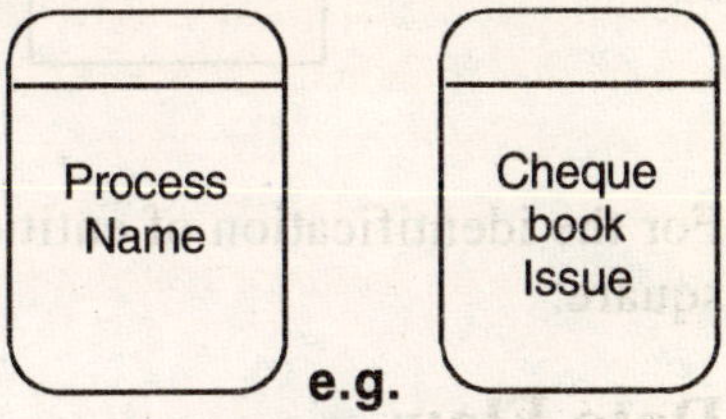


**Figure 11.4**

## Data Store

When an activity takes place in a system, it is likely that it will generate some data. Generally the data is retained in the system, for future reference. Retained data in SSAD terminology is called Data Store. For example, when the bank manager issues the cheque book to the client, he notes down the details in "cheque book issue register". In SSAD terminology, this cheque book issue register will be termed as data store.

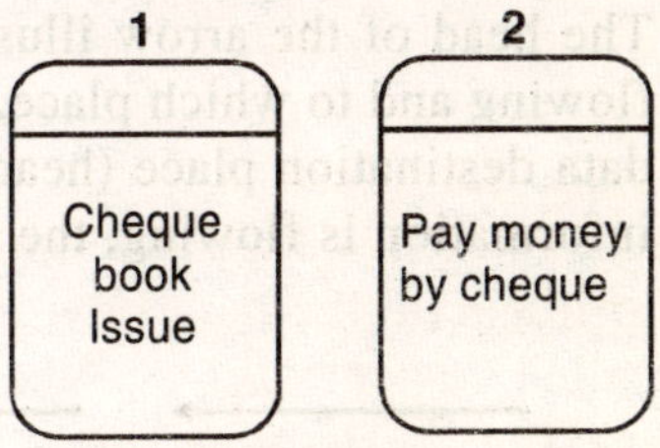


**Figure 11.5**

To depict the data store, following symbol is used.

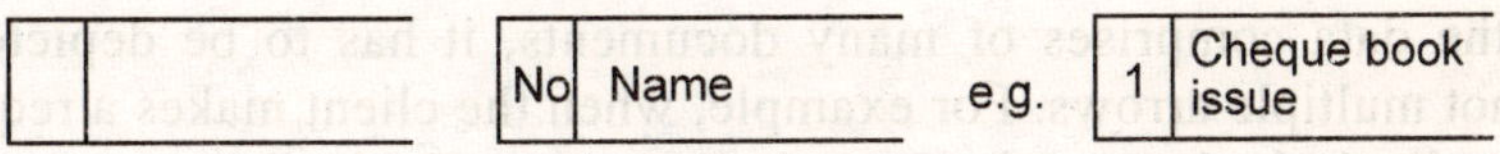


**Figure 11.6**

In the case of data stores also they are assigned a name and a unique number for the reasons mentioned above.

## External Entity

External entities are the outsiders who either supply the data to the system or receive the data from the system. Those entities that supply the data to the system are often referred to as **source entities** or simply **source** and those who receive the data from the system are called **sink entities** or **sink.** External entities can be people, organization or a unit, within the organization, which is not directly using the system but either receives the data or supplies the data. For example, the client who gives application to the bank manager, to issue the cheque book is an external entity. Similarly, head office of the bank, which supplies printed blank check books to the branch bank is an external entity. In a DFD, an external entity is represented as shown in figure 11.7.

**Figure 11.7**

For the identification of entities, each entity is assigned a name, which is written within the square.

## Data Flow

When an activity has to be performed, it requires data as input. After performing an operation on it, it produces data. In this process the data has to move from one place to another.

In DFDs, this data flow is shown with the help of headed arrow, as shown in figure 11.8. The head of the arrow illustrates the direction of flow. To identify from where the data is flowing and to which place, arrow has to start from the data originating place and end at the data destination place (head of the arrow pointing at end point). To identify, which data or information is flowing, the arrow has to be labeled as shown below.

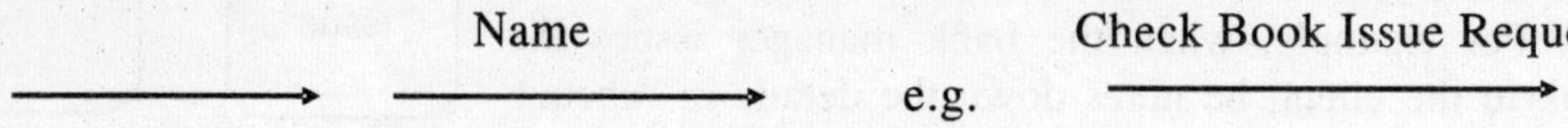


**Figure 11.8**

In addition to the naming conventions, there are other common rules for data flow. These rules are as follows.

1. When the data comprises of many documents, it has to be depicted using single arrow not multiple arrows. For example, when the client makes a request for issuing a cheque book, he has to give two documents to the bank manager viz. request for issue of cheque book and the pass book for verification. To show this data flow, only single arrow will be used not the double arrow.

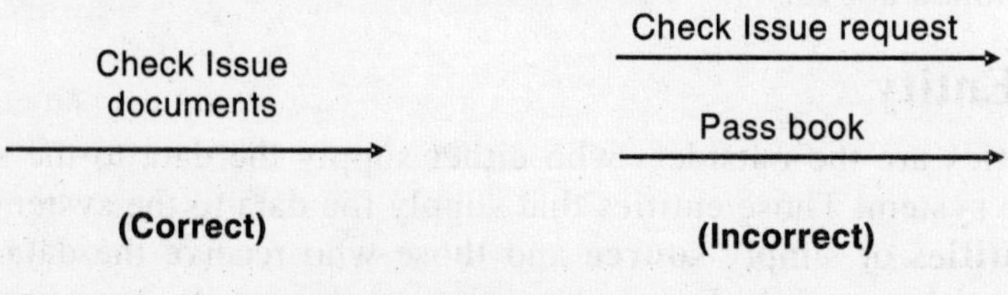


**Figure 11.9**

2. When multiple copies of a report are being produced then also the data flow has to be depicted with the help of single arrow only.

# RULES TO BE FOLLOWED

For drawing the data flow diagrams, following rules have to be followed:

1. No direct data flow can take place between two external entities. For a communication between them there has to be at least one process in between them, as illustrated in figure 11.10.

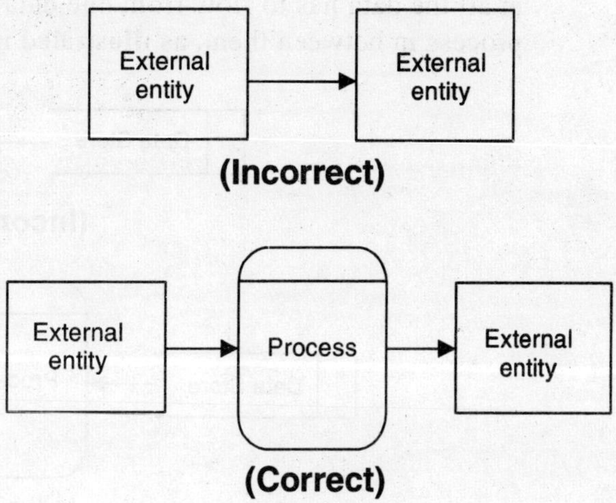


**Figure 11.10**

2. No direct data flow can take place between an external entity and data store. To get the data stored in the data store, there has to be at least one process in between them, as illustrated in figure 11.11

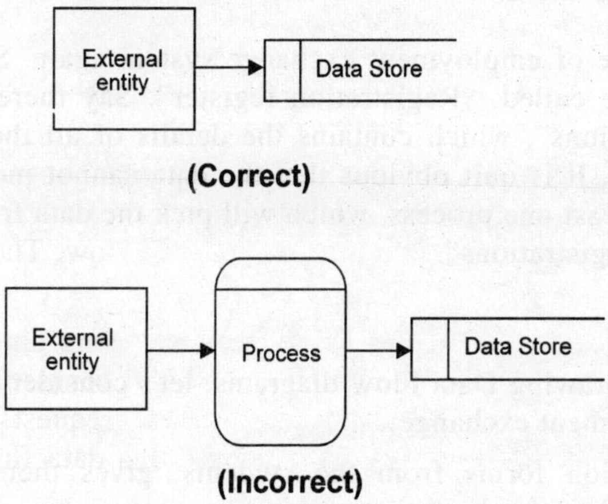


**Figure 11.11**

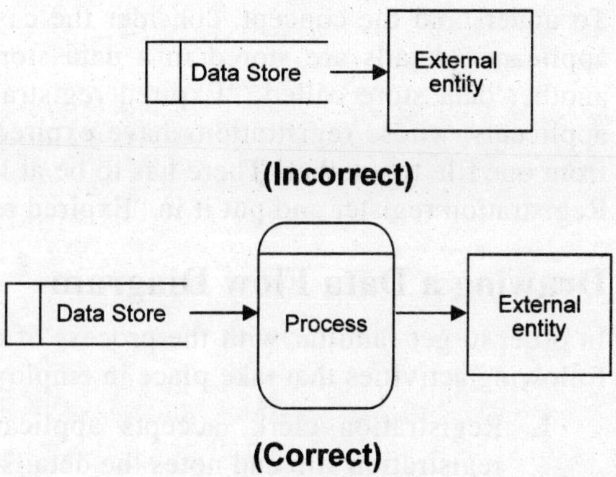


**Figure 11.12**

To understand the concept, consider the case of employment exchange. Applicant cannot directly store the data in registration register. There has to be a process in between them (say registration), which will transfer the data in application form, into registration register.

3. No direct data flow can take place between data store and external entity. If a communication has to take place, there has to be at least one process in between them, as illustrated in figure 11.12.

To understand the concept, consider the case of employment exchange system again. The name of the applicant can not directly move from Job register to employer. There has to be a process, say "Selection" which will transform applicant's details in the form of a letter, which will be posted to the employer.

4. No direct data flow can take place, between one data store to another data store. If at all the data has to flow from one data store to another data store, there has to be a process in between them, as illustrated in figure 11.13.

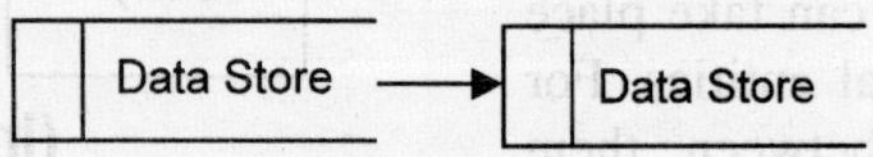


**(Incorrect)**

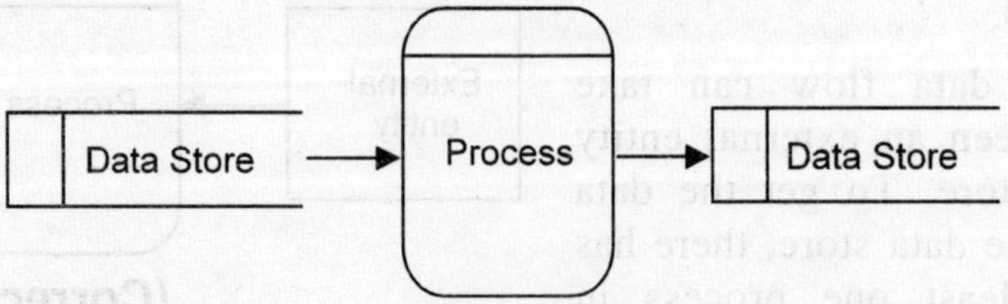


**(Correct)**

**Figure 11.13**

To understand the concept, consider the case of employment exchange system again. Say applicants' details are stored in a data store called, "Registration register". Say there is another data store called, "Expired registrations", which contains the details of all those applicants, whose registrations have expired. It is quit obvious that the data cannot move from one file to another. There has to be at least one process, which will pick the data from Registration register and put it in "Expired registrations".

## Drawing a Data Flow Diagram

In order to get familiar with the process of drawing Data Flow diagrams, let's consider the following activities that take place in employment exchange:

1. Registration clerk, accepts application forms from the students, gives them a registration slip and notes the details in the registration register.
2. Say the posting clerk receives applicant requirement letters from the employers. He writes all the details in job register and finds a suitable match. After that he sends selection information in the form of a letter to the employers, asking them to contact the candidates directly.
3. Posting clerk also sends a call letter to the applicants also. He informs them that they have to report at employer's office for interview.

Data Flow Diagram for the whole process can be drawn as shown below.

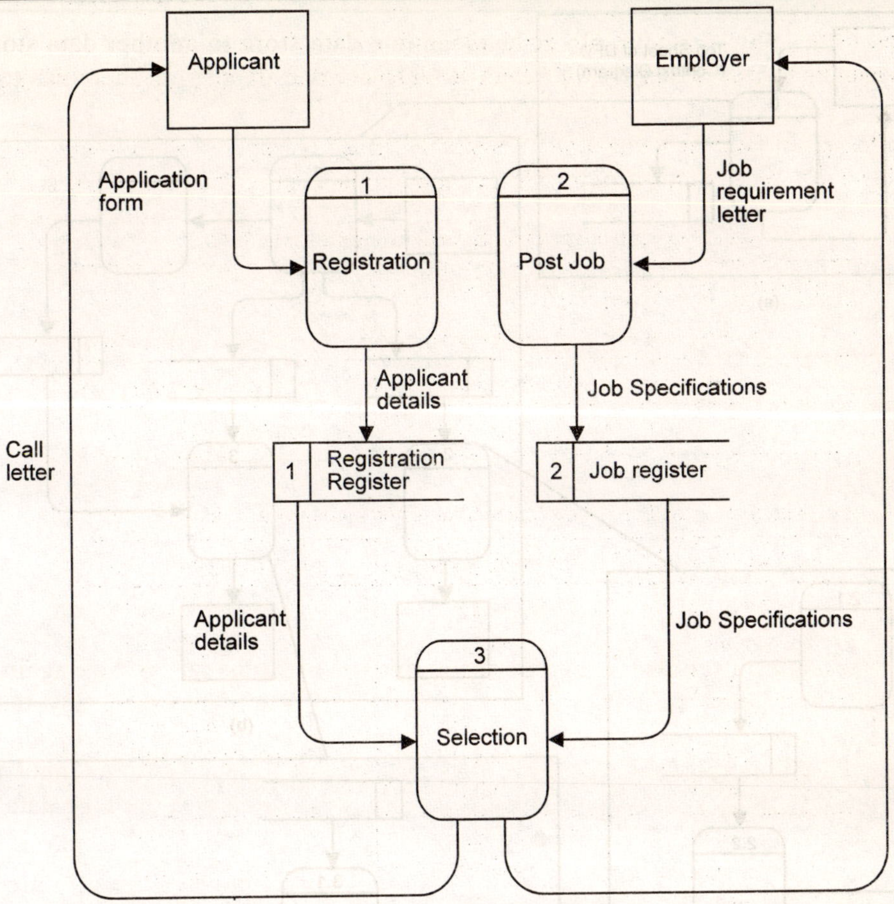


**Figure 11.14**

# CONCEPT OF CONTEXT DIAGRAM AND EXPLODING

An immediate exposure of the complete system, at once, to anybody may not yield desired results. Presenting the complete Data Flow Diagram, to him for understanding the system, is not the right approach. He may get lost into the crowd of symbols and data flows and get confused. So the right approach for presentation is that the person should be gradually exposed to the system. For example, first he should be given a brief idea of overall functions performed by the system then each function should be described in detail.

The diagram that shows one and only one process and defines the scope and boundaries of the system is called Context Diagram. In other words you can say that context diagram shows only those data flows that represent the main objective or most common inputs and outputs of the system.

The process of detailing each activity of context diagram (or some other DFD) through detailed Data Flow Diagrams is called exploding. The concept of exploding is illustrated in figure 11.15.

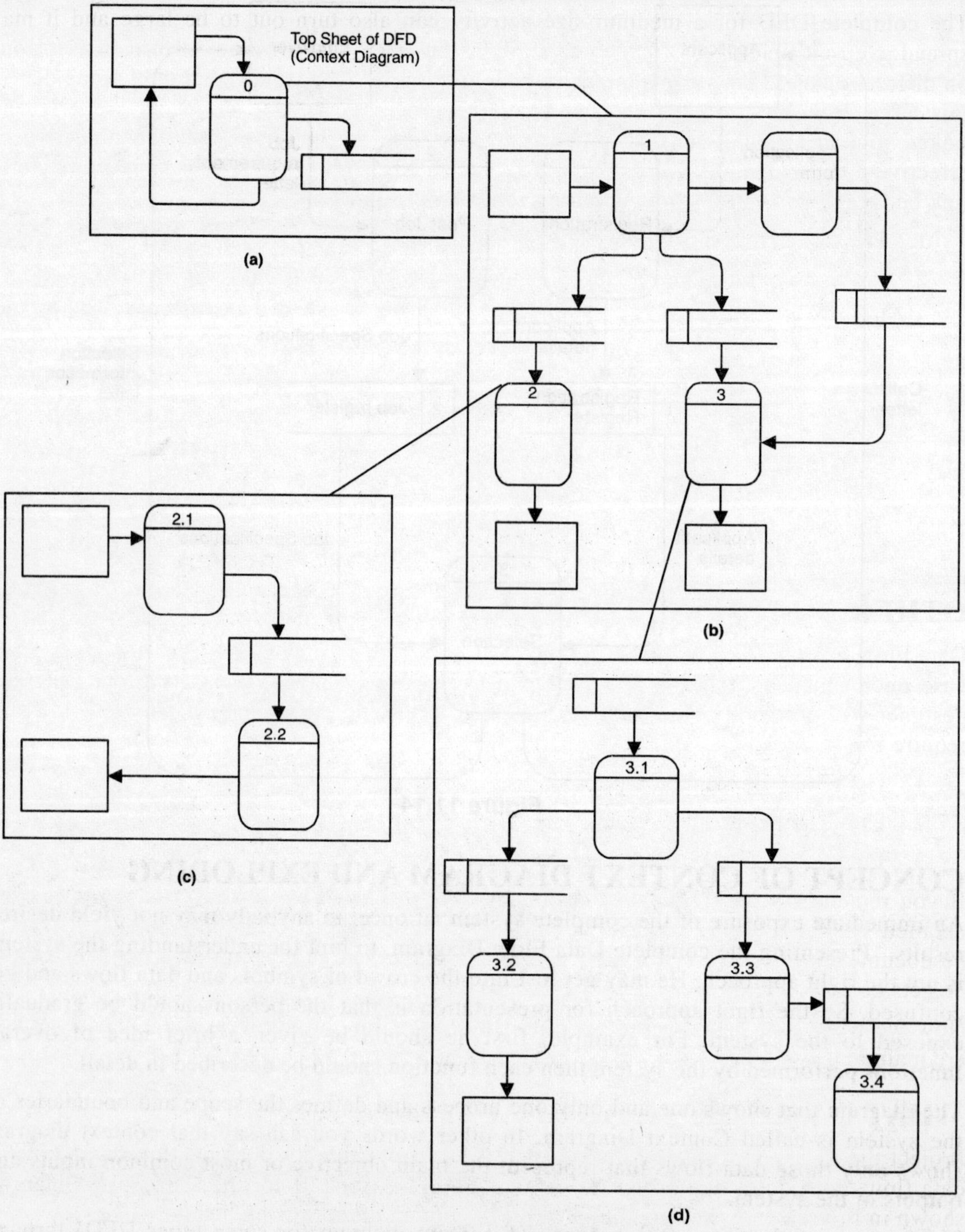


**Numbering the Exploded Process**

**Figure 11.15**

The complete DFD for a medium size activity can also turn out to be large and it may spread over multiple pages. When you explode the chart, you illustrate different explosions on different pages. Thus complete chart becomes a pile of pages. It is quite obvious that in this large crowd of symbols and charts, if the processes are not numbered properly and linked together then it will be difficult to correlate the activities and the objective of effective communication will get lost. So to correlate the processes and maintain a natural link between them, following numbering scheme is followed:

1. Process of context diagram is always numbered as 0 (zero).
2. Processes of second level DFD are numbered as 1, 2, 3, 4, 5 and so on.
3. Explosion of nth process is numbered as n.1, n.2, n.3 and so on. For example, the explosion of third process of second level DFD would result into 3.1, 3.2, 3.3 etc.
4. Explosion of n.m$^{th}$ process would result into the processes that are numbered as n.m.1, n.m.2, n.m.3 etc.

   For example, the explosion of the process, 3.2 will result in a DFD that contains the processes that are numbered as 3.2.1, 3.2.2, 3.2.3 etc.
5. The cycle continues with more points in between the numbers, such as 3.2.3.1, 3.2.3.2 etc.

## OTHER DOCUMENTS OF ANALYSIS PHASE

Data Flow Diagram is a graphical representation of flow of data in a system and it serves as base document for designing the system. But by merely having the Data Flow Diagram, designer will not be able to design the system, because to design the system he would also require following:

1. Contents of each data flow.
2. Contents of each data store.
3. Functional details of all the processes shown in DFD.

If you remember, these details have already been collected by System Analyst during the detailed system study. In order to communicate these details to the designer, they have to be documented properly and linked to DFD so that when he goes through documentation, he gets complete details for designing the system.

Now we will focus our attention on data flow, data store and process logic. These details are maintained in a document called Project Dictionary.

## WHAT IS PROJECT DICTIONARY

Project dictionary is a catalog that System Analyst creates, for documenting the details of data flow, data stores and processes of the DFD. For example, for the Data Flow Diagram, shown in figure 11.16 the data dictionary will contain the following details:

1. Contents of all the data flows i.e. application, accepted application, applicant details and call letter.

2. Contents of registration register.
3. Processing details of the two processes i.e. "Registration" and "Send Call Letters".

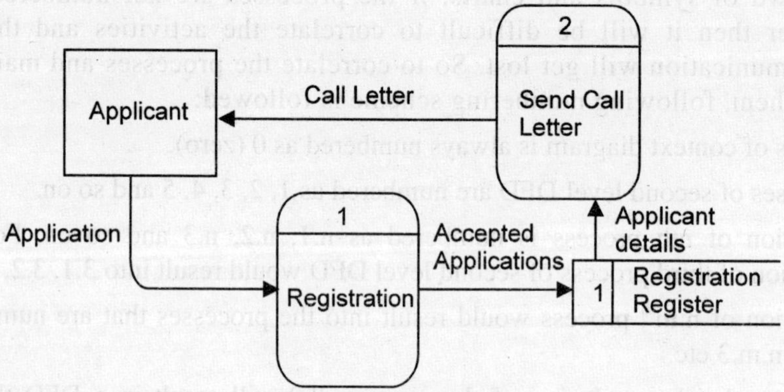


**Figure 11.16**

## Data Flow Entry In Project Dictionary

Data flow entry in data dictionary may consist of four parts as illustrated in the following format:

**Data Flow Name** : *Application*

**Alias** : *Applicant Details*

: *Accepted application*

**Composition** *APPLICATION*

*Name*

*Father's name*

*Date of Birth*

*Post applied for*

*Amount*

**Note** :

**Figure 11.17**

## Data Element Entry In Project Dictionary

Data element entry, in data dictionary may consist of four parts as described below:

**Data element Name** : *Religion*

**Alias**

**Value and Meaning**

| | |
|---|---|
| **H** | *Hindu* |
| **C** | *Christian* |
| **M** | *Muslim* |
| **O** | *Others* |

**Note**

**Figure 11.18**

## Data Store Entry In Project Dictionary

Data stores, physically store the data for future use. To use their data, you have to extract the data from them. When the data volume is large, extracting desired data from them may take long time. To avoid the delays, they are often organized in some special way. For example, they may be sorted or indexed, as the case may be.

Recall that data stores are also described with the help of data structures, so their definition resembles that of data flow, but they contain organization heading also, which describes the data element, on which they are organized. The entry for describing the data store in the data dictionary is described below.

**Data Store Name** : *Application Register*

**Aliases**

**Composition**

*Application*
*Application Number*
*Name*
*Father's name*
*Post applied for Fees*

**Organization** : *Application No*

**Note** :

**Figure 11.19**

## Process Logic Entry In Project Dictionary

Most common method of describing the process logic is that, the System Analyst gathers the details from the client and notes it down in simple English sentences and includes them in the project dictionary. This is the easiest method of explaining the process logic. But it leads to ambiguity problem. To overcome this problem, following tools are used for describing the process logic:

1. Structured English
2. Decision tree
3. Decision table

# ACTIVITIES OF DESIGN PHASE

After analysis phase, comes the design phase, in which the system designer designs the system. Now the question is what we mean by designing the system. Designing the system means, deciding upon the following factors:

- ❑ Codification
- ❑ File Structure
- ❑ Output designs
- ❑ Program Specifications
- ❑ Controls
- ❑ User interface
- ❑ Test Data preparation

# ACTIVITIES OF CONSTRUCTION PHASE

This is the phase, in which most of the programming activity takes place. If it is so then why don't we call it Programming Phase? The answer is that, the contraction is more than programming. It involves many more activities, as described below:

- ❑ Programming
- ❑ Testing & debugging
- ❑ Documentation
- ❑ System implementation
- ❑ System acceptance test
- ❑ Maintenance

At last, we can summarize that Structured System Analysis and Design suggests a complete methodology for system development life cycle. It suggests many activities and few documents that should be produced during the system development life cycle. It also suggests standard symbols and formats that should be used for generating the documents.

By following the guidelines of SSAD not only proper communication is established between client, System Analyst and the developers but a professional working environment is also created that results into the development of efficient applications, cost savings and client's satisfaction.

# EXERCISES CHAPTER 11

## Short Type Questions

### A. Select most appropriate answers for following questions:

1. With which of the following structured system analysis and design is related to?
   (a) Designing of computer (b) Programming
   (c) Designing and development of application software (d) Documentation
2. Which of the following is not an independent phase of system development life cycle?
   (a) Conception phase (b) Analysis phase
   (c) Construction phase (d) Documentation phase
3. In which of the following phases data flow diagrams are drawn?
   (a) Conception phase (b) Construction phase
   (c) Analysis phase (d) Design phase
4. Which of the following report illustrates the advantages of investing money for computerization?
   (a) Feasibility report (b) DFD
   (c) Cost benefit analysis (d) Profit and loss report
5. Which of the following will be shown as data store while drawing a data flow diagram?
   (a) Register (b) Registration process
   (c) Person who is registering (d) Registration form
6. Which of the following will be shown as external entity, while drawing a DFD?
   (a) Applicant (b) Application
   (c) Application process (d) Application register
7. What do you call that document, which contains the details of DFD, process logic, codification schemes etc.?
   (a) Project Register (b) Project Journal
   (c) Project encyclopedia (d) Project dictionary
8. In which phase of system development life cycle, programs are written?
   (a) Analysis phase (b) Design phase
   (c) Testing phase (d) Construction phase

## B. Fill in the blanks:

9. .......................................is the first phase of system development life cycle.
10. SSAD is acronym for..................... ................. ..............................................
11. The report called system proposal is prepared in....................phase of system development life cycle.
12. A rectangular figure is used to shown.......................in data flow diagram.
13. ...............DFD and ...........DFD are the two types of dataflow diagrams.
14. Data flows are shown using ............................figure in data flow diagrams.
15. The person who studies the physical system and analyses it is designated as ......................................
16. The process of writing program is also called ..............................................

## C. State, true or false for following statements:

17. The data flow diagram, which illustrates the movement of data into manual system is called logical data flow diagram.
18. In a DFD data cannot directly flow from one external entity to another. There has to be at least one process in between them.
19. In a DFD, data can directly flow from an external entity to a data store.
20. In a DFD, data cannot directly flow from data store to external entity. There should be at least one process in between.
21. You always explode a process numbered as 2.4, as processes that are numbered as 24.1, 2.4.2, 2.4.3 etc.
22. The first and foremost DFD, showing the main activity of the system is called concert diagram.
23. External entities in a DFD are shown through a circle in DFD.
24. All the processes shown in DFD have to have corresponding description written in project dictionary.
25. Maintenance of developed system starts immediately after finalization of project dictionary.

## D. Answer the following questions in short:

26. Name all the phase of system development life cycle, in correct sequence.
27. In which phase of system development life cycle, feasibility report is made?
28. Draw the symbols that are used for showing external entities, data stores, processes and data flows in a DFD.
29. Name at least three objects that will appear as external entities in a DFD.
30. Name any three objects that will appear as data store in a DFD.
31. Maximum how many processes can be there in a context diagram?
32. What is process logic?

33. Name any three activities of construction phase.
34. Name any three constituent parts of project dictionary.
35. Name any two tools that can be used for describing process logic.

## Detailed Answer Type Questions

**E. Answer the following questions in detail:**

36. What is Structure System Analysis and Design? What is its purpose?
37. What is system development life cycle? How many phases are there in it? Describe the activities of each phase in brief.
38. Name the documents that are generated in each phase of system development life cycle.
39. What is DFD? What is its role in Structured System Analysis and Design?
40. Draw all the symbols that are used while making a DFD.
41. Differentiate between logical DFD and physical DFD.
42. Describe all the rules that apply for drawing a DFD.
43. Draw a data flow diagram for railway reservation process, in which applicant gives a form to the clerk, he gives him a ticket and using reservation data he prepare reservation chart.
44. Draw a data flow diagram to open an account in bank, with deposit and withdraw activities taking place in it.

***

# CHAPTER 12

# *Introduction To Data Structure*

Data structure is basically the arrangement of the data that we store in computer. It is quite obvious that if we arrange the data properly, we will be able to retrieve it easily and the data input output operations will be fast. On the other hand, if the data is not arranged properly, the input output operations will be slow.

To understand the concept, let's take the example of a library, where you have thousands of books. If these books are arranged subject wise, properly, in shelves then on any member's request, librarian will be able to search the book quickly and give it to the person, within very short span of time. On the other hand, if the books are not arranged properly, librarian will not be able to search them easily. He will waste his precious time in going from here to there, while the member would keep waiting in irritated mood at the counter. Similar is the case in computer, the data that we store in it, should be arranged properly so that the program doesn't waste time in searching the required item and throughputs become faster.

One point that is to be understood here is that proper data arrangement not only makes throughputs faster, it also promotes programming simplicity. With the proper arrangement of data, logic of the program becomes easy.

## TYPES OF DATA STRUCTURES

You know that data structure is basically the arrangement of data that is stored in computer. There are various ways of arranging the data. Each arrangement is given a unique name, so you have various types of data structures.

Now the question is why do you have many types of data structure? Well, a data structure is directly linked to the type of search that you would make for getting the desired item from the whole chunk of stored data. It is not necessary that one type of data structure, which suits well in one situation, will equally suit well for other situations also. Truly speaking, it may fail miserably. To understand the concept, let's once again take the example, book library. Say the books are very well arranged subject wise, so when the librarian searches the book on given subject, he is able to hit the right book very quickly. Now what happens, if the subject of the book is not known and the book is to be searched on the basis of its author or publisher. In this situation, subject wise arrangement of book will not be of any great help. Librarian, will certainly take long time in getting the right book from the shelve. Obviously for such type of requirements, library books will require different type of arrangement.

For quite analogous reasons, you have different types of data structures. Following are few commonly used data structure:

1. Arrays
2. Records or Structures
3. Stacks
4. Queues
5. Linked Lists
6. Trees

## ARRAYS

**Array is basically a named list of finite number of variables, of same type, which are placed one after the other and are identified by a common name.** For example, Roll [10] is an Array of 10 variables, which are assigned a common name i.e. Roll. The elements of the array are denoted as follows:

| Roll (1) | Roll (2) | Roll (3) | ............ | ........ | ........ | Roll(10) |
|---|---|---|---|---|---|---|

**Figure 12.1**

Note that the first element of the array is identified as Roll [1] while the 10th element is denoted as Roll[10].

Arrays can be one dimensional, two dimensional or multi dimensional. Details of array and array handling are described in chapter 14 of this book.

## STRUCTURES OR RECORDS

Structure or a Record is basically a group of data items of different types. It is identified by a name. For example, a structure named Student may hold information like his Roll Number, Name, Date of Birth, and Marks. A record is shown in figure 12.2

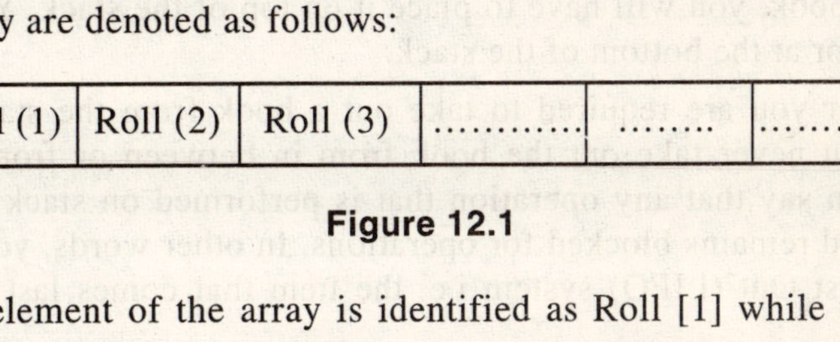


**Figure 12.2**

While dealing with records, the complete record is written or read. Once a record is read, all its members that constitute the record remain available in memory. You need not read the members individually. Similar case happens while writing the records. When you write a record, all its members get written. You need not write the members individually. This approach makes throughputs faster.

# STACK

In general, when you place one thing on top of the other then pile of things gets formed. Often this pile is called stack. To understand the concept of stack, let's take an example. Say you have 10 books with you. You place them on top of one another, as illustrated in figure 12.3 This heap of books, thus formed, will be called stack of books.

Similarly you could have stack of notes, stack of papers or a stack of 100 variables.

**Figure 12.3**

## Characteristics of Stack

Refer stack of books, illustrated in figure 12.3 and note that, whenever you receive a new book, you will have to place it on top of the stack. You never place it in between the books or at the bottom of the stack.

Similarly, whenever you are required to take out a book from the stack, you remove the book from top. You never take out the book from in between or from the bottom of the stack. Thus you can say that any operation that is performed on stack is performed at top end. The bottom end remains blocked for operations. In other words, you can say that stack follows Last In First Out (LIFO) system i.e. the item that comes last into the stack, gets removed first.

## Stack Operations

Adding an item in the stack is called Push operation. When you remove the item from the stack, the operation is called Pop operation.

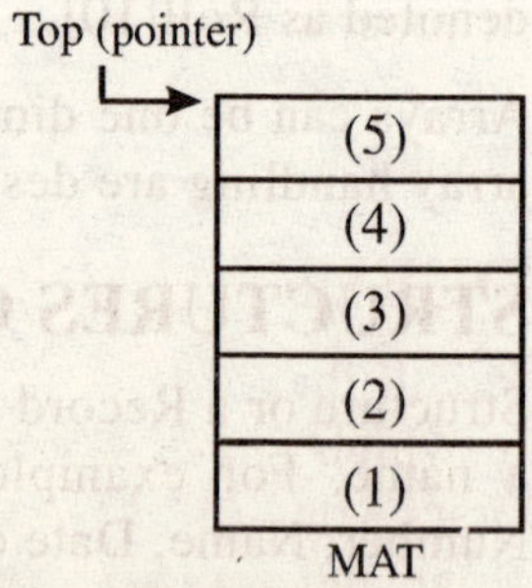


**Figure 12.4**

### Stack of Variables

The examples that we had taken above, pertains to the real world objects. Now the question is, do we have stacks in programming world also? Well, the answer is yes. In programming world, you work with stacks of variables. When you conceptualize the variables to be placed on top of one another, and allow Pop and Push operations to be performed, only at one end (restricting other end to participate in the operation) then a stack is said to be formed. Note that in programming world, stacks are implemented through arrays. To conceptualize the stack of variables, consider the array in figure 12.4.

To implement an array as stack, you have to have a pointer, which should always point to the top position of the array, so that whenever you perform Pop or Push operation, it is performed at current pointer's (Top's) position.

Note that currently the pointer Top is pointing to the fifth position of the stack. To understand the role of pointer in stack operations, let's take an example. Say you want to push an item (6) in the stack. To push this item move the pointer Top to the sixth position and add the item to the stack. This operation is clearly shown in figure 12.5.

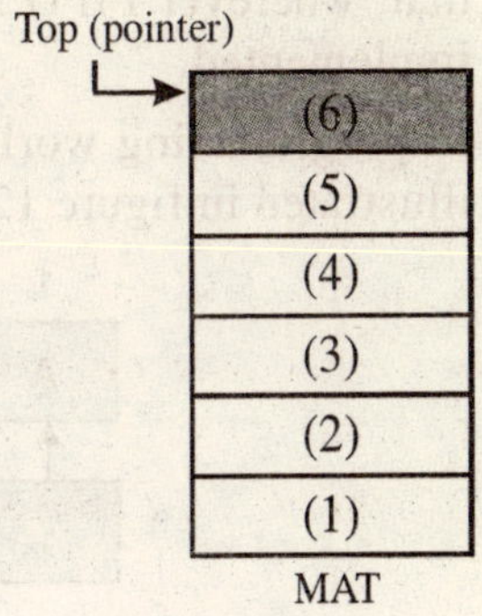


**Figure 12.5**

Now say you wish to perform two pop operations to remove 2 items from the stack. In this case you will have to first pop the item from the top of the stack thus removing item (6) and then move the pointer to the now-top position. (decrementing value by 1). To remove the next item, you will have to perform the same operation again. This operation of two consecutive pops is shown in figure 12.6.

You are now familiar with Stacks, Pop, and Push operations. Note that size of the stack depends upon the number of items currently present in it. When you perform Push operations stack's size increases. When you perform Pop operation its size decreases. Thus you can say that stack is a dynamic data structure with a capacity to enlarge or shrink. You may also note that if you keep on performing Pop operations, then there will come a situation when there will be no item in the stack. This condition is called STACK-EMPTY. On the other hand, when you perform so many push operations that stack can now no more take additional elements then such type of situation is called STACK-FULL.

In programming world, stacks are mostly implemented through arrays.

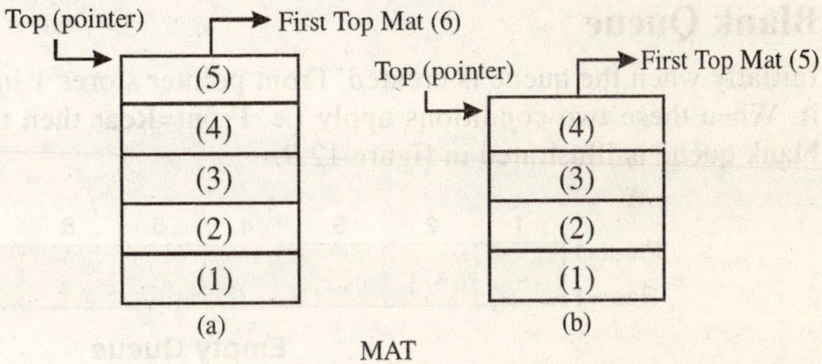


**Figure 12.6**

## QUEUE

Recall that in real life situation queue is basically a line, in which the first member of the line gets the treatment first. After getting the treatment, he leaves the line from front and the next member in the queue now gets the treatment. If a new comer wishes to join the queue, he has to join at the end.

A simple queue of people standing in front of movie ticket counter is illustrated in the figure 12.7.

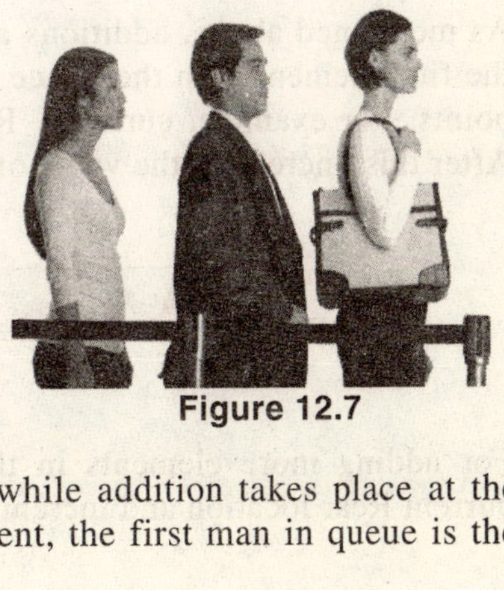

**Figure 12.7**

Refer above illustrated figure and note that queue is an arrangement, in which activities take place at both the ends i.e. removal of the person takes place at front end while addition takes place at the rear end. You may also note that in queue type arrangement, the first man in queue is the first man out. This system is called First In First Out (FIFO). In other words, you can say that, wherever FIFO system of serving has to be maintained, queue arrangement has to be implemented.

In programming world queues are mostly implemented through arrays. An Array queue is illustrated in figure 12.8.

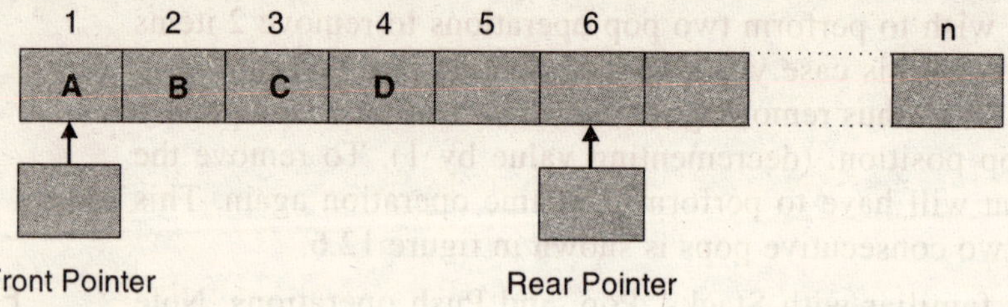


**Figure 12.8**

Above illustrated queue is basically an array of n elements and with two pointers named Front and Rear. Pointer Front stores the index of the first element of the queue. Pointer Rear stores the index of the last element of the queue.

## Blank Queue

Initially when the queue is created, Front pointer stores 1 in it. Rear pointer also stores 1 in it. When these two conditions apply i.e. Front=Rear then the queue is said to be blank. A blank queue is illustrated in figure 12.9.

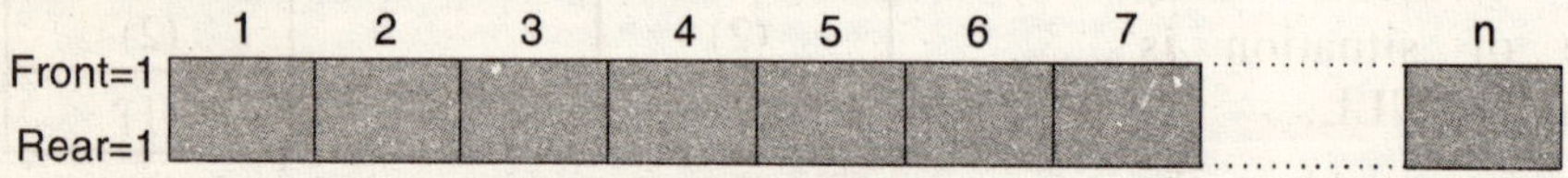


**Empty Queue**

**Figure 12.9**

## Addition In Queue

As mentioned above, additions are done at the rear end of the queue. Say you wish to add the first element A in the queue. For this, insert 'A' at that location where the pointer Rear points. For example, currently Rear is pointing to 1st location so insert 'A' at 1st position. After this, increment the value of Rear by 1. The process is illustrated below.

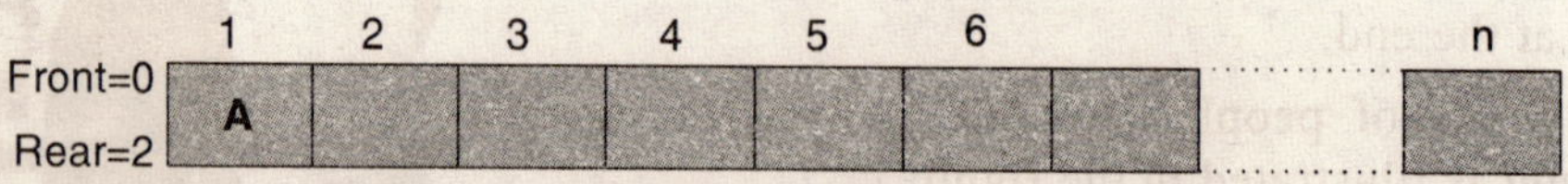


**Figure 12.10**

For adding more elements in the queue, you have to keep on inserting the element at current Rear location and incrementing the value of Rear by 1, as shown in figure 12.1.

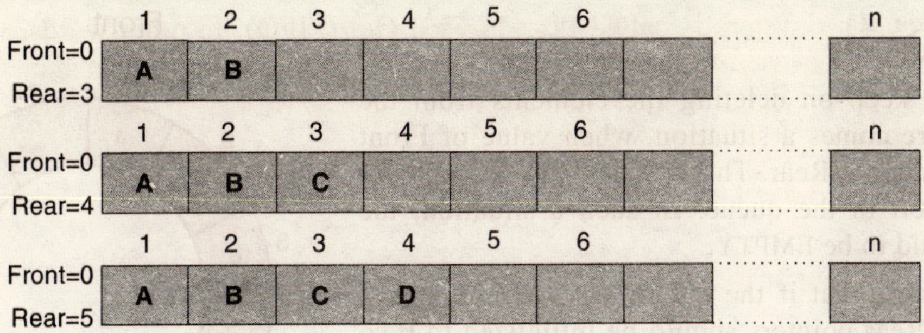


**Figure 12.11**

## Full Queue

When you keep on adding elements in the queue there comes a situation, when Rear>n i.e. no more additions are possible in the queue. In such a situation, queue is said to be FULL.

## Deletion In Queue

As mentioned above deletion of elements in queue is done at the front end of the queue. Say you wish to delete the first element 'A' from the queue. For this, write NULL at the position, where the pointer Front points. At present Front pointer will be pointing to the 1$^{st}$ location of the array so you will write NULL at 1$^{st}$ location thus deleting 'A' from the queue. After this, you will have to increment the value of Front pointer by 1. The process of deletion of first element from the queue is illustrated in figure 12.12

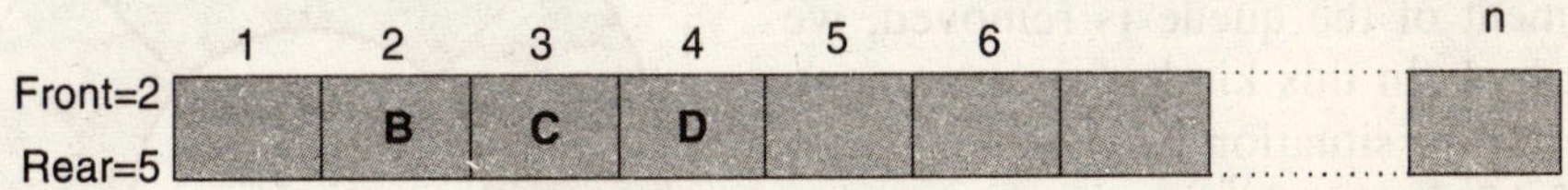


**Figure 12.12**

For deleting more elements from the queue, you have to keep on writing NULL at current Front location and incrementing the value of Front by 1, as shown in figure 12.13.

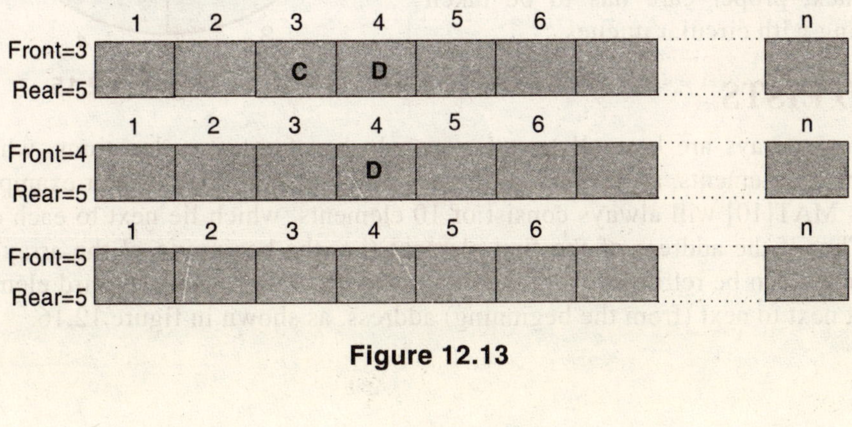


**Figure 12.13**

## Empty Queue

When you keep on deleting the elements from the queue, there comes a situation, when value of Front becomes equal to Rear. That means there are no more elements left in the queue. In such a situation, the queue is said to be EMPTY.

You may note that if the queue has become empty, Front and Rear pointers should be initialized to 0 so that the blank positions of the queue could be reused for inserting the elements.

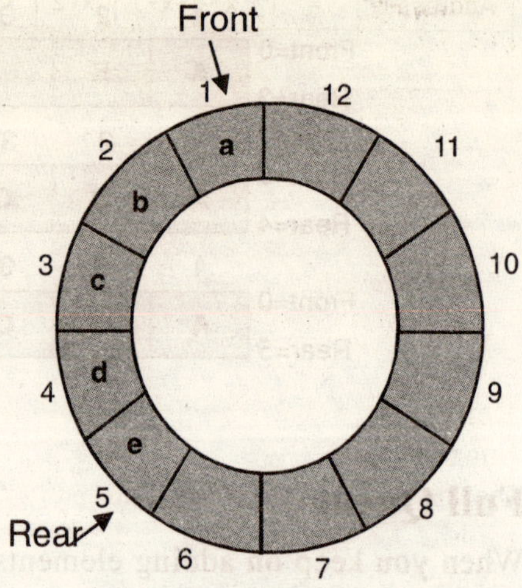

Figure 12.14

## Circular Queue

The pictorial representation of circular queue is shown in figure 12.14

In the above circular queue, the Front pointer points to 1st location (i.e. points to a), whereas Rear pointer points to element 'e', which is stored at 5th location of circular queue. In a circular queue of N elements (maximum), pointer Rear and pointer Front are set to 1 after they cross the limit N. For example, in a circular queue with maximum N elements, when Rear=N and if we insert an element, then this element is assigned to first location i.e. Q[1] i.e. instead of increment Rear to n+1, we reset Rear to 1. Similarly if Front=N and an element of the queue is removed, we reset Front to 1. In this kind of arrangement, a very interesting situation can arise as show in figure 12.15. In this figure, rear is pointing to Q[5] and front is pointing to Q[6]. In this situation, if we try to add an element then Rear will get incremented by 1 and Front will point to same location. To remove this type of draw back, proper care has to be taken while dealing with circular queues.

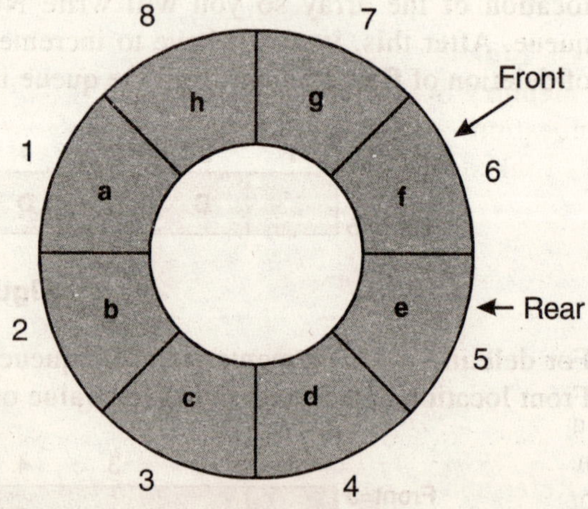

Figure 12.15

## LINKED LISTS

As you know, arrays are basically contiguous blocks of memory locations, which contain fixed number of elements, as declared at the beginning of the program. For example, an array declared as MAT[10] will always consist of 10 elements, which lie next to each other in the memory. Thus if the address of the first element (i.e. the beginning of the array) is known, second element can be retrieved from the next memory address and the third element will be available at next to next (from the beginning) address, as shown in figure 12.16.

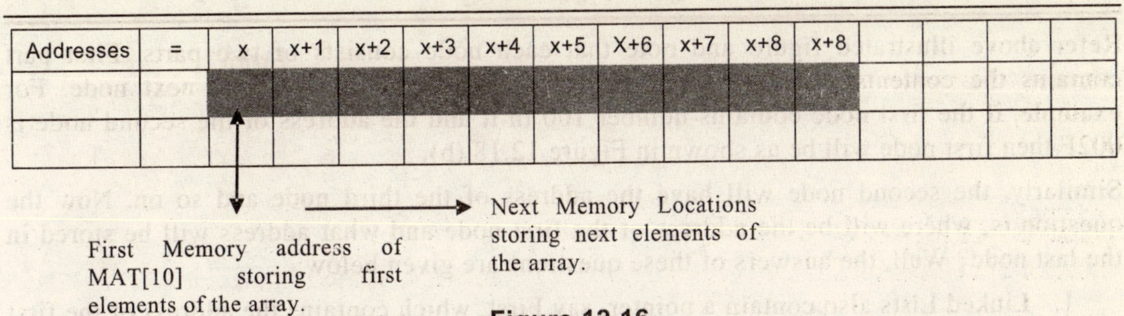


**Figure 12.16**

Refer above illustrated figure and note that management of array in memory is quite easy, but there is a slight problem with it. Say, if you wish to increase the number of elements in the array (say to 11 elements) or decrease the number of elements (say to 8 elements) in the middle of the program then it cannot be done, because memory locations are already created. But truly speaking, such types of requirements do occur, while programming. For example, if you have three numbers in hand then you would like to accommodate three numbers in the memory. On the other hand, if you have 50 numbers, you would like to accommodate 50 numbers in memory. In other words, you can say that you would like to have flexibility of creating as many variables in memory as there are numbers in hand. This obviously cannot be done in arrays, because the array has to have definite number of elements. You cannot create an array as MAT [I] where I is a variable. Note that such types of requirements are handled through different type of data structures, called Linked List.

**In Linked List, elements, called Nodes are created as and when required, at that memory location, which is currently available.** Note that in such process, memory locations allotted to the nodes, may not be contiguous, as shown in figure 12.17. (which is a Linked List of 10 nodes):

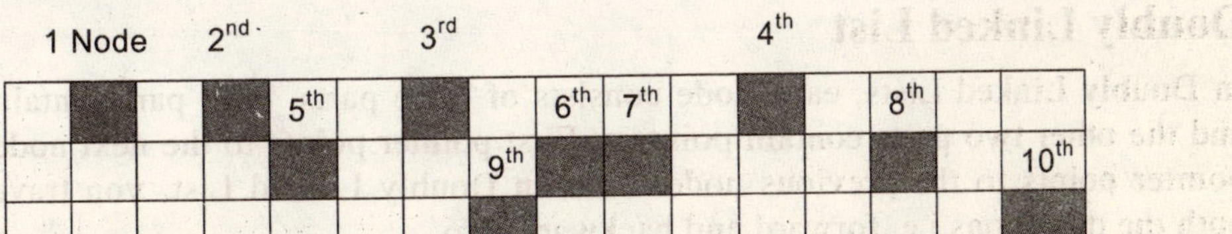


**Figure 12.17**

Refer above illustrated figure and note that, in Linked List, nodes remain scattered in the memory. Now the question is, if they are scattered throughout memory, then how do you retrieve them? Well, each node consists of two parts, as shown in figure 12.18.

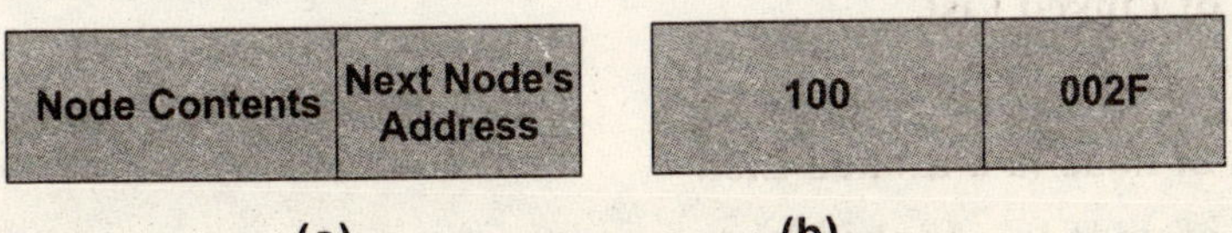


**Figure 12.18**

Refer above illustrated figure and note that each node consists of two parts. First part contains the contents and the second part contains the address of the next node. For example, if the first node contains number 100 in it and the address of the second node is 002F then first node will be as shown in Figure 12.18 (b).

Similarly, the second node will have the address of the third node and so on. Now the question is, where will be the address of the first node and what address will be stored in the last node? Well, the answers of these questions are given below:

1. Linked Lists also contain a pointer, say First, which contains the address of the first node.
2. Since the last node of the Linked List doesn't point to any other node hence it will contain NULL.

Thus a complete Linked List of three nodes will be as shown in figure 12.19.

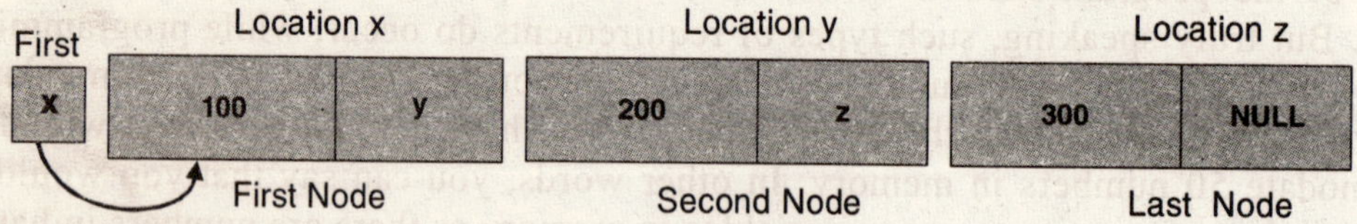


**Figure 12.19**

## Singly Linked List

All those Linked Lists, in which nodes consists of two parts i.e. contents and pointer to next node are called Singly Linked Lists. Thus in Singly Linked List you traverse the list in forward direction.

## Doubly Linked List

In Doubly Linked Lists, each node consists of three parts. First part contains the contents and the other two parts contain pointers. First pointer points to the next node and the other pointer points to the previous node. Thus in Doubly Linked List, you traverse the list in both the directions i.e. forward and backward also.

## Basic Operations On Linked List

Some of the basic operations, which can be performed on Linked List, are summarized below.

1. Creation of Linked List.
2. Traversal in Linked List.
3. Insertion of node in a Linked List.
4. Deletion of node in a Linked List.

These operations are described below:

## Creation Of Linked List

As you know, each element of Linked List is a node, which has two parts: (i) Data part and (2) Next node address i.e. pointer part. Since the Linked List is made of same type of Nodes hence you can define a node as record consisting of:

1. Normal variable
2. Pointer variable

Apart from this, you will also have to create a pointer, say First, which will point to the first node.

Suppose you wish to create a Linked List that contains four nodes named A, B, C and D then you will have to create four different occurrences of record. Thus 4 nodes Node A, Node B, Node C and Node D will have to be created. Such a list is shown in figure 12.24

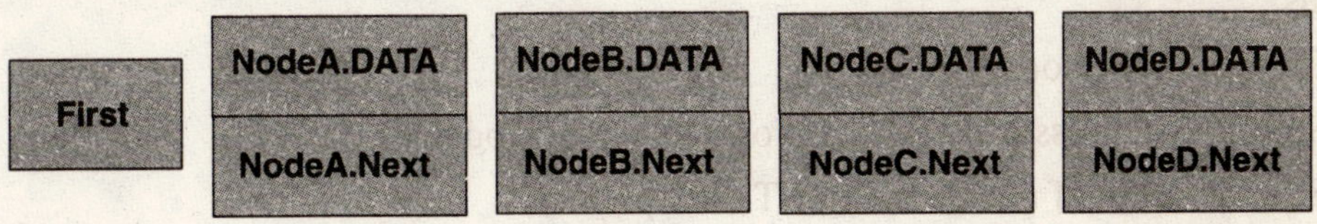


**Figure 12.24**

## Traversal In Linked List

While traversing the Linked List, you start from the pointer that points to the first node. If it is found NULL then Linked List is said to be empty.

If it is not null then you keep on accessing the nodes till pointer value of the node is found NULL.

## Insertion In A Linked List

There are many applications where it is desirable to maintain an ordered linear list. The ordering may be in increasing or decreasing order of the information contained in the field. Such an ordering often results in more efficient processing.

The insertion operation can be classified into three categories.

1. Insertion before first node.
2. Insertion after last node.
3. Insert in between two nodes.

## Insertion Before First Node

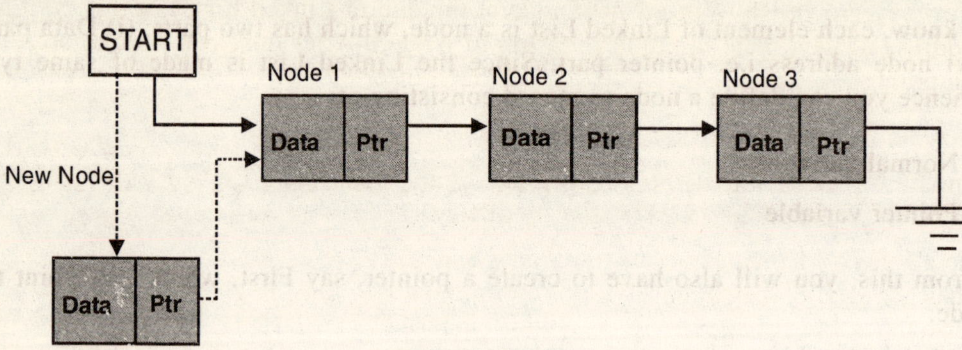


Figure 12.25

Above illustrated figure shows three nodes named as Node 1, Node 2 and Node 3. If you have to insert a new Node before Node 1 then you have to perform following three operations:

1. Create a new Node.
2. Store the address of Previous-Node 1 in new Node.
3. Store address of new Node in START pointer.

## Insertion After The Last Node

To insert a node at the end of the list is very simple. Store the address of new node in the pointer section of previous last node and store NULL value in pointer section of new last Node. This thing is shown in figure 12.26. Here in this figure we have three nodes and we have to insert a new node at the end of Node 3.

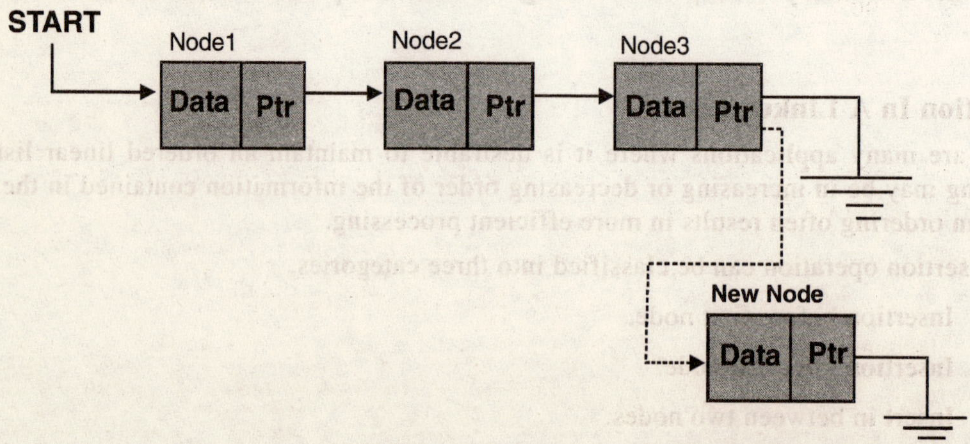


Figure 12.26

As shown in figure12.26, Node3 Pointer contains Null value in it to mark the end of Linked List. Now say, you wish to insert new node at the end of Linked List. For this, you have to store address of new node in pointer section of Node3. After this, you have to store NULL value in pointer section of new node that has been inserted.

### Insert A Node At Any Place In A List

As compared to insertion at the begging and at the end of the Linked List, insertion of a node in between the Linked List is slightly difficult.

Suppose a Linked List comprises of four nodes and you have to insert a new node after a node, which has got data 'C' stored in it. Before inserting the new node in given Linked List, you have to traverse the linked list and search the node, which has data='C' stored in it. Once you find such a node, you can insert new node as shown in figure 12.27

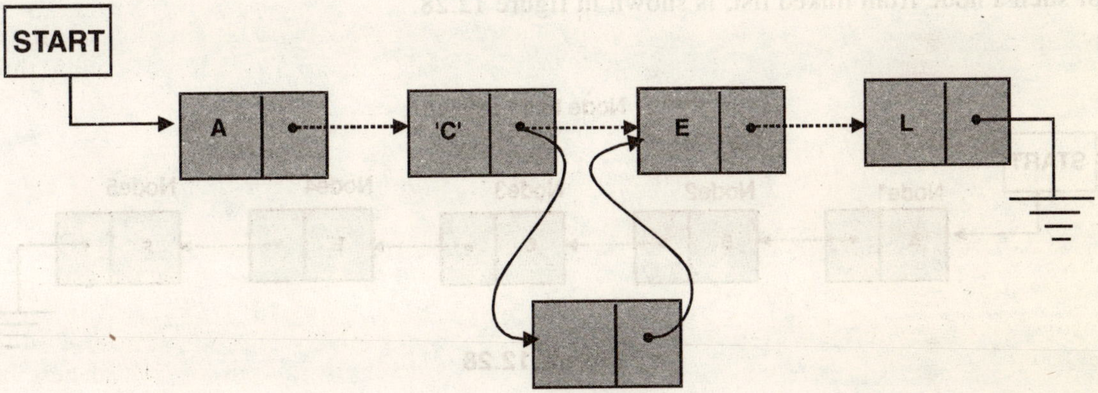


**Figure 12.27**

To insert any node in between two nodes, you have to just manipulate pointers as follows:

1. In pointer section of newly created node, put the address that was written in pointer section of the node, after which the new node has been created.
2. Put the address of newly created pointer in the pointer section of that node, after which the new node has been created.

### Deleting A Node From A Linked List

A delete operation involves the following two points:

1. Search the linked list for the node, which is to be deleted.
2. Delete the node.

Following three types of deletions can take place in linked list:

1. Deleting the first node
2. Deleting the last node
3. Deleting a node in between first node and the last node.

### Deleting The First Node

To delete the first node, you have to simply take the address contained in pointer section of first node ( which, in fact will be the address of second node) and put it in First pointer that points to the first node.

### Deleting The Last Node

To delete the last node, you have to simply put NULL in pointer section of last but one node.

### Deleting A Node In Between First Node And The Last Node

Say you wish to delete a node, which has data = 'C' stored in it. The procedure of deletion of such a node from linked list, is shown in figure 12.28.

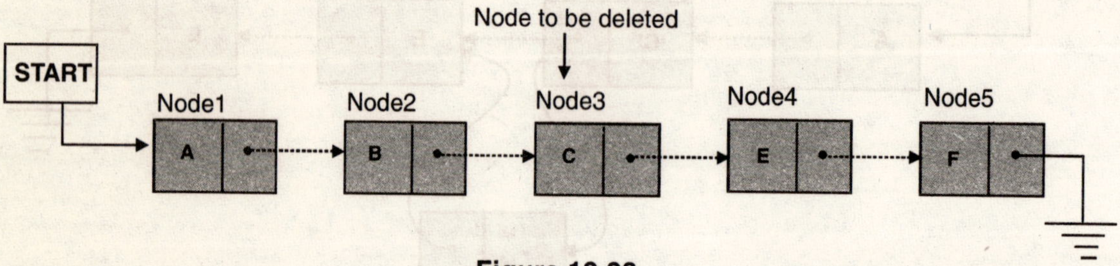


**Figure 12.28**

As shown in figure 12.28, to delete the node having value 'C', you have to perform following two operations:

1. Locate node, which has value C stored in its data section.
2. Store the address of Node 4 in pointer section of Node 2. This will make Node 3 out from the linked list.

## TREE

Tree is basically a multileveled data structure, in which its elements, called nodes are arranged in inverted tree like structure. The node at the top most position is called Root node. The bottommost nodes are called Leaf nodes. Other nodes are called internal nodes. A Tree data structure is illustrated in figure 12.29.

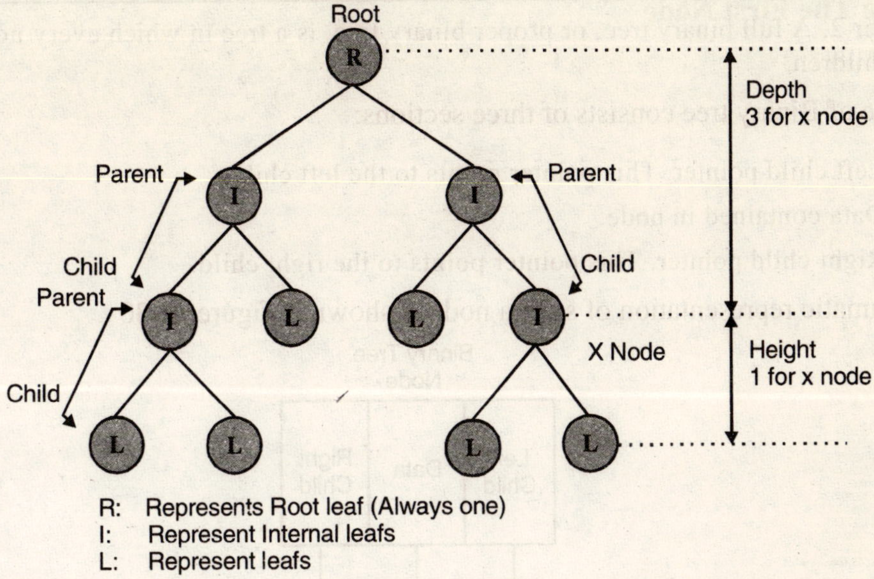


**Figure 12.29**

1. The nodes falling immediately below a node are called the child nodes. Conversely, a node just one level above of any node is called parent node. All children of the same node are siblings. The concept of child and parent node are illustrated in figure 12.29
2. Every tree has a depth and height factor associated with it. These factors are associated with every node in the tree. The depth of a particular node tells you how many generations down, that node is from the Root. Similarly, height of a particular node tells you its distance from the deepest leaf (or leaves) in that branch (or sub tree).

Each node of the tree consists of data and pointer section. Data section contains data and pointer section contains a pointer that points to the node below it.

Following are the operations that can be performed on tree data structures:

- ❑ Enumerating all the items
- ❑ Searching for an item in the tree.
- ❑ Adding a new item at a certain position of the tree.
- ❑ Deleting an item.
- ❑ Removing a whole section of a tree (called **pruning**).
- ❑ Adding a whole section to a tree (called **grafting**).
- ❑ Finding the root for any node.

## Binary Tree

Any tree whose nodes can have at the most two children is called a Binary tree or a tree with order 2. A full binary tree, or proper binary tree, is a tree in which every node has zero or two children.

The node of Binary tree consists of three sections:

1. Left child pointer. This pointer points to the left child.
2. Data contained in node.
3. Right child pointer. This pointer points to the right child.

Diagrammatic representation of such a node is shown in figure 12.30

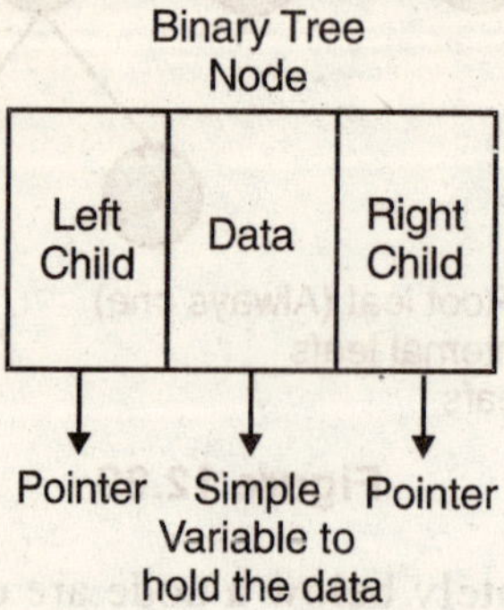


**Figure 12.30**

## Creating A Binary Tree

The binary tree creation follows a very simple principle. For the new element to be added, compare it with the current element in the tree. If its value is less than the current element in the tree, move towards the left side of that element else move towards its right. If there is no sub tree on the left, make your new element as the left child of current element else compare it with the existing left child and follow the same rule. Exactly same procedure has to be followed for the case, when your new element is greater than the current element in the tree but this time with the right child.

## Traversing The Binary Tree

Traversing the tree means, visiting each node of the tree. To traverse the tree, certain method is followed. To be familiar with that method, consider a Binary tree shown in figure 12.31

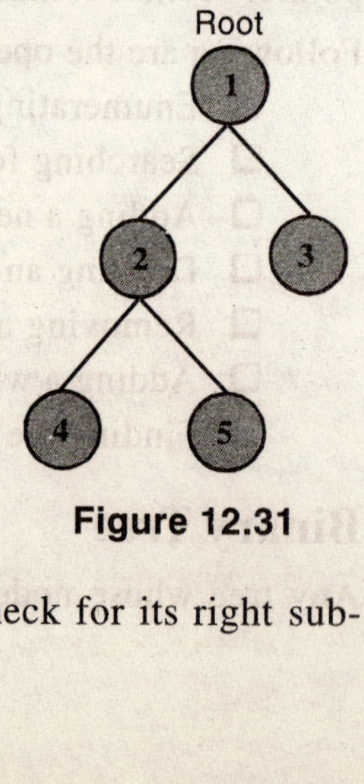


**Figure 12.31**

To traverse the tree, we start from the root i.e. Node 1. We first visit its left sub-tree then visit the node itself and then its right sub-tree. Here, Node 1 has a left sub-tree rooted at Node 2. So, we have to move to Node 2 and check for its left sub-tree (as we are supposed to do it for every node). Again, Node 2 has a left sub-tree rooted at Node 4. So, we check for Node 4's left sub-tree now, but Node 4 doesn't have any left sub-tree so we have to visit Node 4 first and check for its right sub-tree. As Node 4 doesn't have any right sub-tree, we'll go back and visit Node 2; and check for its right sub-tree. Node 2 has a right sub-tree rooted at Node 5 and so we move to Node 5. Note that Node 5 doesn't have any left or right sub-trees so we visit Node 5 and go back to Node 2. Since we have already visited Node 2 hence we go back to Node 1. Since we have not visited Node 1 hence we visit it first. After this we go for checking the right sub-tree of Node 1, which is rooted at Node 3. As Node 3 doesn't have any left or right tree we visit Node 3. So, following the sequence Node4, Node2, Node 5, Node 1, Node 3, we traverse the whole tree.

### Searching The Binary Tree

The search begins from root node of the tree. If the data to be searched, say *x*, matches the data stored in root, the search terminates successfully. If it doesn't then a comparison is made between x and the contents of root. If x is found less, the left sub tree is searched; otherwise *x* must be greater, in which case the right sub tree is searched.

### Deletion Of Node From Binary Tree

1. If the entry to be deleted is not in a leaf, swap it with its successor and then delete the entry from the leaf.
2. If the entry to be deleted is in leaf, delete with no further action.
3. If the node contains 2 entries, consider the two immediate siblings of the node. If one of these siblings has 2 entries then redistribute one entry from the sibling to the parent node and one entry from the parent to the deficient node.
4. If both the immediate siblings have exactly the minimum number of entries then merge the deficient node with one of the immediate sibling node and one entry from the parent node. If this leaves the parent with too few entries then the process is propagated upwards.

## EXERCISES CHAPTER 12

### Short Type Questions

**A. Select most appropriate answers for following questions:**

1. Which of the following is not a data structure?

   (a) Stack (b) Variable (c) Queue (d) Stack
2. Which system of input output is followed in stacks?

   (a) FIFO (b) LIFO (c) LILO (d) FILO
3. Which system of input output is followed in queues?

   (a) FIFO (b) LIFO (c) LILO (d) FILO

4. What do you call the process of removing an item from stack?
   (a) Push (b) Pull (c) Pop (d) Hop
5. Which of the following data structure may not have its elements, arranged contiguously in memory?
   (a) Array (b) Queue (c) Stacks (d) Linked List
6. How many addresses will have to be updated if a node is inserted in between the third and fourth node of a linked list?
   (a) 1 (b) 2 (c) 3 (d) 4
7. What do you call the top node of a tree?
   (a) Root (b) Top (c) Parent (d) Child
8. Maximum how many children can be there for a parent in a binary tree?
   (a) 1 (b) 2 (c) 3 (d) 4
9. What will push operation do in a stack?
   (a) Insert an element between two elements (b) Delete an element
   (c) Add an element at top (d) Add an element at end
10. A tree consists of 10 nodes. You wish to traverse the tree. How many nodes will you visit?
   (a) 10 (b) 5 (c) 1 (d) Less than 10

**B. Fill in the blanks:**

11. In stack, insert and delete operations take place at.................................................
12. In queue additions are done at..................................................................end.
13. Front=N and Rear=N condition for queue denotes that there are.........................item in the queue.
14. The nodes just below a given node in a tree will be called.....................of the given node.

**C. State, true or false for following statements:**

15. Record is not a data structure.
16. In stack all the operations are done at front end.
17. Array is an example of contiguous data structure, in which its elements are arranged one after the other.
18. Pop and Push operations are performed in queues.
19. LIFO is acronym for Last In First Out.
20. When Rear pointer points to higher element than Front pointer, the queue is said to be blank.
21. Each addition in queue increases the value of Rear pointer by one, while Front pointer value doesn't change in this operation.
22. You can have a circular queue.

23. Tree data structure is a special case of linked list.
24. In singly linked list, you can traverse the list, in forward direction only.

## D. Answer the following questions in short:

25. Name any three types of data structures.
26. Which end remains blocked for input output operations in stacks?
27. Which type of linked list you should have so that navigation can be done in both the directions?
28. What are the two types of contents that are stored in singly linked list?
29. Draw a binary tree that has 7 nodes including root.
30. Name a data structure, using which other data structures like stack and queues are implemented.
31. Minimum how many parents can five different children have in a binary tree?
32. What do you call those nodes of a tree, which are at the bottommost position?
33. What activity is performed while traversing a tree?

**Detailed Answer Type Questions**

## E. Answer the following questions in detail:

34. What are data structures? What is their purpose?
35. Why data structures are required?
36. What is stack? Describe its characteristics.
37. Describe pop and push operations in relation to stack.
38. What is queue? Describe its characteristics.
39. What is linked list? Describe it in detail.
40. Write the steps for inserting a node in between two nodes in a linked list.
41. What are circular queue? Describe them.
42. Differentiate between singly linked list and doubly linked list.
43. What is tree? Write the algorithm for traversing the tree.
44. What is binary tree? Describe following terms in relation to a binary tree.
    1. Root
    2. Parent
    3. Child
    4. Leaf
    5. Depth
    6. Height

****

# COMPUTRTS

## CHAPTER 13

# *Introduction To DBMS & SQL*

Database Management System, called DBMS in short, is basically general purpose software, which is specially designed to perform data storage and data retrieval type of activities. It provides a language of its own, using which data can be stored in the computer and retrieved as per requirements. For example, the data of all the students studying in the college may be stored in computer, using database management system and later only the names and marks of those students may be retrieved, who have been securing more that 90 percent marks for last three years.

## FEATURES OF DBMS

Following are the features of database management system:

1. They provide user friendly environment for developing applications.
2. They are easy to learn and can be mastered in short span of time.

   While working in DBMS, you need not write long programs for performing a task. Database language provides English like words. Each word works like a command. Single command of database management system generally performs an equivalent task of approximately 5 to 10 lines of any computer language. Thus database management systems enable fast development of applications.
3. Majority of rudimentary thing are automatically taken care. You don't have to bother about them. For example, when you declare the record layout, data entry screens are automatically generated. Similarly you frame the report layout and the report is automatically generated.
4. Databases are designed in such a manner, that a change done at one place, automatically makes changes at all other required places. Thus the database always remains in consistent state.

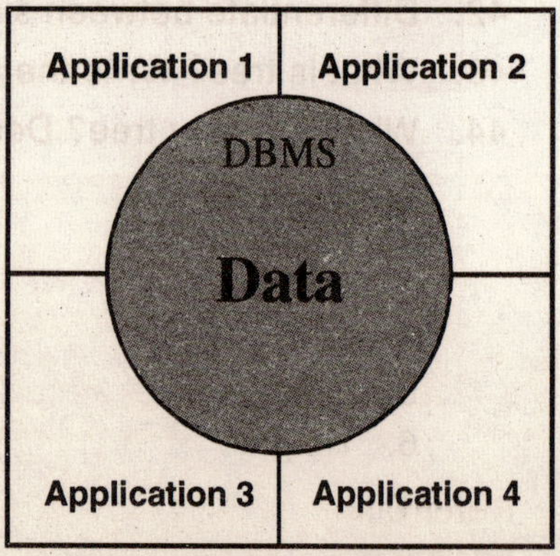


Figure 13.1

5. Data remains centralized in DBMS, so all the applications access the same common pool of data, as shown in figure 13.1.

# INTRODUCTION TO RELATIONAL DATABASE MANAGEMENT SYSTEMS

Relational database management systems are those database management systems, which conceptualize the data to be stored in the form of tables. Tables consist of rows and columns as shown below.

**Table Name: ITEM**

| Item Code | Item Name | Quantity | Rate | Supplier Code |
|---|---|---|---|---|
| ICODE | INAME | QTY | RATE | SCODE |
| 1 | Transistors | 10000 | 5.00 | S1 |
| 2 | Resistors | 25000 | 2.00 | S1 |
| 3 | Transformers | 200 | 500.00 | S2 |
| 4 | Battery | 500 | 12.00 | S3 |
| 5 | Amplifiers | 60 | 2,500.00 | S2 |
| 6 | LED | 2000 | 2.00 | S1 |
| 7 | Plugs | 2500 | 7.00 | S2 |
| 8 | Knobs | 3000 | 5.00 | S4 |
| 9 | Covers | 150 | 175.00 | S5 |
| 10 | Straps | 500 | 85.00 | 55 |

**Table Name: SUP**

| Supplier Code | Supplier Name | Address | City |
|---|---|---|---|
| SSCODE | SNAME | ADD | CITY |
| S1 | Rama Electronics | 16 Park Road | Delhi |
| S2 | ABC Electronics | 25 New Lane | Mumbai |
| S3 | Good Batteries | 20 River Bank Colony | Delhi |
| S4 | Ravi Associates | 30 Shiva Road | Mumbai |
| S5 | Belts & Straps | 22 Vikas Road | Lucknow |

Refer table named ITEM and note that the complete information of one item is called row. For example, there are 10 rows in this table.

Each item of information in the row is called column. For example, the row in ITEM table consists of 5 columns. For example, item code, item name, quantity etc. are columns.

## INTRODUCTION TO INGRES DATABASE MANAGEMENT SYSTEM

Ingres is a relational database management system. It is a robust, state-of-the-art RDBMS. It consists of first-class development tools. INGRES is an acronym for INteractive Graphics REtrieval System.

It was first created in 1970 as a research project at the University of California, Berkeley. It got completed in 1980. As the time passed by, lot of enhancements and modifications were made in the database. New facilities and features were added as per requirement of the time. It was soon commercialized for use. Ingres is still being used today and is quite popular among database users.

## INTRODUCTION TO INGRES SQL

SQL is acronym for Structured Query Language. It is a specially designed language for relational database management systems. Using this language, operations like database creation, data storage, data retrieval etc. are performed in relational databases.

SQL was originally designed by IBM. Later numerous versions of SQL came into existence. In 1986 National Standards Institute (ASNI) published standards for SQL, which were updated in 1992. Now SQL has emerged as standard relational database language and almost every relational database management system makes use of it.

SQL is easy to use language, which comprises of commands, clauses, operators and aggregate functions. Using them you can frame English like sentences. These sentences work as commands and perform operations like:

- ❑ Creating the tables.
- ❑ Establishing relationships between the tables.
- ❑ Defining rules of integrity.
- ❑ Defining security measures for the tables.
- ❑ Entering data into the tables.
- ❑ Modifying and maintenance of data.
- ❑ Retrieving the data from the table.

## CLASSIFICATION OF SQL COMMANDS

SQL commands can be broadly classified into two categories called Data Definition Language (DDL) and Data Manipulation Language (DML). An introduction of the commands of both the categories is given below.

### Data Definition Language (DDL)

All those commands of SQL, using which you create tables are called DDL commands.

## Data Manipulation Language (DML)

All those commands of SQL, using which you manipulate the data of the database are called DML commands.

## INGRES SQL

Ingres SQL provides many commands, using which required data can be retrieved from the database. These commands are described below.

## SELECT Command

SELECT is an SQL command, using which you can retrieve the data from the table. SELECT command has to be written in following format:

```
SELECT * FROM table-name;
```

Here table-name is the name of the table, from which the data has to be retrieved. For example, to display all the details of all the items of ITEM table, SELECT command will have to be executed in the following format:

```
SELECT * FROM ITEM;
```

On execution, SQL will provide following result:

| ICODE | INAME | QTY | RATE | SCODE |
|---|---|---|---|---|
| 1 | Transistors | 10000 | 5.00 | S1 |
| 2 | Resistors | 25000 | 2.00 | S1 |
| 3 | Transformers | 200 | 500.00 | S2 |
| 4 | Battery | 500 | 12.00 | S3 |
| 5 | Amplifiers | 60 | 2,500.00 | S2 |
| 6 | LED | 2000 | 2.00 | S1 |
| 7 | Plugs | 2500 | 7.00 | S2 |
| 8 | Knobs | 3000 | 5.00 | S4 |
| 9 | Covers | 150 | 175.00 | S5 |
| 10 | Straps | 500 | 85.00 | 55 |

### Accessing only Selected Fields

Say, now you wish to select only three columns i.e. ICODE, INAME and RATE from ITEM table. This can be done by using SELECT command in the following format:

```
SELECT [field1], [field2],......... FROM table-name;
```

Here, table-name specifies the name of the table from which the columns have to be selected and field1, field2 etc. are the names of the fields that have to appear in the output. For example, if you wish to view ICODE, INAME, RATE of all the items of ITEM table, SELECT command will have to be executed in the following format:

**SELECT ICODE, INAME, RATE FROM ITEM;**

On execution, SQL will provide following result:

| ICODE | INAME | RATE |
|---|---|---|
| 1 | Transistors | 5.00 |
| 2 | Resistors | 2.00 |
| 3 | Transformers | 500.00 |
| 4 | Battery | 12.00 |
| 5 | Amplifiers | 2,500.00 |
| 6 | LED | 2.00 |
| 7 | Plugs | 7.00 |
| 8 | Knobs | 5.00 |
| 9 | Covers | 175.00 |
| 10 | Straps | 85.00 |

### Selecting Fields Using Criteria

Note that above mentioned SQL commands did not put any restriction on the records of the table. As a result of that, all the records of the table appeared in the output. While writing SQL queries you can also include selection criterion so that only those records appear in the output, which meet the selection criteria.

For example, you may wish to retrieve only those records, in which quantity < 120. For imposing selection criteria in the query, SQL provides WHERE clause that can be used in SELECT command. The general format for including WHERE clause in SELECT command is as follows:

**SELECT [field1], [field2],........**

**FROM table-name**

**WHERE <condition>;**

Here condition specifies the selection criteria, which you wish to apply while selecting the records from the table. For example, if you wish to select ICODE, INAME and RATE columns of only those records in which RATE > 100 then SELECT command will have to be executed in the following format:

**SELECT ICODE, INAME, RATE FROM ITEM WHERE RATE>100;**

On execution, SQL will provide following result:

| ICODE | INAME | RATE |
|---|---|---|
| 3 | Transformers | 500.00 |
| 5 | Amplifiers | 2,500.00 |
| 9 | Covers | 175.00 |

## Use Of AND Operator In Query

Often you would be required to include multiple selection criterion in the query, so as to get the desired record from the table. The objective would be to get only those records in the output, which meet all the given conditions. In such cases, you will have to join two selection conditions (or more, as the case may be) with the help of AND operator. For example, you may wish to retrieve only those records in which RATE is greater than 100 and QTY is also greater than 100. For this, AND operator can be used to buildup the required condition. General format for the usage of AND operator is given below:

```
SELECT [field1], [field2], ........
FROM table-name
WHERE <condition1> AND <condition2>;
```

For above-mentioned example, SELECT command along with WHERE clause will have to be given as follows:

```
SELECT ICODE, INAME, RATE, QTY FROM ITEM
WHERE QTY > 100  AND RATE > 100;
```

On execution, SQL will provide following result:

| ICODE | INAME | RATE | QTY |
|---|---|---|---|
| 3 | Transformers | 500.00 | 200 |
| 9 | Covers | 175.00 | 150 |

## Using OR Operator In The Query

To build the condition for WHERE clause OR operator can also be used in the following format:

```
SELECT [field1], [field2], ........
FROM table-name
WHERE <condition1> OR <condition2>;
```

To understand the usage of OR operator, let's take an example. Say, you wish to view ICODE, INAME and RATE columns of only those records for which either RATE=100 or QTY=100. For this, OR operator will have to be used to buildup the condition as follows:

```
SELECT ICODE, INAME, RATE, QTY FROM ITEM
WHERE RATE =500 OR QTY=500;
```

On execution, SQL will provide following result:

| ICODE | INAME | RATE | QTY |
|---|---|---|---|
| 3 | Transformers | 500.00 | 200 |
| 4 | Battery | 12.00 | 500 |
| 10 | Straps | 85.00 | 500 |

## Ordering The Result Of The Query

SQL provides the facility to sort the output in ascending or descending order on a particular field. For this, you have to make use of ORDER BY clause in SELECT command.

ORDER BY clause has to come after WHERE clause in the query. Note that if no sorting order is specified, by default, ascending order is assumed. General format of ORDER BY clause is given below:

```
SELECT [field1], [field2],.......  FROM table-name
WHERE <condition>
ORDER BY [field1], [filed2],........ Ascending-or-descending-option;
```

If you have to sort the records on more than one field then fields should be separated by comma (,).

If you wish to sort the records in ascending order ASC clause should be specified. On the other hand, for sorting the output in descending order, DESC clause has to be specified.

For example, if you wish to retrieve ICODE, INAME, RATE and QTY from ITEM table and sort the output in ascending order on INAME then following command will have to be executed:

```
SELECT ICODE, INAME, RATE, QTY FROM ITEM
ORDER BY INAME ASC;
```

On execution, SQL will provide following result:

| ICODE | INAME | RATE | QTY |
|---|---|---|---|
| 5 | Amplifiers | 2,500.00 | 60 |
| 4 | Battery | 12.00 | 500 |
| 9 | Covers | 175.00 | 150 |
| 8 | Knobs | 5.00 | 3000 |
| 6 | LED | 2.00 | 2000 |
| 7 | Plugs | 7.00 | 2500 |
| 2 | Resistors | 2.00 | 25000 |
| 10 | Straps | 85.00 | 500 |
| 3 | Transformers | 500.00 | 200 |
| 1 | Transistors | 5.00 | 10000 |

Now let's take another example. Say you wish to select ICODE, INAME, RATE and QTY from

ITEM table for only those records, for which SCODE="S1". Say, you also wish to sort the output on RATE, in descending order. For this, following command will have to be executed:

```
SELECT ICODE, INAME, RATE, QTY FROM ITEM
WHERE SCODE="S1" ORDER BY RATE DESC;
```

On execution, SQL will provide following result:

| ICODE | INAME | RATE | QTY |
|---|---|---|---|
| 1 | Transistors | 5.00 | 10000 |
| 6 | LED | 2.00 | 2000 |
| 2 | Resistors | 2.00 | 25000 |

## Grouping The Result

Often you require that not all the records but the summary of selected fields should appear in the output. For example, you may be interested in knowing the number of items being supplied by the suppliers. Such summaries are in the form of sum of a field, average value of a field, minimum value of a field, count of selected fields and so on. The standard GROUP BY clause of SQL facilitates you to get such results.

This clause is included in the query when you require summary calculation like SUM, COUNT, AVG etc to be done. The general format of SELECT command along with GROUP BY clause is described below:

```
SELECT group-fields-function, AS [name],............... FROM table-name
optional-where-clause
GROUP BY group-field-name, ........;
```

Here, group-fields-function is a field name or an expression made up of fields or group function (described below). Note that at least one group function must be used in such statement. AS keyword followed by name in square braces indicates that you can give any name to the expression, which will appear in the output. Optional WHERE clause is same, which was explained before. Just after GROUP BY keyword you should provide a list of field names, on which grouping is to be done.

Following are the group functions, which are supported by GROUP BY clause.

| *FUNCTION* | *DESCRIPTION* |
|---|---|
| COUNT (*) | Gives count of rows in the group |
| MAX ([field]) | Gives maximum values of specified field, in the group |
| MIN ([field]) | Gives minimum values of specified field, in the group |
| AVG ([field]) | Gives average values of specified field, in the group |
| SUM ([field]) | Gives sum of all the values of specified field, in the group |

Now let's take an example. Say, you wish to know the total number of items being supplied by each supplier. For this, following SQL command along with GROUP BY clause will have to be executed:

```
SELECT SCODE, COUNT(*) AS Total_Items
FROM ITEM
GROUP BY SCODE ;
```

On execution, SQL will provide following result:

| SCODE | Total_Item |
|---|---|
| S1 | 3 |
| S2 | 3 |
| S3 | 1 |
| S4 | 1 |
| S5 | 2 |

To understand the usage of GROUP BY clause more clearly, let's take another example. Say you wish to know the maximum, minimum and average value for RATE field. For this, following command will have to be executed:

```
SELECT ITEM.SCODE, Avg(RATE) AS Avg, Min(RATE) AS Min, Max(RATE) AS Max
FROM ITEM
GROUP BY SCODE;
```

On execution, SQL will provide following result:

| SCODE | Avg | Min | Max |
|---|---|---|---|
| S1 | 3.00 | 2.00 | 5.00 |
| S2 | 1,002.33 | 7.00 | 2,500.00 |
| S3 | 12.00 | 12.00 | 12.00 |
| S4 | 5.00 | 5.00 | 5.00 |
| S5 | 130.00 | 85.00 | 175.00 |

## Having Clause

Refer the results produced by GROUP BY CLAUSE above and note that when SELECT statement is accompanied by GROUP BY clause then all the groups are selected. But there would be times, when you will not be interested in all the groups, but only in those groups, which meet given selection criteria. To deal with such type of requirements, SQL provides HAVING clause. HAVING clause is always used with GROUP BY clause. HAVING clause is followed by a condition. Of course, the condition of this clause must involve the group functions. The general format for including HAVING clause in SQL command is as follows:

```
SELECT group-fields-function, AS [name],................
FROM table-name
optional-where-clause
GROUP BY group-field-name, .........
HAVING <condition> ;
```

To understand the functionality of HAVING clause, let's take an example. Say you are interested in knowing the suppliers, who are supplying more than 2 items. For this, you will have to make use of HAVING clause along with GROUP BY clause in SELECT statement, as follows:

```
SELECT SCODE, Count(*) AS Total_Item FROM ITEM
GROUP BY SCODE
HAVING Count(*)>2;
```

On execution, SQL will provide following result:

| SCODE | Total_Item |
|---|---|
| S1 | 3 |
| S2 | 3 |

## SQL QUERIES FOR MULTIPLE TABLES

If the data items to be stored on the disk are many then often, due to various reasons the data is stored in multiple tables. This data is so distributed in the tables that data of one table can always be linked with data of other table. This is done by maintaining a common field in the tables.

To understand the concept of multiple tables, consider the two tables named ITEM and SUP that were mentioned at the beginning of this chapter. ITEM table contains item related data and SUP table contains supplier related data. Records of these tables are linked together through supplier code, which is SCODE in ITEM table and SSCODE in SUP table.

It is quite obvious that if the data has been stored in multiple tables then there will be times when it will have to be retrieved from multiple tables and presented as single logical output.

Retrieval of the records from multiple tables is also done using SELECT statements, but the syntax gets slightly changed, so as to list the names of all the tables, from which you wish to retrieve the records. General syntax for retrieving the data from multiple tables, using SELECT command is as follows:

```
SELECT [filed1], [field2],........
FROM table-name-1, table-name2, .......
WHERE <condition> ;
```

Here field1, field2 etc. are the column names of table1, table2 etc. that have to appear in the output. Table-name1, table-name2 etc. are the names of the tables, from which the data has to be retrieved. WHERE clause, along with the condition establishes a logical relationship between the data of the tables.

For example, if ICODE and INAME have to be selected from ITEM table and SNAME, ADD and CITY have to be selected from SUP table then following SELECT command will have to be executed:

```
SELECT ICODE, INAME, SNAME, ADD, CITY
FROM ITEM, SUP
WHERE SCODE = SSCODE ;
```

On execution, SQL will provide following result:

| ICODE | INAME | SNAME | ADD | CITY |
|---|---|---|---|---|
| 1 | Transistors | Rama Electronics | 16-Park Road | Delhi |
| 2 | Resistors | Rama Electronics | 16-Park Road | Delhi |
| 6 | LED | Rama Electronics | 16-Park Road | Delhi |
| 3 | Transformers | ABC Electronics | 25 New Lane | Mumbai |
| 5 | Amplifiers | ABC Electronics | 25 New Lane | Mumbai |
| 7 | Plugs | ABC Electronics | 25 New Lane | Mumbai |
| 4 | Battery | Good Batteries | 20 River Bank Colony | Delhi |
| 8 | Knobs | Ravi Associates | 30 Shiva Road | Mumbai |
| 9 | Covers | Belts & Straps | 22 Vikas Road | Lucknow |
| 10 | Straps | Belts & Straps | 22 Vikas Road | Lucknow |

Now let's take another example. Say you wish to know the names of the items that are being supplied by Delhi based suppliers. For this, following SELECT command will have to be executed:

```
SELECT INAME
FROM ITEM, SUP
WHERE CITY="DELHI" AND SCODE=SSCODE;
```

On execution, SQL will provide following result:

| INAME |
|---|
| Transistors |
| Resistors |
| LED |
| Battery |

## Using DISTINCT Clause

DISTINCT clause is used to get unique records in the output. It is used in the following format:

```
SELECT DISTINCT columns FROM tables
WHERE <condition>;
```

To understand the functionality of this clause, let's take an example. Say, you wish to know the names, addresses and city of the suppliers, who are supplying material worth more than Rs. 45,000. For this, DISTINCT clause in SELECT command will have to be used as follows:

```
SELECT DISTINCT SNAME, ADD, CITY
FROM ITEM, SUP
WHERE SCODE=SSCODE AND QTY*RATE>45000;
```

On execution, SQL will provide following result:

| SNAME | ADD | CITY |
|---|---|---|
| ABC Electronics | 25 New Lane | Mumbai |
| Rama Electronics | 16-Park Road | Delhi |

## EXERCISES CHAPTER 13

### Short Type Questions

**A. Select most appropriate answers for following questions:**

1. With which type of activities, DBMS deals with?
   (a) Data storage and retrieval (b) Numerical analysis
   (c) Representing data in the form of charts (d) Statistical analysis
2. In which form, relational DBMS stores its data?
   (a) Queue (b) Stacks (c) Tables (d) Trees
3. Which language is used for performing data storage and retrieval operations in relational database management systems?
   (a) ASP (b) SQL (c) C (d) JAVA
4. What do you call those commands of SQL, using which tables are created?
   (a) DDL (b) DML (c) DEL (d) DSQL

5. Which of the following SQL command will extract data from a database?
   (a) EXTRACT (b) SELECT (c) DISPLAY (d) WHERE

**B. Fill in the blanks:**

6. DBMS is acronym for........................ ........................... ................................
7. SQL is acronym for .......................... ................................ ......................
8. .........clause is used along with SELECT command to extract related records of two tables.
9. HAVING clause is always used along with....................clause in SQL statements.
10. DML is acronym for........................ ........................... ................................
11. DDL is acronym for........................ ........................... ................................
12. ...................columns will get selected, When SELECT* FROM ABC command is executed.
13. In SELECT A FROM B command, A is .........................name while B is the name of ...................
14. You can make use of ........ and ...........operators to join two conditions, which could be used in WHERE clause.
15. .........................clause in SQL can be used for generating sorted output.

**C. State, true or false for following statements:**

16. Data created through DBMS can be used by multiple applications.
17. Relational database management systems store the data in the form of tables.
18. SQL was originally designed by IBM.
19. Ingress is database language.
20. SELECT is a valid SQL command.
21. Those commands of SQL, using which data could be modified are called DML commands.
22. WHERE clause of SQL is used only when records from two tables are to be selected.
23. COUNT (*) function returns total number of selected records.
24. ONLY ONE clause is used for selecting unique records from the table.
25. MIM and MAX functions can only be used along with GROUP BY clause.

**D. Answer the following questions in short:**

26. Which types of functions are performed, using DBMS?
27. Which type of database conceptualize its data to be stored in the form of table?
28. Which language is used for performing data storage and retrieval operations in DBMS?
29. Which type of operations are performed, using DML?

30. Say there is a table named TAB and there are A1, B2 and C3 fields in it. What will you get in the output, if you execute SELECT A1, B1 FROM TAB WHERE C=500 command.
31. Say there is a table, A, which has 30 records and 5 columns viz. A1, A2, A3, A4 and A5 in it. How many records will get selected if SELECT A1, A2 FROM A command is executed?
32. What is the function of ORDER BY clause?
33. Which clause will you use with SELECT command if summary report on a particular field is to be generated?
34. With which clause will you use SUM(PAY) function in SELECT command, where PAY is the name of a column?

## Detailed Answer Type Questions

### E. Answer the following questions in detail:

35. What is database management system? What are its characteristics? For which type of operations it is used?
36. What is SQL? Differentiate between DDL and DML commands?
37. Describe SELECT command of SQL, taking a suitable example.
38. Describe the function, which could be performed by Ingres SQL.
39. Write the SQL commands for operations mentioned from point (a) to (d) and write outputs for operations mentioned in point (e) on the basis of table GRADUATE, given below.

**GRADUATE Table**

| S.No. | Name | Stipend | Subject | Average | Div |
|---|---|---|---|---|---|
| 1. | Karan | 400 | Physics | 68 | 1 |
| 2. | Divakar | 450 | Computer SC. | 68 | 1 |
| 3. | Divya | 300 | Chemistry | 62 | 2 |
| 4. | Arun | 350 | Physics | 63 | 1 |
| 5. | Sabina | 500 | Mathematics | 70 | 1 |
| 6. | John | 400 | Chemistry | 55 | 2 |
| 7. | Robert | 250 | Physics | 64 | 1 |
| 8. | Rubina | 450 | Mathematics | 68 | 1 |
| 9. | Vikas | 500 | Computer SC. | 62 | 1 |
| 10. | Mohan | 300 | Mathematics | 57 | 2 |

(a) List the names of those students who have obtained Div 1 sorted by Name.

(b) Display a report, listing Name, Subject and Stipend received.

(c) To Count the number of students who are either Physics or Computer SC. graduates.

(d) To display the name of all those students, who have secured more than 68 percent marks.

(e) Give the output of following SQL statements based on table named GRADUATE:

1. SELECT MIN(Average) FROM GRADUATE WHERE Subject='Physics';
2. SELECT MIN(Stipend) FROM GRADUATE WHERE Div=2';
3. SELECT AVG(Stipend) FROM GRADUATE WHERE Average>=65;
4. SELECT COUNT(Subject) FROM GRADUATE ;

40. Write SQL commands for operations mentioned in points from (a) to (e) and write the outputs of SQL commands given in point (f) or the table shown below:

**Table: LIBRARY**

| No. | Title | Author | Type | Pub | Qty | Price |
|---|---|---|---|---|---|---|
| 1 | Data Structure | Lipschutz | DS | McGraw | 4 | 217 |
| 2 | Computer Studies | French | FND | Galgotia | 2 | 75 |
| 3 | Advance Pascal | Schildt | PROG | McGraw | 4 | 350 |
| 4 | Dbase dummies | Palmer | DBMS | PustakM | 5 | 130 |
| 5 | Mastering C++ | Gurewich | PROG | BPB | 3 | 295 |
| 6 | Guide Network | Freed | NET | ZPress | 3 | 200 |
| 7 | Mastering Foxpro | Seigal | DBMS | DPB | 2 | 135 |
| 8 | Dos guide | Norton | OS | PHI | 3 | 175 |
| 9 | Basic for beginners | Morton | PROG | BPB | 3 | 40 |
| 10 | Mastering Window | Cowart | OS | BPB | 1 | 225 |

(a) Select all the titles published by McGraw from the table LIBRARY.

(b) Display a list of all books with Price more than 130 and sorted by Qty.

(c) Display all the books sorted by Price in ascending order.

(d) Count the number of books published by PHI.

(e) List all the details of those books, for which multiple copies are present in the library.

(f) Give the output of the following SQL commands on the basis of the table named LIBRARY:

(i) Select MIN(Price) FROM LIBRARY WHERE Price<150;

(ii) Select SUM(Price* Qty) FROM LIBRARY WHERE Qty>3;

****

# COMPUTERS

# CHAPTER 14

## BASIC Language

BASIC is acronym for Beginners All Purpose Symbolic Instruction Code. It is a computer language for beginners. Once upon a time it was quite popular among the students who wanted to learn programming. It was originally designed and developed in 1963, by John George Kemeny and Thomas Eugene Kurtz at Dartmouth College. It is a simple, easy-to-learn, easy-to-use computer programming language with English like commands and mathematical notations. It imposes least possible programming restrictions on programmers and provides easy-to-program environment.

### WRITING A BASIC PROGRAM

Following points are observed while writing programs in BASIC language:

1. Programs are written in upper case or lower case letters.
2. Each line of the program is assigned a unique line number. This number is used for line identification and its reference.
3. All the programs are terminated by END command. This command marks the end of the program.
3. REM command is used for adding comments in the program. As a rule, REM should the first word in comment line. Whatever is written after REM is taken as comments by BASIC compiler or interpreter.

Following is BASIC program, with few comment lines in it:

```
10 REM This program multiplies rate by quantity.
20 LET RATE = 10
25 LET QTY = 50
30 LET AMT = RATE * QTY
40 REM Following line will print the payable amount
40 PRINT "Payable Amount ="; AMT
50 END
```

## BASIC CHARACTER SET

In computer terminology, a letter, digit, or special symbol is called character. All those characters that can be used within BASIC programs are called BASIC character set. BASIC character set is described below:

**Letters:** A-Z, a-z

**Digits:** 0-9

**Special symbols:** + - * / ^ \ () [] {} =! = < > .(dot) ,(comma) " $ ; % ! ? & _(under score) # <= >= @, blank space,

## KEY WORDS

In BASIC language some words are reserved for specific task. They carry special meaning for BASIC complier or interpreter. Thus these words should not be used as identifiers. These reserved words are called keywords. Following is a list of BASIC keyword:

| | | | | | |
|---|---|---|---|---|---|
| ABS | COLOR | DIM | FIX | IOCTL | MERGE |
| AND | COM | DRAW | FNxxxxxxxx | IOCTL$ | MID$ |
| ASC | COMMON | EDIT | FOR | KEY | MKDIR |
| ATN | CONT | ELSE | FRE | KILL | MKD$ |
| AUTO | COS | END | GET | LEFT$ | MKI$ |
| BEEP | CSNG | ENVIRON | GOSUB | LEN | MKS$ |
| BLOAD | CSRLIN | ENVIRON$ | GOTO | LET | MOD |
| BSAVE | CVD | EOF | HEX$ | LINE | MOTOR |
| CALL | CVI | EQV | IF | LIST | NAME |
| CDBL | CVS | ERASE | IMP | LLIST | NEW |
| CHAIN | DATA | ERDEV | INKEY$ | LOAD | NEXT |
| CHDIR | DATE$ | ERDEV$ | INP | LOC | NOT |
| CHR$ | DEF | ERL | INPUT | LOCATE | OCT$ |
| CINT | DEFDBL | ERR | INPUT# | LOF | OFF |
| CIRCLE | DEFINT | ERROR | INPUT$ | LOG | ON |
| CLEAR | DEFSNG | EXP | INSTR | LPOS | OPEN |
| CLOSE | DEFSTR | FIELD | INT | LPRINT | OPTION |
| CLS | DELETE | FILES | INTER$ | LSET | OR |

| OUT | PRINT# | RIGHT$ | SPACE$ | TAB( | VARPTR |
|---|---|---|---|---|---|
| PAINT | PSET | RMDIR | SPC( | TAN | VARPTR$ |
| PEEK | PUT | RND | SQR | THEN | VIEW |
| PEN | RANDOMIZE | RSET | STEP | TIME$ | WAIT |
| PLAY | READ | RUN | STICK | TIMER | WEND |
| PMAP | REM | SAVE | STOP | TO | WHILE |
| POINT | RENUM | SCREEN | STR$ | TROFF | WIDTH |
| POKE | RESET | SGN | STRIG | TRON | WINDOW |
| POS | RESTORE | SHELL | STRING | USING | WRITE |
| PRESET | RESUME | SIN | SWAP | USR | WRITE# |
| PRINT | RETURN | SOUND | SYSTEM | VAL | XOR |

## COMMANDS AND STATEMENTS

A command, such as GOTO is basically a keyword provided in the language. In order to execute it, it should be written in specified format, specifying other required details. For example, to execute GOTO command, it has to be written as GOTO 500. A sentence thus framed is called statement. In other words, it can be said that statement is basically an instruction that is written, using BASIC keywords in specified format. When program runs, statements are executed one by one, in the sequence, in which they are written. As mentioned above, each statement of BASIC program is assigned a unique line number. Following are two statements of BASIC:

```
10 LET A = B + C
20 LET P = Q – R
```

## CONSTANTS

Data items that do not change their valves while the program executes, are called literals or constants. BASIC supports following types of constants:

1. Integer constants
2. Real constants
3. String constant

### Integer Constants

Integer values i.e. whole number values that do not change during the course of the program are termed as integer constants. Following are few examples of valid integer constants:

0, -10, 276, 5968, 32767, - 230

Note that values of integer constants range from -32768 to +32767.

## Real Constants

Those real values (numerical values with decimal points), which do not change during the course of the program are called real constants. Following are few examples of valid real constants:

**5.1 345. -0.77 906.07**

Real constants could either be negative or positive.

## String Constants

String comprises of single or multiple characters. String may be formed by combination of letters or digits enclosed within double quotes. Maximum 255 characters can be assigned to any string. Any string, whose value doesn't change during the program, is called string constant. Following are few valid string constants:

**"India" "Sachin" "R-100" "123"**

# VARIABLES

Variables are basically memory locations, which hold some value in them. They are always assigned a unique name for identification. When you refer a variable, in the program, you basically access its contents. For example, if the variable named SPEED is assigned the value, 100 and you write the statement SPEED = SPEED + 50 in your program then the value of this variable will get incremented by 50 and thus value of SPEED variable will become 150. Note that statements of this kind may appear to be algebraically wrong but they are correct as far as computer languages are concerned. They signify that new value of the variable will now be equal to the old value plus the constant value written in the statement (which is 50 in this example).

## Naming The Variable

Following rules apply for framing a variable name:

1. Variable name should not exceed 40 characters.
2. First letter of the name should always be an alphabet.
3. Keywords should not be taken as variable names.

## Types Of Variables

BASIC variables can be of following three types:

1. Integer variables
2. Real variables
3. String variables

A brief description of all these types of variables is given below:

### Integer Variables

All those variables that hold integer values in them are called integer variables. As a rule of BASIC, a variable whose name ends with a percent sign (%) is taken as integer variable. For example, ROLL% is an integer variable. Similarly PAY% is another integer variable. Note that PAY is not an integer variable.

### Real Variables

All those variables that hold real values in them are called integer variables. As a rule of BASIC, a variable whose name neither ends with % sign nor $ sign is taken as integer variable. For example, MARKS is a real variable. Similarly AMOUNT is another real variable. Note that MARKS% and AMOUNT$ are not real variables.

### String Variables

All those variables that hold string values in them are called string variables. As a rule in BASIC, a variable whose name ends with a $ sign is taken as string variable. For example, NAME$ is an integer variable. Similarly ADDRESS$ is another integer variable. Note that NAME% and NAME are not string variables.

## ARITHMETIC OPERATORS

Following arithmetic operators are available in BASIC language for performing arithmetic operations on integer and real values. In the following table they appear in order of precedence:

| Operator | Operation |
|---|---|
| ^ | Exponentiation |
| - | Negation |
| * | Multiplication |
| / | Floating-point Division |
| + | Addition |
| - | Subtraction |

## RELATIONAL OPERATORS

Relational operators operate upon an expression and evaluate it to find whether it is true or false. In the event of expression being true, they return 1 else they return 0. Following relational operators are available in BASIC language:

| Operator | Relation Tested | Expression |
|---|---|---|
| = | Equality | X=Y |
| <> | Inequality | X<>Y |
| < | Less than | X<Y |
| > | Greater than | X>Y |
| <= | Less than or equal to | X<=Y |
| >= | Greater than or equal to | X>=Y |

Note that equal sign is also used to assign a value to a variable.

# LOGICAL OPERATORS

Logical operators operate upon two relational expressions to build more complex logical expression, which is either true or false. Following are the logical operators, which are available in BASIC language.

## Logical AND Operator

Logical AND operator combines two relational expressions in such a way that the resultant expression gives true only if both the expressions (being combined) are true. In any other condition it gives false.

To understand its usage, assume that S1 and S2 are two relational expressions. The expression S1 AND S2 will be a complex logical expression. It will be true if and only if S1 and S2 are true. In the event of S1 or S2 or both being false, this expression will evaluate to false.

## Logical OR Operator

Logical OR operator combines two relational expressions in such a way that the resultant expression gives false only if both the expressions (being combined) are false. In any other condition it gives true.

To understand its usage, assume that S1 and S2 are two relational expressions. The expression S1 OR S2 will be a complex logical expression. It will be false if and only if S1 and S2 are false. In the event of S1 or S2 or both being true, this expression will evaluate to true.

## Logical NOT Operator

This operator negates the value of the expression. For example, if the expression is true, it will return false and if it is false then it will return true. For example, NOT(100>50) will return false.

# EXPRESSIONS

When two or more variables are joined together, using operators then an expression is said to be formed. BASIC expressions are of following types:

1. Arithmetic expressions
2. Logical expressions
3. Relational expressions

## Arithmetic expressions

When two or more numeric constants or variables are combined together, using, arithmetic operators, they form arithmetic expression. For example (a+b)*100 is an arithmetic expression. Following are sample algebraic expressions and their BASIC counterparts:

| Algebraic Expression | BASIC Expression |
|---|---|
| X–Z/Y | (X-Y)/Z |
| XY/Z | X*Y/Z |
| X+Y/Z | (X+Y)/Z |
| $(X^2)^Y$ | (X^2)^Y |
| X(-Y) | X*(-Y) |

### Rules Of Arithmetic

While writing BASIC arithmetic expressions, following rules must be followed:

1. Two operators should never appear together. For example, A/*B is a wrong expression. Similarly 100^* B is also wrong.
2. No attempt should be made to raise zero to a negative power. For example, in an expression, X^Y, X should never become zero while Y is negative.
3. A negative value should not be raised to real number (fractional number). For example, (-5) ^ 2.5 is not allowed.
4. Denominator of an expression should never be zero. For example, in the expression (A+B) / (C-D), C and D should never become equal.
5. String constants and string variables are not allowed in arithmetic expressions. For example A*3.5="4+2" is an invalid expression.
6. Every left bracket must be matched by a right bracket in the expression.

Any violation of these rules will cause a syntax error and a corresponding error message will get displayed.

### Evaluation of Arithmetic Expressions

Note that computer evaluates arithmetic expressions from left to right. But before that expressions written within brackets are evaluated. To get familiar with the process, consider following expression:

**12 ^ 2 + 25 * 4 – 12 / 4 - (250 - 150) / (20 +30)**

The execution of this expression takes place in the following order:

| | |
|---|---|
| 1. (20+30) = 50 | Parenthesis exp (left most) |
| 2. (250-150) =100 | Parenthesis exp. |
| 3. 12^2 = 144 | Exponentiation |
| 4. 25 * 4 = 100 | Multiplication (left most) |
| 5. 12 / 4=3 | Division (left most) |
| 6. 100 / 50 = 2 | Division |
| 7. 144 + 100=244 | Addition (left most) |
| 8. 244 – 3 = 241 | Subtraction (left most) |
| 9. 241 - 2=239 | Subtraction |

## Relational Expressions

When two or more numeric constants or variables are combined together, using relational operators, they form relational expression. For example, A> (B+C) is a relational expression. Similarly PAY < TOTAL – TAX is another relational expression. To get familiar with the use of relational operators and evaluation of relational expressions, assume A=100, B=50, C=150 and consider the expressions and their results, summarized in the following table:

| Expression | Result |
|---|---|
| A<B | False |
| B<=C | True |
| A>B | True |
| B>=C | False |
| (A+B) > 200 | False |
| (A+B+C) >= 300 | True |

## Logical Expressions

When two or more relational expressions are combined together, using logical operators, a logical expression is said to be formed. For example, (A > 200) AND (B < 500) is a logical expression. Similarly (C > 1000) OR (D < 2000) is another logical expression. To get familiar with the use of logical operators and evaluation of logical expressions, assume A=100, B=50, C=150 and consider the expressions and their results, summarized in the following table:

| Expression | Result |
|---|---|
| (A>B) AND (B>C) | False |
| (A>B) OR (B>C) | True |
| (A>50) AND (B>10) | True |
| (A>100) OR (B>C) | False |
| NOT (B>C) | True |
| NOT (A>B) | False |

## HIERARCHY OF OPERATIONS

A BASIC expression may have many operators in it. Now the question is, in which sequence will they get carried out? The sequence, in which they get carried out, is often referred to as hierarchy of operations. Following is the hierarchy of operations that is followed in BASIC language:

1. Unary minus.
2. Arithmetic multiply and Division.
3. Arithmetic Add and Subtract.
4. Relational operators.
5. Logical NOT
6. Logical AND.
7. Logical OR.

## LIBRARY FUNCTIONS

Library functions are basically prewritten programs. They remain available in inbuilt libraries of BASIC complier or interpreter. This is the reason why they are often called inbuilt functions.

Library functions accept single or multiple values in the form of argument(s). They operate upon those values and return the result. Each function is identified by its unique name and is written in the following format:

**Function-Name(Argument-List)**

Here Function-Name is the name of the function and Argument-List is one or more arguments separated by comma. For example, ABS(X) is a valid function, which provides absolute value of X. For example, if X=(-100) then ABS(X) will return 100.

These functions can either be used independently in the program or can be used in an expression with other variables and constants.

BASIC provides many functions. For the ease of identification they have been categorized in following categories:

1. Numeric functions
2. String functions

Functions falling into these categories are described below.

## Numeric Functions

Numeric functions take numerical values in the form or arguments, operate upon those values and return numeric result. Following are few commonly used numeric functions:

1. **SIN (expression)** : Returns SIN value of the numeric expression.
2. **COS (expression)** : Returns COS value of the numeric expression.
3. **ATN (expression)** : Returns ATN value of the numeric expression.

Note that these functions return the value of the expression in radians.

4. **SQR(expression)** : This function returns square root of a numeric expression. For example, SQR(144) will return 12. Similarly SQR(4) will return 2.
5. **EXP(expression)** : This function is used to calculate the exponential value of the given expression. Note that value of expression must be less than 88.0296 otherwise "Over flow" error will occur. For example, EXP (4) will give 54.59815.
6. **LOG(expression)** : This function is used to calculate natural logarithmic value of given numeric expression. For example LOG (2) will give 0.69314.
7. **INT(expression)** : This function returns integer value, which will be lowest of the adjacent integer values. For example, INT (4.46) will return 4. Similarly INT (-5.6) will return -6.
8. **ABS(expression)** : This function returns absolute value of any numeric expression. For example, ABS(-15) will return 15. Similarly ABS(17) will return 17.
9. **SGN(expression)** : This function returns + 1 if expression is positive, 0 if value of expression is 0 and -1 if value of expression is negative. For example, SGN(100) will return 1 while SGN(-100) will return -1. SNG(100-100) will return 0.
10. **CINT (expression)** : This function rounds off the fractional number to next integer value. For example, CINT(20.73) will return 21. Note that the value of expression that is to be rounded off should be in the range of -32768 to 32767.
11. **RND** : This function generates a random number between 0 and 1.

## String Functions

String functions either take strings as input, in the form of argument or return string as output. Following are few commonly used string functions of BASIC:

1. **LEN(string)** : This function is used for finding the length of the string i.e., the number of characters in the string. For example, LEN("INFOTECH") will return 8. Similarly,

LEN("INDIA") will return 5 because the string, "INDIA" comprises of 5 characters.

**2. INSTR(string1, string2)** : This function is used for finding the position of a substring, string2 within another string, string1. It returns a number, which indicates the position in string1, from where the matching starts. For example, INSTR ("shopkeeper", "keep") will return 5 because the string, "keep" is fully contained in "shopkeeper" from fifth character position.

**3. LEFT(string, n)** : This function takes out left most n characters from the string. For example, LEFT("INFOTECH",4) will take out 4 characters from left of INFOTECH string and return INFO.

**4. RIGHT(string, n)** : This function takes out right most n characters from the string. For example, RIGHT("INFOTECH", 4) will take out 4 characters from right from INFOTECH string and return TECH.

**5. MID$ (string, p, n)** : This function is used to take out n number of characters from the string, starting from position, p. For example, MID$("INFOTECH" 3,4) will give FOTE.

**6. ASC(character)** : This function returns ASCII value (between 0 to 255) of the character that is given to it in the form of argument. For example, ASCII value of A is 65.Thus function ASC("A") will give 65.

**7. CHR$(n)** : This function returns a character whose ASCII value is equal to n. For example, ASCII value of A is 65. Function CHR$(65) will give A. In other words, it can be said that CHR$() function is converse part of ASC() function.

**8. VAL(string-of-numeric-characters)** : This function accepts a string of numeric characters, such as"123", "26" etc. and converts it to equivalent numeric value. For example, VAL("25") will return 25.

**9. STR$(n)** : This function converts a given number into a string. For example, STR$(123) will give "123". Note that 123 and "123" are two different identities. First value is number while second value is a string. In other words, it can be said that STR$() function is converse part of VAL() function.

**10. DATE$** : This function returns current system date. For example, it may return 12-22-2007

**11. TIME$** : This function returns current system time. For example, it may return 10:11:55.

**12. SPC(n)** : This function prints n number of blank spaces. For example, SPC(20) will print 20 blank spaces.

## GETTING DATA INTO MEMORY

You know that variables are basically memory locations, which are identified by their names. When you get the data into variable, it gets stored in the memory. When you refer a variable, you basically refer its value. For example, if you store 100 in a variable, named SPEED and after that when you give the command to print the value of SPEED variable, 100 gets printed.

BASIC provides many commands for getting the data into variable. These commands are described below.

### LET Command

LET command assigns a value to a variable. It is used in the following format:

**<Line-Number> LET variable = expression**

Following are few valid assignment statements:

```
10 LET SPEED% = 100
20 LET DISTANCE =  450.55
30 LET  CITY$ = "DELHI"
40 LET TOT = 100 * RATE
```

Note that while using LET command, the data type of the variable and the type of value that expression returns should always match, otherwise a "TYPE MISMATCH" error will occur. For example, following are few invalid assignment statements:

```
10  A = "MUMBAI"        ( string value is being assigned into real variable).
20  N$ = 200            ( numeric value is being assigned into string variable).
```

### INPUT Command

INPUT command is used for accepting the values of the variables from the keyboard. It is used in following format:

**<Line-Number> INPUT <numeric or string variable>**

e.g. 100 INPUT RATE

When INPUT statement gets executed, it displays a question mark (?) and computer waits for a value to be entered through keyboard. When the value is entered, computer assigns that value to the variable. For example, if you enter 200, when above mentioned statement gets executed then this value (200) will be assigned to RATE variable. Now if you display the value of RATE variable, using a suitable command, 200 will get printed. Following are few valid examples of INPUT command:

```
10 INPUT SPEED, DISTANCE
20 INPUT NAME$, FATHERNAME$, CITY$
30 INPUT PAY, DESG$
```

Above examples clearly indicate that single INPUT statement can be used for accepting values for many variables. For this, each variable name should be separated from other variable name by a comma in INPUT command.

Note that when such statement is executed, the values that we enter from the keyboard should also be separated by comma. If the values are being entered for string variables then they should be enclosed within double quotes.

## Input Command With A Message

If the need be, a message can also be included in INPUT command. This message gets displayed on monitor screen at the time of execution of the command. Format for writing INPUT command, along with message, is given below:

**<Line-Number> INPUT <"message"> ; <list- of-variable-names>**

```
e.g. 100 INPUT "Give Rate "  ;  RATE
     200 INPUT "Roll Number and Name" ; ROLL,  NAM$
```

## DATA & READ Command

DATA and READ are two different commands of BASIC language but they work together. DATA command is used for including data in the program while READ command is used for reading that data. DATA command is used in the following format:

**<Line-Number> DATA data-1, data-2, data-3...........**

Following are few valid DATA commands:

```
DATA 100, 200, 300, 400, 500
DATA "JAN", "FEB, "MAR"
```

Following two points apply for DATA command:

1. Data items written in DATA statement are separated from each other, using comma.
2. String data items are enclosed within double quotes.

To read the data, READ command is used in the following format:

**<Line-Number> READ Variable-Name**

Following are few valid READ commands:

```
40 READ A
50 READ B
60 READ X, Y, Z
70 READ P$, Q$
```

Following points apply for READ command:

1. One READ command can read one or more data items in one go.
2. READ command should include as many variables in the statement as many data items have to be read together. For example, READ A, B statement will read 2 data items at a time from DATA statement.
3. If string data is to be read then string variables should be included in READ command. For example, READ B$ command will read string data items.

To understand the working of DATA and READ commands in relation to each other, consider the following piece of code:

```
05 REM THIS PROGRAM CALCULATES TIME TO COVER THE GIVEN DISTANCE
10 DATA 500, 70
20 READ DISTANCE
30 READ SPEED
40 LET TIM = DISTANCE / SPEED
50 PRINT TIM
60 END
```

On execution of line 20, value 500 (from DATA statement) will be read in DISTANCE variable. Similarly value 70 will be assigned to SPEED variable, on execution of line 30.

Note that instead of having two separate statements for two variables, a single READ statement can be used for reading both the values. For example, above mentioned program can also be written as follows:

```
05 REM THIS PROGRAM CALCULATES TIME TO COVER THE GIVEN DISTANCE
10 DATA 500, 70
20 READ DISTANCE, SPEED
30 LET TIM = DISTANCE / SPEED
40 PRINT TIM
50 END
```

## RESTORE Command

RESTORE command is always used, along with DATA and READ commands. It is used in following format.

**<Line-Number> RESTORE**

e.g. 100 RESTORE

RESTORE command forces READ command to read the data from beginning. To understand the functioning of RESTORE command, consider the following program:

```
10 REM MONTH AND DAYS PROGRAM
```

```
20 DATA "JANUARY", 31, "FEBRUARY", 28, "MARCH", 31
30 READ MTH$, DY
40 PRINT MTH$, DY
50 RESTORE
60 READ MTH$, DY
70 PRINT MTH$, DY
80 READ MTH$, DY
90 PRINT MTH$; DY
100 END
```

This program will display following result on execution:

```
JANUARY 31
JANUARY 31
FEBRUARY 28
```

Note that this program prints "JANUARY 31" twice. This is because first READ command reads JANUARY and 31. PRINT command prints these values.

RESTORE command in line 50 forces the next READ command to read the data from beginning. Thus next READ command will read the same data again, which will be printed by PRINT command. After this, next two values given in DATA statement are read. Thus FEBURARY 28 gets printed. Since there are no more READ statements in the program hence last set of data i.e. MARCH 31 remains unread.

## PRINT COMMAND

PRINT command displays data on monitor screen. It is used in the following format:

**<Line-Number> PRINT <contents-to-be-printed>**

Here contents-to-be-printed can either be variables, constants or an expression. Following are few examples of PRINT command:

| Command | Result |
|---|---|
| PRINT "Hello" | This command will print **Hello** on monitor screen. |
| PRINT | This command will print **a blank line** on monitor screen. |
| PRINT 2 | This command will print **2** on monitor screen. |
| PRINT 2+3 | This command will print **5** on monitor screen. |
| PRINT (20*3)/2 | This command will print **30** on monitor screen. |

## LPRINT COMMAND

LPRINT command prints data on printer. It is used in the following format:

**<Line-Number> LPRINT <contents-to-be-printed>**

Here contents-to-be-printed can either be variables, constants or an expressions. Following are few examples of LPRINT command:

| Command | Result |
|---|---|
| LPRINT "BELL" | This command will print BELL on printer. |
| LPRINT | This command will print a blank line on printer. |
| LPRINT 200 | This command will print 200 on printer. |
| LPRINT 20*3 | This command will print 60 on printer. |
| LPRINT (200*3)/5 | This command will print 120 on monitor screen. |

## PRINT CONTROLS

PRINT and LPRINT commands, mentioned above, were described in their simplest form. Various print controls can be included in PRINT and LPRINT statements so as to have formatted output. Few commonly used print controls of BASIC are described below.

### Comma and Semicolon Controls

Comma and semicolons are used along with PRINT and LPRINT commands. They are used for separating the variables (or values) that need to be printed, using single PRINT or LPRINT command. Use of comma and semicolon is illustrated in the following statements:

```
10 PRINT A, B, C
20 PRINT RATE; QUANTITY
30 LPRINT  NAME$, ADDRESS1$, CITY$
```

Note that comma (,) and semicolon (;) produce different effects in printing. Values separated by semicolon get printed in next column while values separated by comma get printed in next zone. Concept of zone is described below.

### Concept of Zone

For all displaying related tasks, complete monitor screen consists of 80 positions. That means maximum 80 characters can be printed on monitor, in single line.

For all printing (printing on printer) related tasks, a single line consists of 132 positions. That means maximum 132 characters can be printed on printer, in single line.

Total positions available for printing or displaying are logically divided into zones. Each zone comprises of 16 positions. That means first 16 positions constitute first zone. Similarly next sixteen positions constitute second zone and so on. For example, complete monitor screen remains divided into 80 / 16 = 5 zones, as illustrated in figure here.

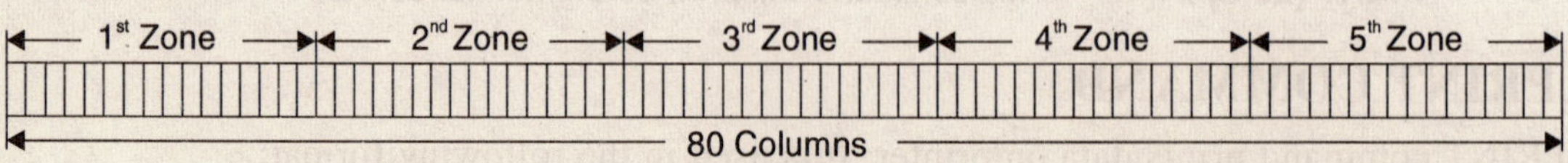


**Figure 14.1**

To understand the effect of comma and semicolon in PRINT statement, consider the following piece of code and the result produced by it:

```
10 LET A=10
20 LET B=20
30 LET C$="DELHI"
40 LET D$="MUMBAI"
50 PRINT A; B
60 PRINT A, B
70  PRINT C$; D$
80 PRINT  C$, D$
90 PRINT A; C$
100 PRINT A, C$
110 PRINT A; B; C$; D$
120 PRINT B; D$
130 END
```

When this program will get executed, it will display the value of the variables in the following manner:

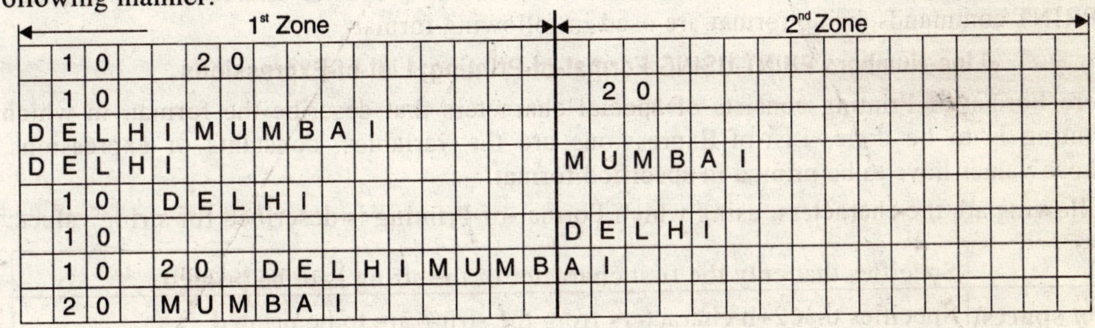


**Figure 14.2**

## TAB Function

TAB function is used as TAB(n). Here n specifies the column number, from where the printing has to start. For example, TAB(15) will let the printing start from fifteenth position, thus leaving 14 blank spaces. Following statement shows usage of TAB function along with PRINT command:

```
10 LET A$="RAHUL"
20 LET B$= "RAVI"
30 PRINT A$; TAB(20); B$
```

On execution, this piece of code will print the values of the variables as follows:

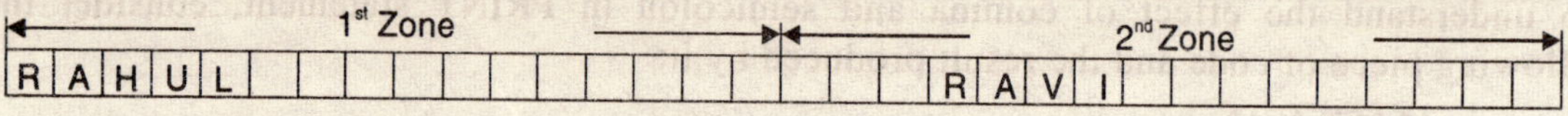


Figure 14.3

Note that if the current print position is already beyond *n* spaces, TAB goes to that position in the next line. For example, if printing has already been done till fifteenth column and TAB(15) command is given, the next printing will start from the fifteenth column of next line. This phenomenon is illustrated in following piece of code:

```
10 PRINT "I LOVE INDIA"; TAB(5); "IT IS MY BIRTH PLACE"
```

On execution, this command will print the words as shown below:

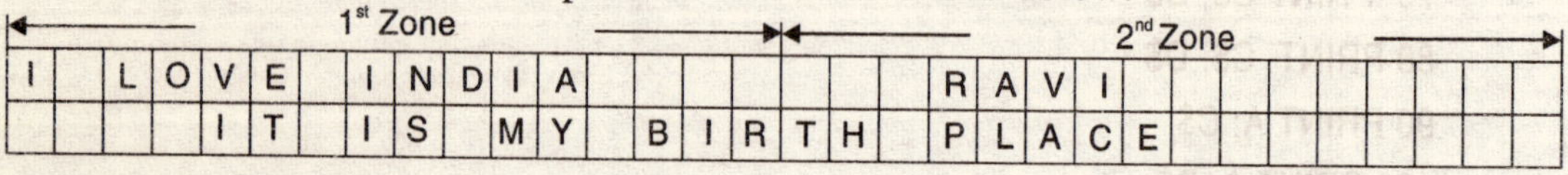


Figure 14.4

## Formatted Printing

In all the examples mentioned above, printing was unformatted printing. PRINT and LPRINT commands, both, allow the use of format along with them so that the values of the variables, constants or expressions that are being printed through them are printed in specified format. These formats are described using few special characters. PRINT and LPRINT commands, using format are used in following format:

**<Line-Number> PRINT USING Format-of-Printing; List-of-Expressions**

Here Format-of-Printing consists of special characters that describe the format, in which printing is to be done. List-of-Expressions are the variables, constants or expressions, whose values have to be printed in specified format.

Following are the characters, using which Format-of-Printing is described for string values:

| | |
|---|---|
| ! | Specifies that only the first character of the string is to be printed. |
| ***n* spaces** | Specifies that 2+n characters from the string are to be printed.<br>If the backslashes are typed with no spaces, two characters are printed. If the backslashes are typed with one space, three characters are printed, and so on.<br>If the string is longer than the field, the extra characters are truncated. If the field is longer than the string, the string is left justified in the specified space and it is padded with spaces on the right. For example:<br>10 A$="BABU": B$="RAM"<br>30 PRINT USING "!"; A$; B$<br>40 PRINT USING"\  \"; A$; B$<br>50 PRINT USING"\   \"; A$; B$;"**"<br>On execution, this piece of code will generate following output:<br>BR<br>BABURAM<br>BABU RAM** |

<table>
<tr><td>&</td><td>Specifies a variable length string field. When the format is specified, using &, complete string is printed. The phenomena is illustrated in the following example:<br><pre>10 A$="BABU": B$="RAM"
20 PRINT USING "!"; A$
30 PRINT USING "&"; B$</pre>On execution, this piece of code will generate following output:<br><pre>B
RAM</pre></td></tr>
</table>

Following are the characters, using which format-of-Printing is described for numeric values:

<table>
<tr><td>#</td><td>Hash sign is used to represent that one digit has to be printed at this position. When this digit is specified the digit positions are always filled. If the number to be printed has fewer digits than positions specified, the number is right justified (preceded by spaces) in the field.<br>A decimal point may be inserted at any position in the format constituted by #s. If the format string specifies that a digit is to precede the decimal point, the digit always is printed (as 0 if necessary). Numbers are rounded as necessary. For example:<br><pre>PRINT USING "##.##";.60</pre>On execution this line will give following output:<br><pre>0.60
PRINT USING "###.##";123.456</pre>On execution this line will give following output:<br><pre>123.45
PRINT USING "##.##" ;11.3,7.9,34.246,.123</pre>On execution this line will give following output:<br><pre>11.30 7.90 24.79 0.12</pre></td></tr>
<tr><td>+</td><td>A plus sign at the beginning or end of the format string causes the sign of the number (plus or minus) to be printed before or after the number.</td></tr>
<tr><td>-</td><td>A minus sign at the end of the format field causes negative numbers to be printed with a trailing minus sign. For example:<br><pre>PRINT USING"+##.##";-44.95,6.2,11.6,-8</pre>On execution this line will give following output:<br><pre>-44.95 +6.20 +11.60 -0.80
PRINT USING"##.##-";-66.95,55.449,-7.01</pre></td></tr>
</table>

| | |
|---|---|
| | On execution, this line will give following output:<br>66.95 55.45 7.01- |
| ** | A double asterisk at the beginning of the format string causes leading spaces in the numeric field to be filled with asterisks. The ** also specifies two more digit positions. For example:<br>PRINT USING "**#.#";22.29,-0.9,765.1<br>*22.3* -09765.1 |
| $$ | A double dollar sign at the beginning of the format string causes a dollar sign to be printed to the immediate left of the formatted number. The $$ specifies two more digit positions, one of which is the dollar sign. The exponential format cannot be used with $$. Negative numbers cannot be used unless the minus sign trails to the right. For example:<br>PRINT USING "$$###.##";123.38<br>$123.38 |
| **$ | The **$ at the beginning of a format string combines the effects of the above two symbols. Leading spaces are filled with asterisks, and a dollar sign is printed before the number. **$ specifies three more digit positions, one of which is the dollar sign. For example:<br>PRINT USING "**$##.##";1.94<br>***$1.94 |
| , | A comma to the left of the decimal point in the format string causes a comma to be printed to the left of every third digit to the left of the decimal point. A comma at the end of the format string is printed as part of the string.<br>PRINT USING "#,###.##";1234.5<br>1,234.50 |
| ^^^^ | Four carets may be placed after the digit position characters to specify exponential format. The four carets allow space for E+xx to be printed. Any decimal point position may be specified. The significant digits are left justified, and the exponent is adjusted. Unless a leading + or trailing + or - is specified, one digit position is used to the left of the decimal point to print a space or a min us sign. For example:<br>PRINT USING "##.##^^^^";934.56<br>9.35E+02<br>PRINT USING ".####^^^^-";888888<br>PRINT USING "+.##^^^^";123<br>+.12E+03 |

| | |
|---|---|
| | Note that in the above examples the comma is not used as a delimiter with the exponential format. |
| _ | An underscore in the format string causes the next character to be output as a literal character. For example:<br>PRINT USING "_!##.##_!";77.22<br>!77.22!<br>The literal character itself may be an underscore by placing "_" in the format string. |

## IF...THEN COMMAND

IF...THEN command is basically a decision making command. It involves a condition and an executable statement. On its execution, it checks the condition. If the condition is found true, it executes the associated statement otherwise it bypasses it. IF...THEN statement is used in the following format:

**<Line-Number > IF <Condition> THEN <Statement>**

Here Condition is a relational expression that is to be evaluated and statement is an executable statement that is to be executed in the event of condition being true. Following statement is a valid IF...THEN statement:

```
10 IF AGE > 18 THEN PRINT "ADULT"
```

This statement will print ADULT only if the value of age variable is greater than 18. If it is found less than or equal to 18, nothing will be printed by this statement.

Following are few more examples of IF...THEN statement:

```
10 IF A=B THEN C=D
20 IF A > B THEN BIG = A
30 IF MARKS > =50 THEN PRINT "YOU HAVE PASSED"
```

## IF...THEN ...ELSE COMMAND

IF...THEN...ELSE command is basically decision making command. It involves one condition (say e1) and two courses of actions (say S1 and S2). When this statement gets executed, given condition (e1) is evaluated. If it is found true then first statement (S1) gets executed. On the other hand, if the given condition evaluates to false, second statement (S2) gets executed. Characteristic behavior of IF...THEN...ELSE statement is illustrated here in figure 14.5.

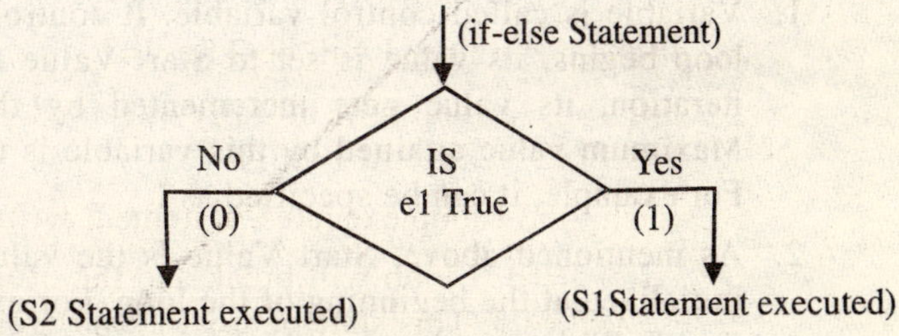


**Figure 14.5**

To understand the usage of IF…THEN…ELSE statement, assume that:

1. Expression e1 is (AGE >= 60)
2. Statement S1 is PRINT "SENIOR CITIZEN"
3. Statement S2 is PRINT "NOT AN OLD PERSON"

At the time of execution of it, if AGE=75 then expression e1 will be true. Thus S1 statement will get executed and the program will print "SENIOR CITIZEN". On the other hand if AGE = 45 then e1 will be false and S2 statement will get executed. Thus "NOT AN OLD PERSON" will get printed.

IF…THEN…ELSE statement is executed in the following format:

**< Line-Number > IF < Condition > THEN < Statement1 > ELSE < Statement2 >**

Here Condition is a relational expression that is to be evaluated and Statement1 is the statement that is to be executed if condition is found true. Statement2 is the statement that is to be executed when condition is false. Following are few valid IF…THEN…ELSE statements:

```
10 IF MARKS > 49 THEN PRINT "PASS" ELSE PRINT "FAIL"
20 IF A > B THEN BIG = A ELSE BIG = B
```

## FOR...NEXT LOOP

In reference to programming, loop means, executing a group of contiguous lines (or a line) repeatedly. For example, if line 100,110,120 and 130 are being executed continuously 5 times then it will be said that program runs in a loop that iterates 5 times.

BASIC language provides FOR…NEXT command for forming loop within the program. It also provides means for controlling the number of iterations that the loop will make. FOR…NEXT command is used in the following format:

```
<Line-Number>  FOR Variable = <Start Value> TO <Stop Value>  [STEP <Value>]
    Statements
    NEXT [variable]
```

Here, in this statement meanings of various clauses are as follows:

1. Variable is called control variable. It controls the iteration of the loop. When the loop begins, its value is set to Start Value mentioned in FOR statement. In each iteration, its value gets incremented by the value specified by STEP clause. Maximum value attained by this variable is the value stated by Stop Value clause. For example, it can be specified as I.
2. As mentioned above, Start Value is the value, with which the control variable is initialized at the beginning of the loop. For example, if Start Value is specified as 1 then I will be equal to 1 at the beginning of the loop.
3. As mentioned above, Stop Value is the last value, which the control variable attains during the execution of the loop. After attaining this value, the loop breaks. For

example, if Stop Value is specified as 10 then loop will come to an end after I=10. When the loop ends, statement written immediately after NEXT statement gets executed.

4. STEP <value> governs the steps, in which the value of control variable will be incremented. For example, if STEP 2 is specified then value of I will be incremented in steps of 2. That means, in first iteration it will be 1, in second iteration it will be 3, in next iteration it will be 5 and so on. Note that STEP clause is optional. If it is not specified, value of control variable is incremented by 1 in each iteration of the loop.
5. Statements are BASIC commands that are to be executed repeatedly, till the loop continues.
6. NEXT Statement marks the end of the loop. It sends program control back to FOR statement. Variable mentioned in this statement is control variable. For example, if I was taken as control variable in FOR statement then NEXT statement will also have I as its variable.

Following is a valid FOR loop that prints numbers from 1 to 100:

```
10 FOR I = 1 TO 100
20 PRINT I
30 NEXT I
40 END
```

Following is a valid FOR loop that prints all odd numbers from 1 to 100:

```
10 FOR I = 1 TO 100 STEP 2
20 PRINT I
30 NEXT I
40 END
```

***Problem***

Let's write a program that asks for two numbers (say X and N) and then prints sum of series of following type, for that number:

```
1+X+X^2+X^3+ ....+X^n.
10 S=1
20 INPUT "ENTER VALUE OF X" ;  X
30 INPUT "ENTER VALUE OF N "; N
40 FOR I=1 TO N
50 S=S+X^I
60 NEXT I
70 PRINT "SUM = " ; S
80 END
```

## Nested FOR Loop [Loop within a Loop]

A FOR loop can have one or more FOR loops inside it. Using a loop within another loop, is called nesting of loops. For nesting the loops, following rules apply:

1. Each loop that is placed within a loop, must have its own FOR and NEXT statements.
2. The inner loop must fully lie within the outer loop. It means that loops must not overlap.
3. When loops are nested, each loop must have a unique variable name.
4. The NEXT statement for the inner loop must appear before the NEXT statement of outer loop, as shown below:

```
10 FOR I=1 TO 5
20 FOR J=1 TO 10
30 PRINT I, J
40 NEXT J
50 NEXT I
60 END
```

## Execution Of Nested FOR Loops

When FOR loops are nested then inner loop gets fully executed for each iteration of outer loop. To understand the concept, consider the following program:

```
10 FOR I = 2 TO 20
20 FOR J = 1 TO 10
30  PRINT  I; "X"; J; "="; I*J
40 NEXT J
50 NEXT I
60 END
```

Execution of nested loops will proceed as follows:

1. In the beginning, I will be 2 i.e. I=2.
2. For I=2, J loop will be fully executed. That means it will make 10 iterations i.e. J=1, J=2 .........till J=10.
3. Each time J loop executes, statement written in line 30 will get executed.
4. After J=10, J loop will break and the program control will come to line number 50, i.e. NEXT I. This statement will send program control back to line number 10.
5. Now I will become 3, i.e. I=3.
6. For I=3, J loop will once again get fully executed and the statement written in it will get executed 10 times.

7. This way, program will proceed till I=20. After this, program will come to an end.

From the explanation, given above, it is clear that statement written in line number 30 will get executed 19 * 10 = 190 times. If you closely follow the logic of the program and the contents being printed, you will find that this program prints tables of numbers, starting from 2 to 20.

## WHILE...WEND LOOP

WHILE...WEND loop consists of an expression and few commands associated with it. WHILE statements marks the beginning of the loop and WEND statement marks the end of the loop. Condition is places with WHILE statement and all the statements that are to be executed repeatedly in a loop are placed between WHILE and WEND statements. WHILE...WEND loop is written in following format:

```
<Line-Number> WHILE < Condition >
    Statements
    WEND
```

At the time of execution, condition is evaluated. If it is found true all the statements, written within WHILE and WEND are executed one after the other. After executing the last statement of the loop, program control returns back to the beginning of the loop. Now again, the expression is evaluated, if it is found true the body of the loop is executed again else the loop terminates. When the loop terminates, the statement placed next to the WEND command gets executed.

Following points need to be taken care for proper execution of WHILE loop:

1. The expression should be so framed that its value changes in each iteration of loop.
2. If the expression is ill framed and its value doesn't change within the loop, loop will never terminate. The loops, which do not terminate are called infinite loops. As a general rule of programming, infinite loops have to be avoided at any cost.
3. Initially if the expression is found false, loop will not execute even once.

To understand the functioning of WHILE loop, let's write program for following problem:

***Problem***

Program should accept Roll Number, Physics, Chemistry and Maths marks of the student. It should then calculate the percentage and print it, assuming that subject marks have been awarded out of 100 marks. The program should work in a loop till the time user wants.

***Program***

```
10 REM PRINTING PERCENTAGE
20 LET Q$="Y"
30 WHILE (Q$="Y")
40 INPUT "GIVE ROLL NUMBER"; ROLL$
```

```
50 INPUT "GIVE PHYSICS MARKS"; PHY
60 INPUT "GIVE CHEMISTRY MARKS"; CHEM
70 INPUT "GIVE MATHS MARKS"; MATHS
80 LET PER = (PHY+CHEM+MATHS) * 100 / 300
90 PRINT "PERCENTAGE FOR ROLL NO "; ROLL$;  "IS "; PER
100 INPUT "DO YOU WINSH TO CONTINUE ? GIVE Y or N "; Q$
110 WEND
120 END
```

***Program Analysis***

To understand the dynamics of the loop note the following points:

1. In line 20 value of Q$ has been initialized to "Y".
2. Loop has been set to iterate till the time Q$="Y"
3. In line number 100 program asks, if the user wishes to continue. If he enters Y against this question, value of Q$ continues to be equal to Y. Thus condition in WHILE loop continues to be true and the loop executes for other student. On the other hand, if "Y" is not entered, condition becomes false and loop breaks. When the loop breaks, statement written just after WEND command gets executed. Note that this command is END statement, so program terminates.

## GOTO COMMAND

GOTO command makes program-control jump to another line of the program. It is an unconditional jump command. That means, as and when it is executed, it makes the program control jump to other location without any condition. It is used in the following format:

**<Line-Number> GOTO line number**

e.g. GOTO 900

Note that line number should be a valid line number and should exist within the program. If line number refers to a non executable statement like REN or DATA etc., the execution continues from the first executable statement after the given line number.

To understand the functioning of GOTO command, consider the following program, which prints the area of square only if its side length is positive and greater than 0.

***Program***

```
10 REN TO CALCULATE AREA OF A SQUARE
20 REN THIS IS DEMONSTRATION OF GOTO STATEMENT
30 INPUT "ENTER THE SIDE OF A SQUARE"; X
40 IF X =< 0 GOTO 80
```

```
50 PRINT "AREA OF SQUARE=";A
60 LET A=X^2
70 GOTO 100
80 PRINT " GIVE POSITIVE NUMBER "
90 GOTO 30
100 END
```

## GOSUB AND RETURN (SUBROUTINES)

GOSUB and RETURN statements are two statements that work together. The structure that they form within the program is called subroutines. GOSUB statement transfers the program control to some other part of the program, like GOTO command. Thus statements written there start getting executed one by one. When RETURN statement is encountered there, program control returns back to the statement, written immediately after GOSUB statement. GOSUB and RETURN statements are written in the following format:

```
< Line Number > GOSUB <Transfer-Line-Number >
                      Statement-1
                      Statement-2
                      Statement-3


< Transfer-Line-Number >    Statement-A
                            Statement-B
                            Statement-C
 RETURN
```

To understand the functioning of GOSUB statement, go through the following program, which has four logical sections or subroutines:

1. Main section, in which all other sections are called, using GOSUB command.
2. Entry section, in which length and width of rectangle are accepted from the keyboard.
3. Calculation section, in which area and perimeter of the rectangle are calculated.
4. Printing section, in which area and perimeter of the rectangle are printed.

***Program***

```
10 REM MAIN SECTION
20 GOSUB 100
30 GOSUB 200
```

```
40 GOSUB 240
50 END

100 REM ENTRY SECTION
110 INPUT "Length"; L
120 INPUT "Enter width", W
130 RETURN

200 REM CALCULATION SECTION
210 AREA = L*W
220 PERIMETER=(L+W)*2
230 RETURN

240 REM DISPLAY SECTION
250 PRINT "AREA OF RECTANGLE", AREA
260 PRINT "PERIMETER OF RECTANGLE", PERMETER
270 RETURN
```

## ON...GOSUB COMMAND

ON GOSUB command is extension of GOSUB command. In this command, one of the line numbers is chosen from various options available, depending upon the value of a variable. It is used in the following format:

**<Line Number > ON < Variable> GOSUB Line-Number1, Line Number-2,.......**

e.g. 100 ON N% GOSUB 1000, 2000, 3000

Here Variable is a variable or expression that attains values like 1, 2, 3 etc. Line-Number1 is that line number, to which the control has to be passed when the value of the variable is 1. Similarly Line-Number2 is that line number, to which the control has to be passed when the value of the variable is 2 and so on.

Note that there should be as many line numbers as possible values of the variable. For example, if variable can attain values up to 12 then there should be 12 different line numbers mentioned in ON GOSUB statement.

Note that if at run time, value of variable is 0 or greater than the number of line numbers mentioned in GOSUB statement, program control automatically gets transferred to the statement, written immediately after GOSUB statement.

To understand the usage of ON GOSUB command, let's write a program for the following problem:

***Problem***

Let's write a program that accepts month number from the keyboard (i.e. a number from 1 to 12) and prints the name of the corresponding month. For example, if 3 is given, it should print March.

***Program***

```
10 INPUT "ENTER MONTH IN NUMBER"; N%
20 ON N% GOSUB 100,200,300,400,500,600,700,800,900,1000,1100,1200
30 END

100 PRINT "JANUARY"
110 RETURN

200 PRINT "FEBRUARY"
210 RETURN

300 PRINT "MARCH"
310 RETURN

400 PRINT "APRIL"
410 RETURN

500 PRINT "MAY"
510 RETURN

600 PRINT "JUNE"
610 RETURN

700 PRINT "JULY"
710 RETURN

800 PRINT "AUGUST"
810 RETURN
900 PRINT "SEPTEMBER"
910 RETURN
```

```
1000 PRINT "OCTOBER"
1010 RETURN

1100 PRINT "NOVEMBER"
1110 RETURN

1200 PRINT "DECEMBER"
1210 RETURN
```

## USER DEFINED FUNCTIONS

You are already familiar with library functions and know that they remain inbuilt in the compiler/interpreter. Apart from this, BASIC language provides facility for writing your own functions. The functions that you write, as per your requirements, are called user defined functions. User defined functions are written in following format:

**DEF Function-Name(Argument1, Argumant2, ......)=Expression**

Here DEF is a keyword, which is written as such. Function-Name is name of the function. Note that Function-Name should always start with FN. After FN, any character(s) can be suffixed. For example, FNCALC is valid function name, while CALC is an invalid name. Argument1, Argument2 etc. are the variables, in which function receives values from main program.

To get familiar with the process of writing user defined functions, let's write a program for the following problem:

***Problem***

Say we have to write a function, which calculates cube of the given number.

***Program***

```
10 DEF FNCUBE (Z)= Z*Z*Z
20 FOR I=1 TO 5
30 INPUT "GIVE NUMBER"; N
40 PRINT FNCUBE(N)
50 NEXT I
60 END
```

To understand the dynamics of above given program, go through the following points:

1. Function has been defined in line 10. As per rules, name FNCUBE has been assigned to the function. Note that the name starts with FN.
2. FOR loop accepts five numbers from the keyboard, one by one.

3. In each iteration, the number accepted from the keyboard is passed to FNCUBE function for the calculation of cube. Cube value is displayed on the monitor screen. Note that both these activities are being performed in line number 40.
4. After repeating the process of accepting the number, calculating the cube through user defined function and printing it, the loop breaks and the program ends.

## SUBSCRIPTED VARIABLES OR ARRAYS

Subscripted variables are also called array. Array, in general, can be defined as similar type of items placed side by side. For example, an array of students, an array of books or an array of cars. In reference to programming, array comprises of group of variables that are of same data type (Integer, Real, String etc.). Such an array is illustrated in figure 14.6(a). Since it spreads in one direction hence it is called one dimensional array.

An array is identified by its name and the number of variables it consists of. For example, the array shown in figure 14.6(a) comprises of 5 variables and its name is x. Thus this array will be identified as x(5).

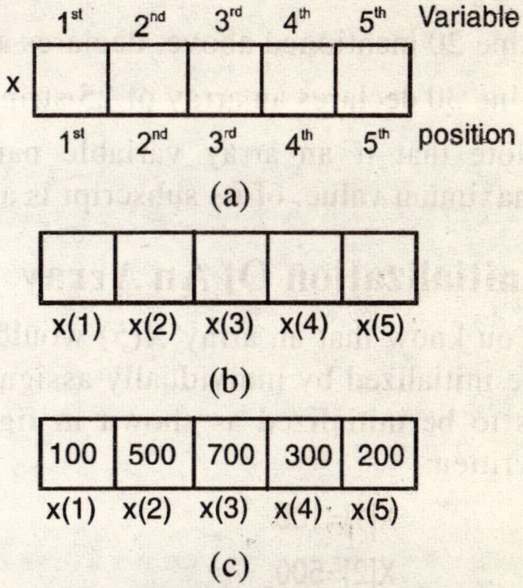


**Figure 14.6**

Individual element in the array is identified by the array name and its position in the array. Position in the array starts from 1 i.e. the first element will be at 1st position, second element will be at 2nd position and so on. The name of the element is written in the following format:

**Array Name(element's position in the array)**

Following this format, all the element of the array x, will be named as illustrated in figure 14.6(b).

Note that the first element will be at 1st position and will be named as x(1). Similarly the last element will be at fifth position and will be named as x(5).

After all, all the elements of the array are variables and variables are utilized for storing some value. Thus an array of integer variable will be able to store integer values. Similarly a string array will be utilized for storing string values. Any mismatch in the array type and the data being stored in it, leads to error. Now the question is how do you decide the array type? Well, that is done at the time of defining the array. As a general rule, all the arrays must be declared before they are used in the program. The procedure for declaring the array is described below.

### Declaring Single Dimension Arrays

As a rule, all the array variables must be declared at the beginning of program. It is done by writing DIM statement in the following format:

**<Line-Number> DIM Array-Name (integer)**

```
e.g. 10 DIM   MARKS%(100)
     20 DIM   PAY(15)
     30 DIM SNAME$(25)
```

Refer above mentioned statements and note that first statement (line number 10) declares MARKS% array of 100 elements. It is in an integer array so it will be able to store 100 integer values in it.

Line 20 mentioned above, declares an array named PAY for storing 15 real numbers.

Line 30 declares an array of 25 strings. It has been named as SNAME$.

Note that if an array variable name is used without using DIM statement for it, the maximum value, of its subscript is assumed to be 10.

## Initialization Of An Array

You know that an array X(5) would consist of 5 elements of real data type. The array can be initialized by individually assigning values to all its elements. For example, if the array is to be initialized as shown in figure 14.6(c) then following statements will have to be written:

```
X[1]=100
X[2]=500
X[3]=700
X[4]=300
X[5]=200
```

## FOR Loops And Arrays

Arrays and FOR loops go hand in hand. For all practical purposes, control variable of FOR loop is used as subscript variable in array. In each iteration, control variable, as subscript variable, refers to the next element of the array. To understand the usage of FOR loop along with arrays, let's write programs for the following problem:

***Problem***

Let's write a program that initializes all the elements of an integer array to 0. Say the array consists of 50 elements.

***Program***

```
10 REM INITIALIZING ELEMENTS OF ARRAY, USING FOR LOOP
20 DIM A%(50)
30 FOR I% = 1 TO 50
```

```
40 A%(I%) = 0
50 NEXT I%
60 END
```

## Reading Data Into Array, Using DATA And READ Statement

DATA and READ statements can be used for reading the data into array. For this, the data to be read is stated in DATA statement and READ statement is put within the loop. The variable used in READ statement is taken as array element, with varying subscript. Loop is defined in such a way that it iterates as many number of times as data items have to be read. For example, if 5 data items are to be read then loop is set to iterate 5 times.

To understand the process of reading the data through DATA and READ statements, into array, let's write a program for the following problem:

***Problem***

Let's write a program, which reads names of first six months into a string array, through DATA and READ statements and then prints them.

***Program***

```
10 REM READING DATA INTO ARRAY, USING DATA AND READ STATEMENTS
20 DIM N$(6)
30 DATA "JAN", "FEB", "MAR", "APR", "MAY", "JUNE"
40 FOR I% = 1 TO 6
50 READ N$(I%)
60 NEXT I%
70 REM DISPLAYING DATA
80 FOR I% = 1 TO 6
900 PRINT N$(I%)
100 NEXT I%
110 END
```

## Reading Data Into Array Using, INPUT Statement

Data can be read into array, using INPUT statement. For this, INPUT statement is put within the loop. In each iteration of the loop, a new item will be read into array. To understand the process of reading the data through INPUT statements, into array, let's write a program for the following problem:

***Problem***

Let's write a program that accepts Roll numbers and marks of 10 students from the keyboard and prints roll numbers and names of those students, who secured more than average marks.

**Program**

```
10 REM ACCEPTING DATA INTO ARRAY, USING INPUT STATEMENT
20 DIM ROLL%(10), MARKS(10)
30 FOR I% = 1 to 10
40 INPUT "GIVE ROLL NUMBER AND MARKS " ; ROLL(I%),MARKS(I%)
50 LET SUM = SUM + MARKS(I%)
60 NEXT I%
70  LET AVRG = SUM / 10
80 PRINT "AVERAGE MARKS ARE =  " ; AVRG
90 PRINT
100 PRINT "FOLLOWING STUDENTS GOT MORE THANAVERAGE MARKS"
110 FOR I% = 1 to 10
120 IF MARKS(I%) > SUM THEN PRINT ROLL(I%), MARKS%
130 NEXT I%
140 END
```

## DOUBLE DIMENSION ARRAYS

You are now familiar with single dimension arrays and know that they can be visualized as multiple elements of same type, placed side by side. Its details were described in the previous section. Now conceptualize many such one dimensional arrays placed below each other. Such an arrangement is shown in figure 14.7(a). Such an arrangement is called two dimensional array. Horizontal arrangement of elements, in such an array is called row and vertical arrangement is called column. Refer figure 14.7 (a) and note that this array consists of 3 rows and four columns.

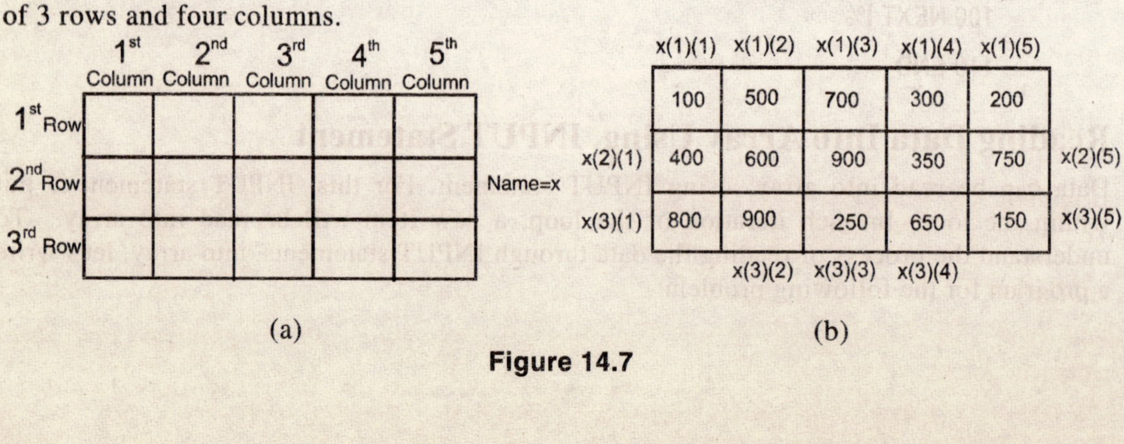


(a) (b)

**Figure 14.7**

Two dimensional arrays are identified by their names and the number of rows and columns in them. For example, X(3)(5) is a valid two dimensional array. In this arrangement, first number within the bracket denotes number of rows in the array, while the second number denotes number of columns in the array. Thus X(3)(5) denotes an array, X, which comprises of 3 rows and 5 columns. Similarly Y(10)(7) will denote an array Y that consists of 10 rows and 7 columns.

As illustrated in figure 14.7 (a), the first row is denoted by 1, second row is denoted by 2 and so on. Similarly the first column is denoted by 1, second column is denoted by 2 and so on.

Individual elements in the array are identified by the array name and their position in terms of row and column.

For example, the first element of the array X will be X(1)(1), second element in the first row will be X(1)(2). Similarly first elements in the second and third row will be denoted as x(2)(1), x(3)(1) respectively. Few other elements of X(3)(5) array are listed in figure 14.7(b), for your reference.

## Declaring Two Dimensional Arrays

Two dimensional arrays are declared in the following manner:

```
<Line-Number> DIM   Array-Name(rows)(columns)
e.g. 10 DIM   AMT%(3)(5)
     20 DIM   BUDGET(4)(5)
     30 DIM CITY$(10)(5)
```

Refer above mentioned statements and note that first statement (line number 10) declares an array named AMT%. It consists of 3 rows and 5 columns. It is in an integer array so it will store 3 x 5=15 integer values in it.

Line 20 mentioned above, declares an array named BUDGET consisting of 4 rows and five columns. Since it is real array hence it will store 4 x 5=20 real numbers in it.

Line 30 declares an array of strings. It has been names as CITY$. It consists of 10 rows and 5 columns. Thus it will store 10 x 5=50 elements in it.

## Reading Data Into Two Dimensional Array

Nested loops can be used for reading the data into two dimensional arrays. Control variable of one loop can be used as subscript variable for row and control variable of another loop can be used as subscript variable for column. The data can either be read within the program, using DATA and READ statements or it can be read from keyboard, using INPUT statement.

To understand the usage of nested FOR loops for handling two dimensional arrays, let's write programs for the following problems:

***Problem***

Let's write a program, which accepts data for two dimensional array of integers and then prints row totals and column totals.

***Program***

```
10 SUM=0
20 DIM NUMB (4,5)
30 FOR I=1 TO 3
40 PRINT "Enter the Row " ; I "Value"
50 FOR J=1 TO 4
60 PRINT "Enter Value";
70 INPUT NUMB(I,J)
80 NEXT J
90 NEXT I
100 PRINT
110 FOR I=1 TO 3
120 FOR J=1 TO 4
130 SUM = SUM+NUMB(I,J)
140 NEXT J
150 NUMB(I,5)=SUM
160 SUM=0
170 NEXT I
180 SUM=0
190 FOR J= 1 TO 4
200 FOR I= 1 TO 3
210 SUM = SUM+NUMB(I,J)
220 NEXT I
230 NUMB (4,J)= SUM
240 SUM=0
250 NEXT J
260 FOR I=1 TO 4
270 FOR J= 1 TO 5
280 PRINT NUMB(I,J),
290 NEXT J
300 NEXT I
310 PRINT
320 END
```

***Problem***

Let's write a program, which accepts data for two dimensional square array (an array, in which number of rows are equal to number of columns) and prints its diagonal elements.

***Program***

```
10 SUM=0
20 ROWPOS=0
30 X=0
40 DIM NUMB (4,4)
50 FOR I=1 TO 4
60 PRINT "Enter the Row " ; I "Value"
70 FOR J=1 TO 4
80 PRINT "Enter Value";
90 INPUT NUMB(I,J)
100 NEXT J
110 NEXT I
120 PRINT
121 PRINT "MATRIX DISPLAY"
130 FOR I=1 TO 4
140 FOR J= 1 TO 4
150 PRINT NUMB(I,J),
160 NEXT J
170 PRINT
180 NEXT I
190 PRINT
191 PRINT "DISPLAY LEFT DIGNAL OF MATRIX"
200 FOR I=1 TO 4
210 FOR J=1 TO 4
220 IF I=J THEN  PRINT NUMB(I,J)
230 NEXT J
240 NEXT I
250 PRINT
251 PRINT "DISPLAY RIGHT DIGNAL OF MATRIX"
260 FOR I=1 TO 4
```

```
270 FOR J=1 TO 4
280 ROWPOS=4-X
290 IF J=ROWPOS THEN  PRINT NUMB(I,ROWPOS)
300 NEXT J
310 X=X+1
320 NEXT I
330 PRINT
340 END
```

## EXERCISES CHAPTER 14

### Short Type Questions

**A. Select most appropriate answers for following questions:**

1. Which of the following doesn't hold good for BASIC program?
   (a) Each statement of the program should bear a unique number.
   (b) Constants, variables and operators can be used for forming an expression.
   (c) Keywords can be used as variable names.
   (d) RE M command can be used to include comments in the program.
2. Which of the following fact doesn't hold good for GOSUB?
   (a) GOSUB is a keyword.
   (b) It transfers the control to some other part of the program.
   (c) It always has an associated RETURN statement.
   (d) It forms a loop within the program.
3. Which of the following is a subscribed variable?
   (a) A$(10) (b) A$
   (c) A% (d) / A
4. Which of the following is wrong assignment statement?
   (a) A%=5 (b) NAME= "INDIA"
   (c) A$(10)="RAM" (d) X=100.7
5. Which of the following statement is wrong in relation to AMT=AMT+100?
   (a) Initially AMT will be 100 and 100 will be added to it.
   (b) After execution of this statement AMT value will increase.
   (c) New value of AMT will be old value of AMT+ 100.
   (d) It is a correct statement.

6. What are AND and OR?
   (a) Commands (b) Logical operators
   (c) Expressions (d) Relational operators
7. What will be obtained by function RIGHT$ ("ACDEH", 2)?
   (a) AC (b) EN
   (c) CD (d) ED
8. Which of the following is valid name for user-defined function?
   (a) DEF (b) DEFR
   (c) FNROOT (d) ROOT
9. Which of the following command is not directly related to obtaining data in a variable?
   (a) INPUT (b) LET
   (c) DATA & READ (d) FOR
10. Which command will print the value of the variable on printer?
    (a) PINT (b) PRINTER
    (c) LPRINT (d) GOTO

## B. Fill in the blanks:

11. BASIC is acronym for ............... ....................... ............... . ......................
12. Those quantities, which change their values during the program are called .........................
13. .....................and...................are the two commands, using which loops are formed within the program.
14. All integer variable names should end with ...................sign.
15. A loop, which is defined as FOR I = 1 TO 24 STEP 3 will iterate...............times.
16. In BASIC.............................command is used for accepting data from the keyboard.
17. ................command will have to be used for making the program control jump to some other location of the program.
18. Array A$ (7)(3), will store...............value, which are of ...........data type.
19. ......................command sets the data read counter to the beginning in DATA and READ command.

## C. State, true or false for following statements:

20. All real variable names must to be terminated with % sign.
21. X= X+100+500 is a correct BASIC statement.
22. (A+B) AND (X-C) is a valid logic expression.
23. All BASIC expressions are evaluated from left to right.

24. ASCII(66) will give B.
25. All FOR statements must have a corresponding NEXT statement.
26. All DATA and READ statements must have a corresponding RESTORE command.
27. Use of ; in PRINT command will print the item in next printing zone.
28. ON.....GOSUB command involves more than one line numbers.
29. IF few commands are to be executed till the time a given condition is false then while loop can be used.

**D. Answer the following questions in short:**

30. How will you differentiate between integer variable and real variable?
31. If Y=200 then what will get printed when you print the value of Y, after executing Y= Y+ Y/2 command?
32. In which situation will the expression (A1 AND A2) give true?
33. In which situation will the expression (B1 OR B2) give false?
34. Which command will you use to display a message and accept a value from the keyboard?
35. Which function will you use to convert the number 124 6 into a string?
36. Which function will you use to find out the length of a given string?
37. In a nested loop, outer loop iterate 4 times, inner loop iterates 5 times. How many times the PRINT statement will get executed that has been written within inner FOR loop?
38. How many blank spaces will be left before printing the value of X, when PRINT TAB(10); X command is given?
39. How many value will be stored in A%(10)(5) array?

## Detailed Answer Type Questions

**E. Answer the following questions in detail:**

40. Differentiate between following:
    (a) Constant and variable.
    (b) Integer variable name and Real variable name.
    (c) GUSUB and GOTO command.
    (d) FOR loop and WHILE loop.
    (e) IF----THEN and IF-----THEN-------ELSE command
41. What are subscripted variables? How do you declare them?
42. What do you understand by nested FOR loops? Explain them giving suitable piece of code.

43. FOR what purpose, following are used?
    (a) LET command.
    (b) INPUT command
    (c) WHILE-----WEND command
    (d) ON GUSUB command.
    (e) User defined function.
44. Differentiate between one-dimensional and two-dimensional array. Write a program to create an array A(10)(5) in the program and reading values in it through keyboard. After storing the values in array, program should also print the elements of the array.
45. Write a program in BASIC to accept 10 numbers from the keyboard and print the largest number among them.
46. Write a program to accept value of n from the keyboard and generate the series of following type:

    **1+4+9+16+25+ -------(n×n)**
47. Write a program to accept roll number and marks of 10 students and print the roll number of those students who secured more than average marks.
48. Write a program to print the sum of two arrays.
49. Write a program that takes temperature in centigrade and converts it the Fahrenheit.

****

# COMPUTER

## CHAPTER 15

# *Computer Graphics*

For quite long time after first computer came into existence, the emphasis was on numerical calculations. Computers were mainly used for arithmetic and logical calculations. All the developments and enhancements were made from the point of view of doing more complex calculations at much faster speed. For quite long time, the scenario remained same. As a result of this, computer was more or less treated as number crunching machine.

Computers of the olden days were thin in resources. They had few kilobytes of memory and a CPU of limited capabilities. Display systems were also poor. They could not display high resolution images. All put together, computer resources of that time were not sufficient to take up graphics applications. But gradually the scenario changed. With the advent of powerful Integrated Circuit Technology, computers became more powerful and their cost also came down. These two factors encouraged a shift in focus from number applications to graphical applications.

## BIRTH OF COMPUTER GRAPHICS

In 1961, Ivan Sutherland, a student of MIT created a computer drawing program called Sketchpad. Sketchpad program allowed its users to draw simple shapes on the computer screen, using a light pen. These shapes could be saved and even recalled later. This event is marked as origin of computer graphics.

In 1961 itself, another student at MIT, Steve Russell, created the first video game, Spacewar, written for the DEC PDP-1 computer. Spacewar was an instant success and copies of this program flowed to other PDP-1 computers.

In 1963 E. E. Zajac, a scientist at Bell Telephone Laboratory created a film called "Simulation of a two-giro gravity attitude control system". In this computer generated film, Zajac showed how the altitude of a satellite will be altered when it orbits the Earth. He created the animation on an IBM 7090 mainframe computer.

Other scientists Ken Knowlton, Frank Sindon and Michael Noll at Bell Telephone Laboratory started working in the computer graphics field. Sindon created a film called Force, Mass and Motion, illustrating Newton's laws of motion in operation. At Lawrence

Radiation Laboratory, Nelson Max created the films, "Flow of a Viscous Fluid" and "Propagation of Shock Waves in a Solid Form." Boeing Aircraft created a film called "Vibration of an Aircraft."

In mid sixties IBM released the first commercially viable graphics computer with IBM 2250 graphics terminal.

Around 1971, microprocessors came into existence and computer graphics took a great leap. Lot of animation companies came into existence and video games were their main focus area.

In 1990s computer graphics became so popular that it became the need of the time. Graphical User Interface (GUI) based operating systems came into existence. Windows Operating system is an example of this. Such operating systems changed the computer working scenario completely. With the advent of GUI, working on computer became quite easy. Coupled with price falling scenario, computer became the tool of masses.

With the popularity of computer graphics, special graphics capabilities were built into hardware. MMX technology, high resolution monitors, LCD screens etc. are the result of graphics demand. High power video chips are now available for use.

Applications have gradually shifted their focus from 2D to 3D. Fields like Multimedia and Virtual reality came into existence because of graphics. These are the fields which go beyond graphics, but graphics plays a major role in them.

## COMPUTER GRAPHICS DISPLAY SYSTEM

Visual Display Unit (VDU) of computer is the basic hardware, where graphics is displayed. VDU basically consist of many assemblies, which put together display the graphics that is given to it for display by the CPU. In present context, we are concerned with following two assemblies of VDU:

1. Three electron guns
2. Phosphor quoted display screen

As the name indicates, the electron guns produce electrons. When the electron hits phosphor quoted display screen, the hit spot get illuminated. Complete arrangement is made in such a way that one electron gun produces red colored spot; second gun produces green color spot while the third gun produces blue colored spot. When the spots of varying intensity of these colors are illuminated very near to each other, other color impressions are created. For example, colors like orange, yellow, indigo, cyan etc. are created with the help of these three colors only.

After some time the illuminated dot tends to loose its illumination. To keep the illumination alive, arrangement is made to refresh the brightness.

## VDU PERFORMANCE

Some of the VDUs display sharp and impressive images, while others don't. Following are the factors, which contribute to the quality of the image that is displayed in VDUs:

1. Video adapter card.
2. Monitor resolution.
3. Refresh rate.
4. Dot pitch.

## Video Adaptor Card

**Figure 15.1**

Video adaptor card is the basic circuit that provides data needed to create a display on the monitor. Most commonly used video adaptor cards are either VGA cards or SVGA cards.

VGA is short form of Video Graphics Array. VGA adapter cards provide either 16 colors with resolution of 640x480 or 256 colors with resolution of 320x200.

SVGA is short form of Super Video Graphics Array. SVGA card provides resolution of 800 x 600 or 1024 x 768 pixels. It supports palette of 16 million colors, but the number of colors that actually get displayed are governed by the amount of video memory present in the computer. Modern video adaptor cards contain local memory so that computer's memory is not used for display purpose. In addition to this, most video adaptors these days, have a graphics coprocessor for performing graphics calculations. These adaptors are called graphics accelerators.

It is quite obvious now that the quality of the picture will depend upon the type of video adaptor card being used in the system, Better the card, better would be the image displayed on the monitor.

## Monitor Resolution

You know that image on the monitor is made by illuminating the dots on the screen. These dots can either be very near or far away from each other. Nearer they are, finer is the image formed on the monitor. Each such dot is called pixel.

Monitor resolution is measured in terms of how many pixels are there per inch. When monitors resolution is expressed as 800 x 600, that mean when you are moving along width of the monitor, there are 800 dots per inch and when you are moving along length, there are 600 dots per inch.

It is quite obvious that higher resolution monitors will display better images than low-resolution monitors.

Note that monitor is not the only factor that governs the resolution of the display. Video adapter card also contributes to it. So even if you have a high-resolution monitor and if it is coupled with lower graded video adapter card, the image that you would see will be of inferior quality.

## Refresh Rate

You know that when a spot on the screen is illuminated, it fades after some time. To keep the image unchanged on the screen, it needs to be illuminated again. This process of re illuminating is called refreshing. Refresh rate decides the number of time this process will be performed in a second. Its unit of measurement is Hertz (Hz) or cycle per second. For example, if a monitor refresh rate is 70 Hz that means the refreshing process will be performed 70 times per second.

If the refresh rate is low, the image will not be stable in intensity. It will keep on fading and brightening. Thus the monitor will cause lot of eyestrain. A refresh rate of 75 Hz or higher produces stainless images.

## Dot Pitch

You know that in color monitors, screens have 3-color arrangement, either in the form of dots or strips to generate color display. Dot pitch is the distance between two adjacent dots of the same color. For example, distance between two blue color dots will be called dot pitch. It is measured in millimeter. It is quite obvious that monitors with smaller dot pitch will generate better quality image than those having higher dot pitch value. For personal computer monitors, it ranges from 0.15 mm to 0.30mm. It is also called phosphor pitch.

# PIXEL

Pixel is the most common word that is used in reference to graphics. The word pixel has been derived from two other words i.e. **Pic**ture and **El**ement (c of Pic is taken as x). Pixels are dots, used for displaying an image on a screen or on printed matter.

In graphics context, the word Pixel is used as the smallest unit of an image. It is important to recognize that pixels are not a measure of size.

A VDU is made up of many millions of pixels arranged in the form of grid. VDUs also have a bit depth, which controls how many grays or colors each pixel is capable of displaying. In an 8-bit display, for example, each of the pixels can display 8 bits worth of colors i.e. equal to two to the power of eighth, or 256, colors. With 24-bit displays, each of the pixels can display more than sixteen-million colors.

While pixels are the smallest complete element of an image, they comprise of even smaller elements. In a standard RGB monitor, for example, each pixel has three dots within it: a red, blue, and green dot. In theory, these dots all converge at the same point, making them visibly seamless, but in practice some fuzziness is often apparent. The measure of total pixels in an image is referred to as its resolution, and the higher the resolution in a small area, the more difficult it is to distinguish between individual pixels.

## Physical Pixels

Physical pixels are the actual dots displayed on an output device. Each pixel takes up a small amount of space on the surface of the device. Physical pixels are manipulated directly by the display hardware and form the smallest indenendently programmable physical elements on the display surface. Using a program they can either be illuminated or left untouched.

Because physical pixels cover a fixed area on the display surface, there are practical limits to how close together two adjacent pixels can be. Asking a piece of display hardware to provide too high a resolution, too many pixels on a given display surface will create blurring and other detoriation of image quality if adjacent pixels overlap or collide.

## Logical Pixels

Logical pixels are coordinates. They specify a location, but are assumed to occupy no area. Thus following logical pixels, physical pixels can be illuminated and the image be displayed on VDU. Note that in this process of logical to physical mapping, a dense and brightly colored bitmap may lose its vibrancy when displayed on too large a monitor, because the pixels will be spread out to cover the surface. This is illustrated in figure 15.2.

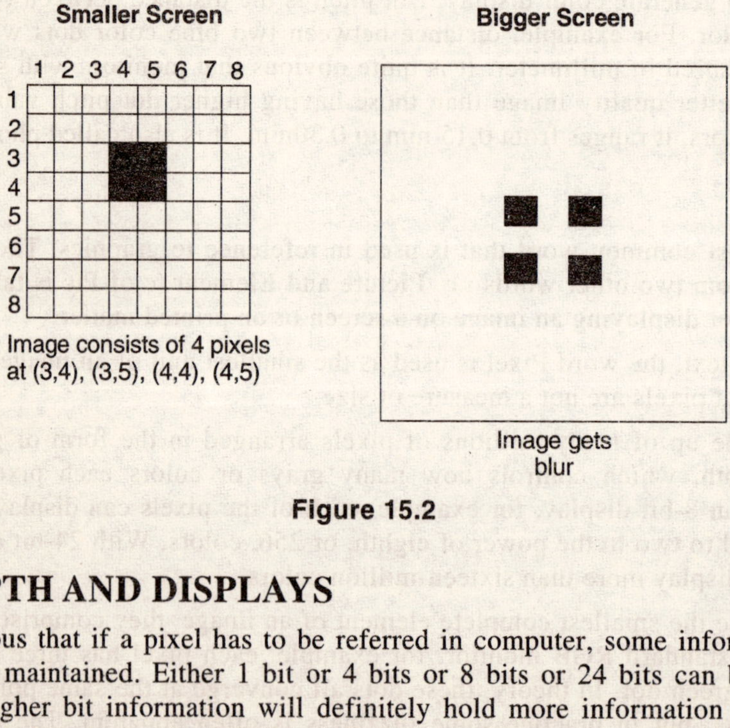


**Figure 15.2**

## PIXEL DEPTH AND DISPLAYS

It is quite obvious that if a pixel has to be referred in computer, some information about it will have to be maintained. Either 1 bit or 4 bits or 8 bits or 24 bits can be used for this purpose. The higher bit information will definitely hold more information about the pixel while lesser bit information will hold less information. In graphics terminology, number of bits associated with pixel is called Pixel Depth. Thus Pixel depth governs the number of colors that pixel can exhibit. Higher the pixel depth, greater is the number of possible colors.

Note that more bits per pixel also means more space is needed to store the information of the pixels that constitute the graphical picture.

A bit can either be 0 or 1. So following relationship exists between the Bit Depth and number of colors that can be displayed:

| Bit Depth | Formula | Possible Number Of Colors |
|---|---|---|
| 1 | $2^1$ | 2 |
| 4 | $2^4$ | 16 |
| 8 | $2^8$ | 256 |
| 24 | $2^{24}$ | 16,777,216 |

## TRUE COLORS

It is said that human eye can discriminate between 16,777,216 colors. These colors are called true colors. A device capable of matching or exceeding the color-resolving power of the human eye under most conditions is said to display true colors. In practice, this means 24 Bit Depth displays true colors.

## COLOR DISPLAYING ISSUES

The image that is stored on the disk, in the form of logical pixels and the image that is displayed on the output device are two different identities. It is the responsibility of the rendering application that displays the image to maintain greatest possible similarities between the two. Often it so happens that pixel values stored in a file differ from those that can be displayed on the surface of an output device. Some times their value is lower and sometimes higher.

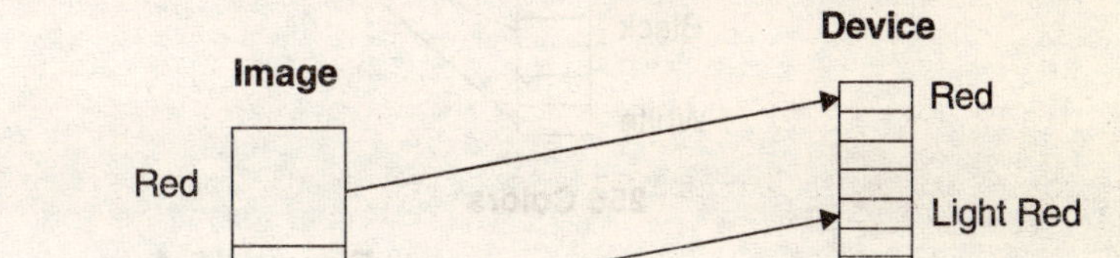


**Figure 15.3**

There is generally no problem when the number of colors defined in the file (source) is much less than the number that can be displayed on the output device (destination). The rendering application in this case is able to choose among the destination colors to provide a match for each source color. This is shown in the figure 15.3.

A problem occurs when the number of colors defined in the file exceeds the number that can be displayed on an output device. Rendering program tries to match nearest possible color available in display system. This is the reason why new colors appear in the image when it is displayed on a different system than the one, on which it was produced. This is shown in figure15.4.

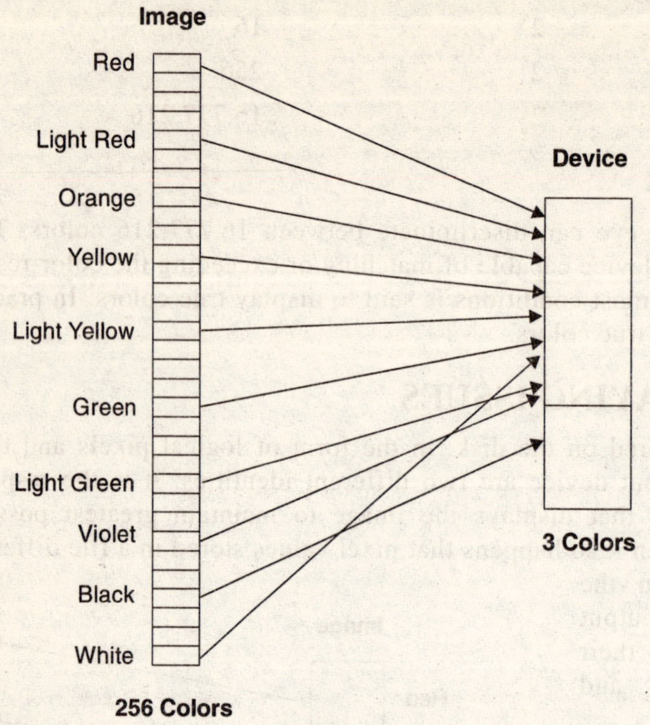


**Figure 15.4**

## COLOR PALETTE

Color Palette is basically a lookup table that specifies colors for different combination of bits. To understand the concept, take the case of 4 bit depth. That means, with each pixel 4 bit data will be attached. That means $2^4$ i.e. 16 colors can be associated with the pixel.

The 4 bits associated can take any of the following values:

| | | | |
|---|---|---|---|
| 1. | 0000 | 9. | 1000 |
| 2. | 0001 | 10. | 1001 |
| 3. | 0010 | 11. | 1010 |
| 4. | 0011 | 12. | 1011 |
| 5. | 0100 | 13. | 1100 |
| 6. | 0101 | 14. | 1101 |
| 7. | 1110 | 15. | 1110 |
| 8. | 0111 | 16. | 1111 |

Now the question is which value mentioned above will represent which color? Well, color palette mentions these values and the color represented by them in the form of a table. Note that higher the Bit Depth, bigger will be the color palette.

# COLOR SYSTEMS

Several mathematical systems exist, which are used to describe colors. Color systems can be separated into following two categories:

1. Additive color systems
2. Subtractive color systems

## Additive Color Systems

Colors in additive systems are created by adding colors to black to create new colors. More the color added, resulting color tends more towards white. The presence of all the primary colors in sufficient amounts creates pure white, while the absence of all the primary colors creates pure black. Additive color environments are self-luminous. For example, colors on monitors, are additive.

## Subtractive Color Systems

Color subtraction works in the opposite way. Conceptually, primary colors are subtracted from white to create new colors. More the color subtracted, the resulting color tends more towards black. Thus, the presence of all the primary colors theoretically creates pure black, while the absence of all primary colors theoretically creates pure white. Another way of looking at this process is that black is the total absorption of all light by color pigments. Subtractive environments are reflective in nature, and color is conveyed to us by reflecting light from an external source. A colored image reproduced on paper is an example of use of a subtractive color system.

# COMMON COLOR SYSTEMS

There are many color systems. Few common color systems are described below.

## RGB (Red-Green-Blue) Color System

RGB is the most widely used color system. It is an additive system. In this system Red, Green and Blue colors are taken as primary colors. Other colors are formed by adding varying amounts of red, green, and blue colors. Graphics files using the RGB color system represent each pixel as a color triplet. For example (255,70,80) is a color triplet in RGB color system and it represents a color. Each value in the triplet can take any value between 0 to 255. For example, in 24-bit color, the triplet (0,0,0) represents black, and the triplet (255,255,255) represents white. When the three RGB values are set to the same value, for example, (63,63,63) or (127,127,127), or (191,191,191), the resulting color is a shade of gray.

## CMY (Cyan-Magenta-Yellow) Color System

CMY is a subtractive color system. It is generally used in printing. When illuminated, each of the three colors absorbs the light of its complementary color. Cyan absorbs Red. Magenta absorbs Green and Yellow absorbs Blue. By increasing the amount of Yellow ink, for example, the amount of blue, in the image, is decreased.

As in all subtractive systems, we say that in the CMY system colors are subtracted from white light by pigments to create new colors. The new colors are the wavelengths of light reflected, rather than absorbed, by the CMY pigments. For example, when cyan and magenta are absorbed, the resulting color is yellow. The yellow pigment is said to "subtract" the cyan and magenta components from the reflected light. When all of the CMY components are subtracted, or absorbed, the resulting color is black.

## CMYK (Cyan-Magenta-Yellow Black) Color System

CMYK system is variant of CMY system. In this system, K stands for black color. To compensate for inexpensive and off-specification inks, the color black is taken into the color system and treated something like an independent primary color variable. For this reason, use of the CMYK scheme is often called 4-color printing, or process color.

In CMYK color system, each color is represented as a set of four values. In many real-world color composition systems, the four CMYK color components are specified as percentages in the range of 0 to 100. Thus their values vary from 0 to 100. The value (0,0,0,0) represents white and (100,100,100,100) represents black in CYMK color system.

## HSV (Hue, Saturation, and Value) Color System

HSV is one of many color systems that vary the degree of properties of colors to create new colors, rather than using a mixture of the colors themselves. Hue specifies "color" in the common use of the term, such as red, orange, blue, and so on. Saturation (also called chroma) refers to the amount of white in a hue; a fully (100 percent) saturated hue contains no white and appears pure. By extension, a partly saturated hue appears lighter in color due to the admixture of white. For example, Red hue with 50 percent saturation appears pink. Value (also called brightness) is the degree of self-luminescence of a color that is, how much light it emits. A hue with high intensity is very bright, while a hue with low intensity is dark.

HSV (also called HSB for Hue, Saturation, and Brightness) most closely resembles the color system used by painters and other artists, who create colors by adding white, black, and gray to pure pigments to create tints, shades, and tones. A tint is a pure, fully saturated color combined with white, and a shade is a fully saturated color combined with black. A tone is a fully saturated color with both black and white (gray) added to it.

# TYPES OF COMPUTER GRAPHICS

There are two kinds of computer graphics. They are Raster graphics and Vector graphics. They are rendered in different ways and have their own characteristics. Both of them are described in detail below.

## Raster Graphics

Raster images are also called bitmap images. Bitmap graphical pictures are composed of pixels They can be conceptualized as a grid of individual pixels where each pixel can be of a different color or shade. Image shown below is a Bitmap image or Raster graphics.

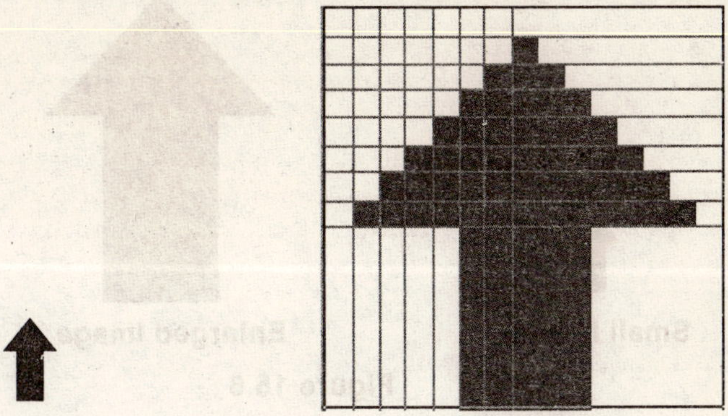

**Small Image** **Enlarged Image**

**Figure 15.5**

Refer this figure and note that the figure is shown composed of small squares. Consider these squares as pixels. If somehow each pixel is described in terms of location and color information then using that information, image can either be saved on the disk or reformed on output device. Following this strategy, location and color information of the image is stored as a collection of information of individual pixels.

Bitmap images are edited at the pixel level. In other words, the color of any one pixel can be changed. Additional attributes of bitmap images include:

1. Bitmap images are usually created and edited in "photo" or "paint" programs such as Adobe Photoshop.
2. Bitmap images are mapped to a grid.
3. The size of the image is based on the image's resolution.
4. Bitmap images are not easily scalable.

When a raster image is created, the image on the screen is converted into pixels. Each pixel is assigned a specific value which determines its color. The raster image uses RGB color system.

When a raster image is viewed, the pixels usually smooth out visually for the user, who sees a photograph or drawing. When blown up, the pixels in a raster image become apparent. Depending on resolution, some raster images can be enlarged to very large sizes, while others quickly become difficult to see. The smaller the resolution, the smaller the digital image file. For this reason, people who work with computer graphics find a balance between resolution and image size.

## Vector Graphics

Vector graphics are basically algorithms and data. In other words it can be said that they are drawing instructions and the color values for the pixels. Graphics shown in figure 15.6 is a vector graphics.

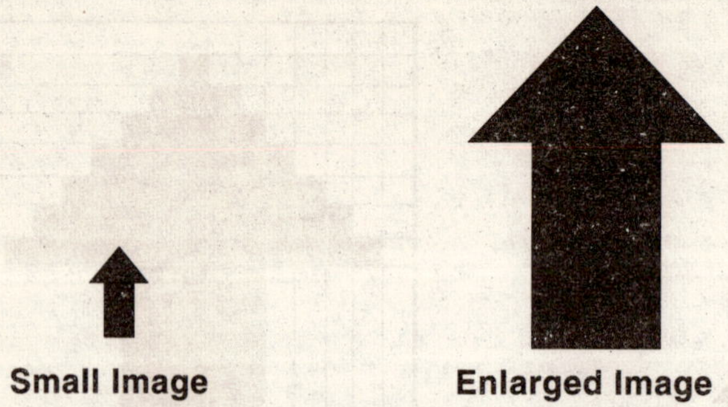


**Figure 15.6**

Following are the attributes of vector graphics:

1. Vector-based images are usually created and edited in "draw" or "illustrate" programs such as Adobe Illustrator.
2. Vector-based images have smooth edges.
3. Vector-based images create curves or shapes.
4. Vector-based images are good for precise illustrations, but are not as good for photorealistic images.
5. Vector-based images are easily scalable, due to their use of mathematical formulas.

Since vector graphics are stored as drawing instructions so they occupy less space on the disk and can be easily enlarged without distorting the image. Bitmap art, on the other hand, becomes jagged when enlarged because the squares used to create the image are enlarged.

Another advantage of vector art is the fact that each line within the image is representative of a single object. This makes it easier to re-edit vector art as and when necessary. The same is not true of bitmap art.

## GRAPHIC FILE FORMATS

Drawings that we make, need to be stored in a file. The format, in which they are saved is called file format. Although there are wide variety of file formats for graphics but few commonly used file formats are described below:

| File Format Name | Graphics Type | File Extension | Description |
|---|---|---|---|
| Bitmap | Raster | .bmp | It is commonly used by Microsoft Windows. |

| File Format Name | Graphics Type | File Extension | Description |
|---|---|---|---|
| CorelDraw | Vector | .cdr | It is a proprietary file format invented by Corel Corporation. Drawings made in CorelDraw package make use of this format. |
| Flash Source File | Vector | .fla | It is a file format used by Macromedia, Flash authoring software. |
| Graphic Interchange Format | Raster | .gif | This format is widely used on Web. Apart from still graphics, it also supports animated graphics. It supports only 255 colors. |
| Joint Photographic experts Group | Raster | .jpeg & .jpg | This file format is extensively used for photos on web. |
| Image Photo Pack CD | Raster | .pcd | It is proprietary Kodak format. It supports 24-bit colors. |
| Portable Document Format | Raster and Vector | .pdf | Supported by Acrobat Reader |
| Portable Network Graphics | Raster | .png | It is a file format, which gives minimum loss of colors when file is compressed. It is gradually becoming very popular on Web. It supports bit depths from 1 to 48. |
| Photoshop Document | Raster | .psd | This file format is supported by Photoshop program, which is commonly used for photo editing. |
| Tagged Image File Format | Raster | .tiff & .tif | It is extensively used for print graphics |
| Windows Metafile | Vector | .wnf | Stores vector graphics and raster graphics as a sequence of commands to be issued to the graphics layer of the Microsoft Windows operating system. |

## FONTS

Fonts that we use on computer are after all graphics. Some of them are raster graphics and some of them are vector graphics.

True type fonts are vector graphics. Thus true font file contains directions to draw the character. This is the reason why they can be enlarged to any practical extent.

Raster fonts being pixel based fonts cannot be scaled to very large size. The problem of jagging starts appearing. But for all practical purposes, bitmaps for mostly used point sizes are generally made available.

## COMPRESSION AND DECOMPRESSION

Compression is a process, using which file size is reduced to a considerable extent. Since graphics files are generally large in size hence portability becomes a problem. The problem gets enhanced when they are to be transported over network or Internet. Transportation of large files takes time and blocks the traffic for others. Thus compressing the files and then transporting them solves the problem to a large extent.

Compression is done through a set of algorithms. These algorithms compress the data and thus the file size gets reduced. Concept of file compression is illustrated in figure 15.7

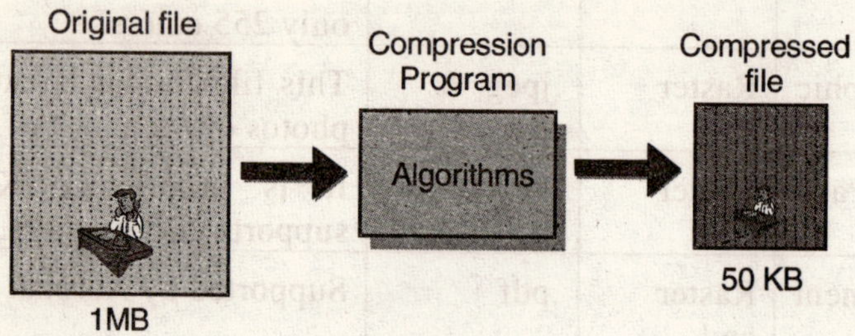


**Figure 15.7**

Decompression is reverse of compression process. It is a process, in which compressed file is operated upon and original file is retrieved. Thus reduced file size becomes large. Concept of file decompression is illustrated in figure 15.8.

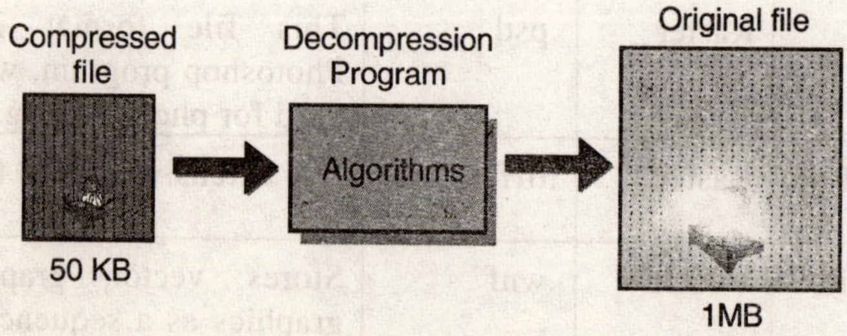


**Figure 15.8**

### Compression Loss

When a data file is compressed and then decompressed to attain its original state, often some data gets lost in the process. This happens due to the algorithms used for compression and decompression. If the algorithms are not efficient, data loss will be greater. On the other hand, few efficient algorithms ensure lossless data compression.

In the case of computer graphics, often this data loss is such that it doesn't become visible to the eyes hence the compression process becomes acceptable and serves as good solution for file transportability.

## Compression Techniques

Although there are various compression techniques available these days but following techniques are very common:

1. Shannon Fano algorithm
2. Huffman algorithm
3. Extended Huffman algorithm
4. Arithmetic coding
5. LZ77 compression technique
6. LZ78 compression
7. LZW compression algorithm
8. MHEG (Multimedia Hyper Expert Group)
9. JPEG (Joint Photographic Expert Group)

# CREATING COMPUTER GRAPHICS

In computers, following means can create graphics:

1. Through graphic input devices.
2. Through graphics softwares.
3. Through programming.

## Graphics Through Input Devices

Special graphic input devices are available these days at very nominal cost. They can be connected to the computer and graphics can be brought in the computer. For example, scanner, digital camera, light pen etc. are the devices that can be used for this purpose.

## Graphics Through Graphics Softwares

There are various drawing packages available for creating drawings on computer. For example, CorelDraw is one of the most popular drawing packages in computer drawing industry. Paint program of Windows is also used for making drawings on computer. These packages provide coloring facility also. Using their color systems and color palettes, drawings can be filled with attractive colors.

3D-Studio MAX is another popular package that provides facility to create 3-D models.

Photoshop is a popular photo editing package. Using this package, raster graphic images can be edited, color corrections can be made and special effects can be introduced in the picture. For example, an image captured through camera can be subjected to a filter and a special effect of pastel color be introduced in it, as if it was made using pastel colors.

## Graphics Through Programming Languages

Programming languages like C, C++, Java etc. provide exclusive libraries of graphic routines. Using these libraries calls, graphics programming can be done. Such drawings are very much required in engineering applications.

## EXERCISES CHAPTER 15

### Short Type Questions

**A. Select most appropriate answers for following questions:**

1. What do you call those small spots on the monitor screen that form an image, on illumination?

   (a) Pic sell (b) Pixel (c) Pic Exel (d) Picl

2. Which factor governs the number of colors displayed by a pixel?

   (a) Pixel Width (b) Pixel depth (c) Pixel height (d) Pixel length

3. How many primary colors are used in RGB color system?

   (a) 1 (b) 2 (c) 3 (d) 4

4. To which category does CMYK color model belong?

   (a) Additive (b) Subtractive (c) Divisive (d) Multiplicity

5. What rang of values is used in RGB color model?

   (a) 0 to 225 (b) 1 to 225 (c)1 to 256 (d) 0 to 256

**B. Fill in the blanks:**

6. Computer graphic pictures have been categorized into two major groups. They are...........graphics and................graphics.
7. The graphics that deforms on enlargement definitely belongs to .........category of graphics.
8. Graphic files with .bmp extension are ....................graphics pictures.
9. Vector graphics are stored in the form of ................while restore graphics are stored in the form of pixel.
10. The drawings made in CorelDraw package is examples of .............graphics.
11. True fonts are examples of ...........graphics.
12. .........is a process, using which file size is reduced.
13. ............................is an example of compression technique.
14. .......................is a device, which is used for displaying image in computer.

**C. State, true or false for following statements:**

15. Scalar is one of the categories of computer graphics.
16. Vector graphics doesn't get deformed, when it is enlarged.
17. The dots on VDU, using which graphics is displayed on the screen are examples of logical pixels.
18. 8-pixel width is capable of displaying 8 colors only.
19. In CMYK model, K represents black color.

20. RGB model holds good for those devices, which function on color additive principle.
21. IN HSV color system S stands for situation.
22. The word jagging is related to vector graphics.
23. Decompression is a process, which converts vector graphics to raster graphics.
24. Scanner can bring external graphical pictures into computer.

### D. Answer the following questions in short:

25. Name any two-output device, using which graphic pictures could be brought into computer.
26. Name any two output devices that could be used for outputting graphical pictures.
27. Name two softwares, using which pictures could be drawn in computer.
28. Name a software, using which graphical pictures could be edited in computer.
29. Name any four factors, which govern the quality of the image displayed on VDU.
30. Which colors are used in RGB color system to generate other colors?
31. Name any three-color system, which are used for filling colors in computer graphics.
32. Which type of graphics doesn't get deformed when enlarged?
33. Which software generates files with .fla extension?
34. Which file format is used for web figures?

## Detailed Answer Type Questions

### E. Answer the following questions in detail:

35. What is pixel? How do you define pixel depth?
36. Explain the relationship between the pixel depth and number of colors displayed in pixel.
37. What are true colors? Approximately how many true colors are there?
38. Differentiate between additive color system and subtractive color system.
39. What is raster graphics? Describe its characteristics.
40. What is vector graphics? Describe its characteristics.
41. What is file compression? Why it is required? Name any three-file compression techniques.
42. What is file decompression? Why it is required?
43. What is data loss in relation to file compression? Why it is acceptable during decompression? What is lossless compression? Which file format results into lossless compression?
44. Why file compression is required? Name any 5 file compression techniques.
45. Explains the concept of Color Palette.
46. Describe RGB color system.

****

# CHAPTER 16

# Computer Applications in Pharmaceutical and Clinical Studies

You know that pharmacy deals with drugs and clinical studies is planned examination of the effectiveness of a new drug or treatment for a disease as compared to a controlled group not receiving the treatment. Computers find extensive use in both the areas.

In recent past there has been a rapid growth of use of computers in the field of biological science, pharmaceuticals, medical information, disease diagnostics, doctoral treatment and medication, all across the world. The main reason for this is that power of computer has increased to a great extent with its prices slashed down. Use of computer has become so easy that it is being treated as an assisting tool for every field. Networking and Internet have completely changed the olden style of functioning. Now the emphasis is not on data hiding but it is on data sharing. Ocean of information, on any desired topic, is now available on single click of computer mouse.

Many useful web sites offering outcomes of clinical researches, experimentations, diseases symptoms, medicine details are on the web. Any person willing to use that information can access these web sites.

There is hardly any scholar, who is not making use of computer these days. However few important uses of computer in pharmaceuticals and clinical studies are described in this chapter.

## NEED TO USE COMPUTER APPLICATIONS IN PHARMACY

You know that computer is a powerful tool. Modern computers possess tremendous computing capabilities. In computers, large and complex data can be organized and stored in easily accessible formats. Activities like data retrieval, data processing and data analysis are very efficiently done through computers. Thus computer suits well for complex data analysis that involves bulk data.

Pharmaceutical and clinical studies are the fields that involve lot of complex data. Data related to these fields requires lot of analysis and correlation with previous knowledge and outcomes of other experiments and their findings, before it is used for any practical purpose. A single mistake at any stage can lead to disaster. For example, a new drug comes into existence after lot of studies and experimentation under strict observation. Any mistake at any level can cause lot of problems and side effects.

With the increasing number of researches and studies, pharmaceutical data has soared multifold. First of all it becomes impossible to get required data by manual methods. If somehow this data becomes available, using this data and analyzing it manually for some specific purpose, is a difficult, time consuming and error prone task. Here computer comes handy. Networking and Internet have emerged as best possible medium to share the information electronically. Research work done in one country can be easily accessed by the people of other countries. Latest outcomes and developments are easily available on Internet these days so there is no need to reinvent the wheel.

Once the data is available, computer can be used as data processing and analyzing tool to perform the task efficiently and in error free manner.

Knowledge base systems and decision support systems coupled with artificial intelligence are talk of the time. Numerous pharmaceutical applications based on these technologies are available for use. Scientists are using them and are taking benefit of them for saving their time and efforts and utilizing them for better productivity.

## APPLICATION OF COMPUTER IN PHARMACEUTICAL AND CLINICAL STUDIES

In the field of pharmaceutical research and development, computers are mainly used for:

1. Data storage and retrieval
2. Scientific information handling and enhancing productivity
3. Statistical modeling in pharmaceutical research and development
4. Rigorous data analysis
5. Discovering, developing and marketing new medicines
6. Predicting human response to drugs
7. For predicting drug metabolism
8. Clinical developments
9. Understanding diseases

### Using Computer For Data Storage And Retrieval

As mentioned above, biological and chemical data need to be organized properly in categorized form so that it could be accessed in various ways, as per requirement. Computer based database management systems are the best tools available for this purpose. In fact Bioinformatics is one such field that has emerged recently that makes biological database available for studies and analysis.

Bioinformatics is the application of computer technology for the management of biological information. Computers are used to gather, store, analyze and integrate biological and genetic information, which can then be applied to gene-based drug discovery and development.

The science of Bioinformatics, which is melding of molecular biology with computer science, is essential for the use of genomic information in understanding human diseases and in the identification of new molecular targets for drug discovery. In recognition of this, many universities, government institutions and **pharmaceutical firms have formed bioinformatics groups, consisting of computational biologists and bioinformatics computer scientists.** Such groups will be key to unraveling the mass of information generated by large scale sequencing efforts underway in laboratories around the world.

## Use Of Computers In Bio And Chemoinformatics Data Management

Now a days, computers are extensively used for bio and chemoinformatics data management. These systems enable drug discovery organizations to capture, store, secure and share their drug development operations.

They integrate research data in a scalable, extensible environment that provides secure management of a drug discovery organization's principle asset - data. Using these softwares scientists can easily access, visualize and analyze information throughout an entire organization, communicating and processing information faster and more accurately.

By maximizing the value of research data, these applications improve efficiency and protect intellectual property across the spectrum of discovery activities, providing a rapid return on investment.

## Using Computer Software In Understanding Protein-Protein Docking

You know that most of the proteins have interacting partners in a cell. Characterization of these interactions is critical to an exact understanding of biological process and can be used to design new therapeutic agents, such as drugs, biomarkers, or diagnostics, to treat a wide variety of diseases. As the structures of many protein-protein complexes have not been experimentally determined, computational methods are required that can provide fast alternatives to establish the important geometric relationships.

Computer softwares like ZDOCK and RDOCK are now available for studying protein-protein docking. These softwares enable more efficient life science research by helping scientists to understand the action of therapeutically interesting proteins by predicting the manner in which two proteins interact at the atomic level.

As our awareness of the complexities of the human body increases, pharmaceutical and biotech companies are finding that they need to account for the multitude of interactions that a protein makes in the cell in selecting new protein targets or designing drugs". These softwares provide the ability to predict what interactions a protein makes with other proteins. Using these softwares scientists can now prioritize, which of these interactions to study further experimentally. This saves significant time to market and cost of development.

## Use Of Computer In Drug Discovery

There are many powerful software modeling tools, which have the potential to reduce drug development costs to a large extent. These tools are basically knowledge based applications that have revolutionized the modern drug discovery process. They possess the ability to extract meaningful information from data sets, design new biologically active structures and to predict their properties that are essential for competitive drug discovery in the 21st century.

For example, 'MolSAR is a software designed and developed by CSIRO, using which lot of work to find and develop new drugs can be done on a computer.

Such softwares can be used to extract useful information from large quantities of data, for example from pharmacological experiments, which can then be used in the drug discovery process. It can also search through existing collections of chemicals to find new drug leads and help in designing new chemical collections, which have a higher likelihood of containing good drug leads.

These packages can be used for a wide range of problems related to the human pharmaceutical discovery process, as well as other commercially important areas such as agrochemical and veterinary drug discovery.

## Use Of Computer In Diagnosing Diseases

These days almost all the serious illnesses and diseases require computer to diagnose them. Inputs for diagnosing these serious diseases are so complex and varied in nature that it is impossible for any human being to reach to any definite conclusion without rigorous pathological tests. All the pathological tests involve use of computer and computer based dedicated equipments.

Now a days many expert systems are available, which take patient's symptoms as input, they diagnose the disease and suggest medicines. Such systems being artificially intelligent become more expert as the time passes by.

## Use Of Computer For Clinical Care

Computers are also being used for clinical care. Patient's records are maintained on computer. Analyzing softwares are deployed, which generate automatic reports and suggest caring treatment.

# ADVANTAGES OF USING COMPUTERS

Following are the advantages of using computers in pharmaceutical and clinical studies;

1. Data handling becomes easy.
2. Data analysis and correlation becomes an easy task.

3. It increases efficiency.
4. Chances of making error get reduced.
5 Drug testing becomes easy.
6. Complex issues that would have had otherwise remained unhandled, with use of computer are easily handled. This opens new directions for researches and development.
7. Cost of development of new drugs reduces.

## EXERCISES CHAPTER 16

### Short Type Questions

**A. Answers for following questions:**

1. Why computers are required for pharmacy and clinical studies?
2. Describe 3 areas of pharmaceutical operations that could be computerized.
3. Describe 2 areas of clinical studies that could be computerized.
4. Describe the role of computer in inventing new drugs.
5. How computer helps in data storage and retrieval of pharmaceutical data?
6. What is protine-protine docking? For what purpose computer can be utilized in this area? Name the software available for this purpose.
7. What is MOLSAR? What it is used for? What are its advantages?
8. How computer can be used in diagnosing deceases?
9. What are the advantages of computerization? Describe any two softwares that are related to pharmaceutical and clinical studies.

****

# PRACTICAL

# CHAPTER 17

## *UNIX Lab Assignments*

You know that UNIX is an operating system. Theoretical aspects of UNIX were covered in chapter 7 of this book. UNIX security features and commands were also described in that chapter. To get more familiar with them perform following tasks in computer lab sessions:

1. Get your login name and password from the system administrator.
2. Ask the name of your home directory from the system administrator, when he creates it for you.
3. To login into the computer, give correct login name and wrong password. System will display an error message. Read that message.
4. Try to login by giving wrong login name and correct password this time. System will again display an error message. Read that message.
5. Try to login again by giving correct login name and correct password this time. You will be able to login now and you will be placed in your home directory.
6. Execute "pwd" command to display the current working directory. Match the name with the name of your home directory that system administrator had told you.
7. Now execute "who am i" command and note the response. Match the output of this command with your login name.
8. Now execute "man" command.
9. Execute "man <some-command-name>" command. Read the output of both the above-mentioned commands. You will learn to take help from UNIX itself.

10. Now execute vi.command and enter the following lines in it:

> Depression cure summary: Here are some changes you can make to become a happier person:
>
> - Be true to yourself
> - Look to the present, drop the future worry and the past regret
> - Focus on positives not negatives
> - Give yourself permission to move forward
> - Work on your self confidence and self esteem.

11. Perform following operations on this text in vi.
    - (i) Add few more lines at the top, bottom and in the middle.
    - (ii) Copy first three lines at the end.
    - (iii) Move fifth lines at seventh position.
    - (vi) Join third line with next lines.
    - (v) Break line 4 into two lines.
    - (vi) Find the word "look" in the text.
    - (vii) Substitute the word forward with ahead.
    - (viii) Get line numbers in the text.
    - (ix) Remove the line numbers.
    - (x) Save the text in a file named happy and come out of vi.
12. Display the contents of my text file on VDU.
13. Make a duplicate copy of happy file with another name, happyback.
14. Add the contents of happy and happyback file into another file named happysum, using cp command.
15. Display the contents of happysum file on VDU.
16. Make another copy of happysum file with name happysum2.
17. Execute ls command and then ls -l command and note the difference between the output of both the commands.
18. Now execute ls -al command and note the additions made in the output.
19. Make a directory with name mydir.

20. Move happysum and happysum2 files in mydir directory.
21. Copy happy file from current directory to mydir directory.
22. Now change your current position from present directory to mydir directory.
23. Execute a command to print current working directory.
24. Delete happysum file from current directory.
25. Try to remove mydir directory. You will get an error message. Analyze that message.
26. Go to root directory.
27. Execute ls -R command and observe the output. Try to answer the question, What does -R switch do in ls command?
28. Go to your home directory now.
29. Try to remove mydir directory. You will get an error message. Note that message and perform combination of steps to remove the directory.
30. Now change the file permissions for my text file, using chmod command.
31. Change the ownership of your text file to some other user.
32. Try to delete or edit your text file.
33. Execute other UNIX commands like date, time, who etc. given in chapter 7 of this book.
34. Change your password and ensure that it has changed by logging out and logging in again.
35. Move the root directory and give ls -l command and note the names of directories present in it.
36. End current session by logging out.

****

# PRACTICAL

# CHAPTER 18

## MS-DOS Lab Assignments

You know that MS-DOS is an operating system. Theoretical aspects of MS-DOS were covered in chapter 7 of this book. MS-DOS architecture, features and commands were also described in that chapter. To get more familiar with them perform following tasks in computer lab sessions:

1. Switch on the computer and boot it with MS-DOS.
2. Make a directory with the name TEST.
3. Go into TEST directory.
4. Make a file, INDIA.TXT and enter following text into it:

> 1. India is the world's largest, oldest, continuous civilization.
> 2. India never invaded any country in her last 10000 years of history.
> 3. India is the world's largest democracy.
> 4. Varanasi, also known as Benares, was called "the ancient city" when Lord Buddha visited it in 500 B.C.E, and is the oldest, continuously inhabited city in the world today.
> 5. India invented the Number System. Zero was invented by Aryabhatta.
> 6. The World's first university was established in Takshashila in 700BC. More than 10,500 students from all over the world studied more than 60 subjects. The University of Nalanda built in the 4th century BC was one of the greatest achievements of ancient India in the field of education.
> 7. Sanskrit is the mother of all the European languages. Sanskrit is the most suitable language for computer software - a report in Forbes magazine, July 1987.
> 8. Ayurveda is the earliest school of medicine known to humans. Charaka, the father of medicine consolidated Ayurveda 2500 years ago. Today Ayurveda is fast regaining its rightful place in our civilization.

5. Perform following operations on this file.
   (a) Add few lines at top, bottom and in middle.
   (b) Delete few lines from the text.
   (c) Copy some portion of the text to some other part.
   (d) Move some portion of the text to other location.
   (e) Save the text in the file and exit.
6. Execute DIR command and observe the output.
7. Make a duplicate copy of INDIA.TXT file with another name, INDIA1.TXT.
8. Type the contents of INDIA.TXT file.
9. Rename the file INDIA.TXT to BHARAT.TXT.
10. Make another directory with the name JUNK.
11. Copy BHARAT.TXT file into JUNK directory with same name.
12. Copy BHARAT.TXT file into JUNK directory with another name BHATRAT.BAK.
13. MOVE to JUNK directory.
14. Execute DIR commands to see how many files are there.
15. Now execute CD .. command first and then print the current directory name so as to know the effect of CD .. command.
16. Remove the file BHARAT.BAK.
17. Try to remove JUNK directory, you will get an error.
18. Perform all necessary steps and remove JUNK directory.
19. Take a floppy and format it.
20. Make the floppy bootable.
21. Copy few files from TEST directory to floppy.
22. Move one file from hard disk to floppy.
23. Remove the file that you had copied on floppy.
24. Display the names of the files present on floppy.
25. Delete all the files from floppy.
26. Reboot the system from floppy.
27. Give C: command to make hard disk as default disk.
28. Print the name of working directory.
29. Go to root directory.
30. Go to TEST directory.
31. Take another floppy that has few files in it. Using DISKCOP command, copy all the contents of previous floppy to this floppy.

32. Move around the system. Go to other drives.
33. Perform DEL A:*.* command for floppy and note the message that appears against this command.
34. Execute other DOS command given in chapter 7.
35. Confirm the presence of COMMAND.COM, IO.SYS and MSDOS.SYS files on bootable floppy.
36. Make an exit from the system and switch it off.

****

# PRACTICAL

## CHAPTER 19

# BASIC Language Lab Assignments

You are already familiar with BASIC language and BASIC programming. Details of BASIC language were described in chapter 14 of this book.

In practical sessions, you can enter following programs compile and execute them:

## PART-1

***Problem -1***

Let's write a program that asks for two numbers (say X and N) and then prints sum of series of following type, for that number:

***Program-1***

```
1+X+X^2+X^3+ ....+X^n.
10 S=1
20 INPUT "ENTER VALUE OF X" ;  X
30 INPUT "ENTER VALUE OF N "; N
40 FOR I=1 TO N
50 S=S+X^I
60 NEXT I
70 PRINT "SUM = " ; S
80 END
```

***Problem-2***

Let's write a program that accepts Roll Number, Physics, Chemistry and Maths marks of the student. It then calculates the percentage and prints it, assuming that subject marks have been awarded out of 100 marks. The program works in a loop till the time user wants.

***Program-2***

```
10 REM PRINTING PERCENTAGE
20 LET Q$="Y"
30 WHILE (Q$="Y")
40  INPUT "GIVE ROLL NUMBER"; ROLL$
50  INPUT "GIVE PHYSICS MARKS"; PHY
60  INPUT "GIVE CHEMISTRY MARKS"; CHEM
70  INPUT "GIVE MATHS MARKS"; MATHS
80 LET PER = (PHY+CHEM+MATHS) * 100 / 300
90 PRINT "PERCENTAGE FOR ROLL NO "; ROLL$;  "IS "; PER
100 INPUT "DO YOU WINSH TO CONTINUE ? GIVE Y or N "; Q$
110 WEND
120 END
```

***Problem-3***

Let's write a program, which prints the area of square only if its side length is positive and greater than 0.

***Program-3***

```
10 REM TO CALCULATE AREA OF A SQUARE
20 REM THIS IS DEMONSTRATION OF GOTO STATEMENT
30 INPUT "ENTER THE SIDE OF A SQUARE"; X
40 IF X =< 0 GOTO 80
50 PRINT "AREA OF SQUARE=";A
60 LET A=X^2
70 GOTO 100
80 PRINT " GIVE POSITIVE NUMBER "
90 GOTO 30
100 END
```

***Problem-4***

Let's write a program that makes use of GOSUB command for:

1. Taking length and width of a rectangle.
2. Calculating the area and perimeter of the rectangle.
3. Printing area and perimeter of the rectangle.

***Program-4***

```
10 REM MAIN SECTION
20 GOSUB 100
30 GOSUB 200
40 GOSUB 240
50 END

100 REM ENTRY SECTION
110 INPUT "Length"; L
120 INPUT "Enter width", W
130 RETURN

200 REM CALCULATION SECTION
210 AREA = L*W
220 PERIMETER=(L+W)*2
230 RETURN

240 REM DISPLAY SECTION
250 PRINT "AREA OF RECTANGLE", AREA
260 PRINT "PERIMETER OF RECTANGLE", PERMETER
270 RETURN
```

***Problem-5***

Let's write a program that accepts month number from the keyboard (i.e. a number from 1 to 12) and prints the name of the corresponding month. For example, if 3 is given, it should print March.

***Program-5***

```
10 INPUT "ENTER MONTH IN NUMBER"; N%
20 ON N% GOSUB 100,200,300,400,500,600,700,800,900,1000,1100,1200
```

```
30 END

100 PRINT "JANUARY"
110 RETURN

200 PRINT "FEBRUARY"
110 RETURN

300 PRINT "MARCH"
310 RETURN

400 PRINT "APRIL"
410 RETURN

500 PRINT "MAY"
510 RETURN

600 PRINT "JUNE"
610 RETURN

700 PRINT "JULY"
710 RETURN

800 PRINT "AUGUST"
810 RETURN

900 PRINT "SEPTEMBER"
910 RETURN

1000 PRINT "OCTOBER"
1010 RETURN

1100 PRINT "NOVEMBER"
```

```
1110 RETURN

1200 PRINT "DECEMBER"
1210 RETURN
```

***Problem-6***

Say we have to write a function, which calculates cube of the given number.

***Program-6***

```
10 DEF FNCUBE (Z)= Z*Z*Z
20 FOR I=1 TO 5
30 INPUT "GIVE NUMBER"; N
40 PRINT FNCUBE(N)
50 NEXT I
60 END
```

***Problem-7***

Let's write a program that initializes all the elements of an integer array to 0. Say the array consists of 50 elements.

***Program-7***

```
10 REM INITIALIZING ELEMENTS OF ARRAY, USING FOR LOOP
20 DIM A%(50)
30 FOR I% = 1 TO 50
40 A%(I%) = 0
50 NEXT I%
60 END
```

***Problem-8***

Let's write a program, which reads names of first six months into a string array, through DATA and READ statements and then prints them.

***Program-8***

```
10 REM READING DATA INTO ARRAY, USING DATA AND READ STATEMENTS
20 DIM N$(6)
30 DATA "JAN", "FEB", "MAR", "APR", "MAY", "JUNE"
40 FOR I% = 1 TO 6
50 READ N$(I%)
```

```
60 NEXT I%
70 REM DISPLAYING DATA
80 FOR I% = 1 TO 6
900 PRINT N$(I%)
100 NEXT I%
110 END
```

***Problem-9***

Let's write a program that accepts Roll numbers and marks of 10 students from the keyboard and prints roll numbers and names of those students, who secured more than average marks.

***Program-9***

```
10 REM ACCEPTiNG DATA INTO ARRAY, USING INPUT STATEMENT
20 DIM ROLL%(10), MARKS(10)
30 FOR I% = 1 to 10
40 INPUT "GIVE ROLL NUMBER AND MARKS " ; ROLL(I%),MARKS(I%)
50 LET SUM = SUM + MARKS(I%)
60 NEXT I%
70  LET AVRG = SUM / 10
80 PRINT "AVERAGE MARKS ARE =  " ; AVRG
90 PRINT
100 PRINT "FOLLOWING STUDENTS GOT MORE THANAVERAGE MARKS"
110 FOR I% = 1 to 10
120 IF MARKS(I%) > SUM THEN PRINT ROLL(I%), MARKS%
130 NEXT I%
140 END
```

***Problem-10***

Let's write a program, which accepts data for two dimensional array of integers and then prints row totals and column totals.

***Program-10***

```
10 SUM=0
20 DIM NUMB (4,5)
30 FOR I=1 TO 3
40 PRINT "Enter the Row " ; I "Value"
```

```
50 FOR J=1 TO 4
60 PRINT "Enter Value";
70 INPUT NUMB(I,J)
80 NEXT J
90 NEXT I
100 PRINT
110 FOR I=1 TO 3
120 FOR J=1 TO 4
130 SUM = SUM+NUMB(I,J)
140 NEXT J
150 NUMB(I,5)=SUM
160 SUM=0
170 NEXT I
180 SUM=0
190 FOR J= 1 TO 4
200 FOR I= 1 TO 3
210 SUM = SUM+NUMB(I,J)
220 NEXT I
230 NUMB (4,J)= SUM
240 SUM=0
250 NEXT J
260 FOR I=1 TO 4
270 FOR J= 1 TO 5
280 PRINT NUMB(I,J),
290 NEXT J
300 NEXT I
310 PRINT
320 END
```

***Problem-11***

Let's write a program which accepts data for two dimensional square array (an array, in which number of rows are equal to number of columns) and prints its diagonal elements.

***Program-11***

```
10 SUM=0
20 ROWPOS=0
30 X=0
```

```
40 DIM NUMB (4,4)
50 FOR I=1 TO 4
60 PRINT "Enter the Row " ; I "Value"
70 FOR J=1 TO 4
80 PRINT "Enter Value";
90 INPUT NUMB(I,J)
100 NEXT J
110 NEXT I
120 PRINT
121 PRINT "MATRIX DISPLAY"
130 FOR I=1 TO 4
140 FOR J= 1 TO 4
150 PRINT NUMB(I,J),
160 NEXT J
170 PRINT
180 NEXT I
190 PRINT
191 PRINT "DISPLAY LEFT DIGNAL OF MATRIX"
200 FOR I=1 TO 4
210 FOR J=1 TO 4
220 IF I=J THEN  PRINT NUMB(I,J)
230 NEXT J
240 NEXT I
250 PRINT
251 PRINT "DISPLAY RIGHT DIGNAL OF MATRIX"
260 FOR I=1 TO 4
270 FOR J=1 TO 4
280 ROWPOS=4-X
290 IF J=ROWPOS THEN  PRINT NUMB(I,ROWPOS)
300 NEXT J
310 X=X+1
320 NEXT I
330 PRINT
340 END
```

# PART 2

1. Write a program that asks the user to select one option among following options:

   (i) Calculate area of a rectangle

   (ii) Area of a circle

   (iii) Area of a triangle

   After accepting the choice it takes inputs according to the shape selected and prints the area. Make use of GOSUB statement in the program.

   **Write, edit, compile and execute the program for above mentioned problem.**

2. Write a program to print prime number between 1 to 100. Recall that prime numbers are the numbers that are divisible by 1 and themselves only.

   **Write, edit, compile and execute the program for above mentioned problem.**

3. Write a program that accepts a number and tells if the given number is an even number or not.

   **Write, edit, compile and execute the program for above mentioned problem**

4. Write a program that accept 20 numbers from the keyboard and prints their average.

   **Write, edit, compile and execute the program for above mentioned problem.**

5. Write a program that accepts a number, say n from the keyboard and prints the series of following type.

   **n+(n+1)+n+2)+.......... +(n+n)**

   **Write, edit, compile and execute the program for above mentioned problem.**

6. Write a program that accepts a number say n from the keyboard and prints factorial of that number. Recall that factorial 5 is:

   **1 x 2 x 3 x 4 x 5**

   **Write, edit, compile and execute the program for above mentioned problem.**

7. Write a program in BASIC to accept 10 numbers from the keyboard and print the largest number among them.

   **Write, edit, compile and execute the program for above mentioned problem.**

8. Write a program to accept value of n from the keyboard and generate the series of following type:

   **1+4+9+16+25+ -------(n×n)**

   **Write, edit, compile and execute the program for above mentioned problem.**

9. Write a program to accept roll number and makes of 10 students and print the roll number of those students who secured more than average marks.

   **Write, edit, compile and execute the program for above mentioned problem.**

10. Write a program to print the sum of two arrays.

    **Write, edit, compile and execute the program for above mentioned problem.**

11. Write a program that takes the temperature in centigrade and converts it the Fahrenheit.
    **Write, edit, compile and execute the program for above mentioned problem.**

***

# PRACTICAL

# CHAPTER 20

# *MS-WORD & Lab Assignments*

MS-Word is word processing software that has been designed and developed by Microsoft Corporation of USA. It is mainly used for entering the text in computer and formatting it in presentable form. Its working details are described below.

## OPENING A NEW DOCUMENT AND TYPING THE TEXT

For getting started with MS-Word, perform following steps:

1. Click the mouse on "Start" button. When you do so, a menu, as illustrated in figure 20.1 (a) will appear on the screen.
2. Select "All Programs" option from this menu. When you do so, a submenu, as illustrated in figure 20.1 (b) will appear on the screen.
3. Select "Microsoft Word" option from this submenu.

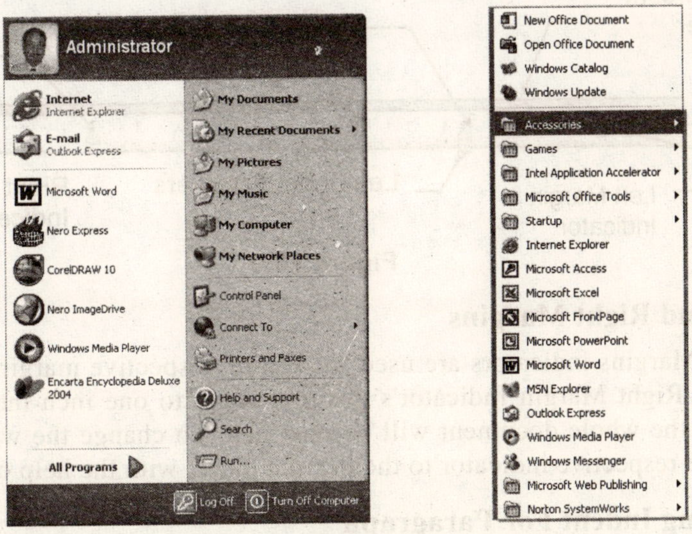


(a) (b)

**Figure 20.1**

When you do so, a window, as illustrated in figure 20.2 will appear on the screen. This window is called Word Window.

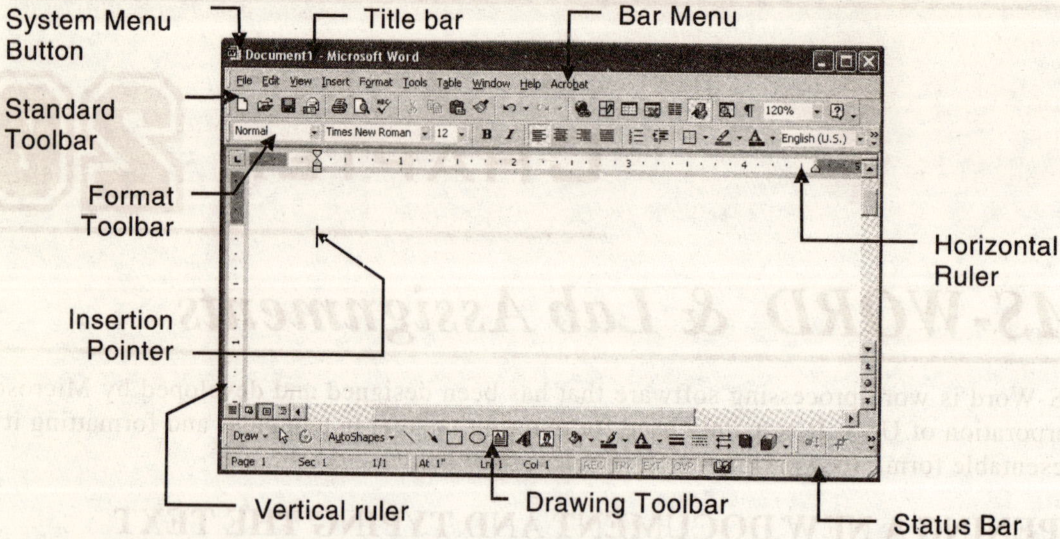


**Figure 20.2**

# PAGE LAYOUT SETTING

The process of setting the margins and paragraphs etc. is called page setting. Ruler is the best medium for setting the page layout. An exclusive view to ruler is shown in figure 20.3

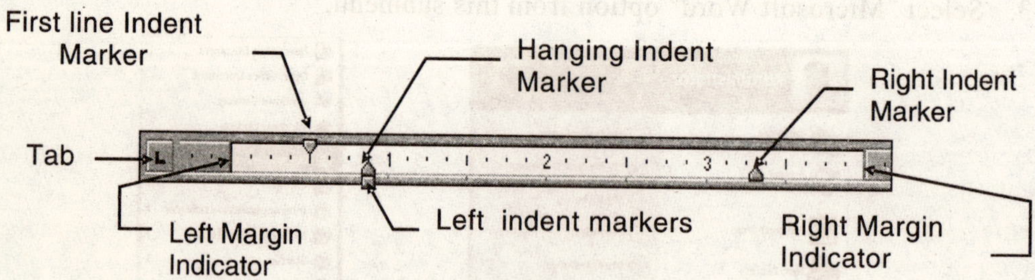


**Figure 20.3**

## Setting Left and Right Margins

Left and Right Margins indicators are used for setting respective margins of the document. For example, if Right Margin Indicator's position is set to one inch then the width of the left margin, for the whole document will be one inch. To change the width of the margin, you can drag the respective indicator to the desired place, with the help of mouse.

## Setting Hanging Indent For Paragraph

Usually all the paragraphs of the document start from left margin but Hanging Indent Marker provides the facility to start the current paragraph from a different position. It could be dragged to a new position to define the starting place for the current paragraph. For example, if the right margin has been set to 1 inch and Hanging Indent Marker has been positioned at 1.5 inch then all the paragraphs other than the current paragraph will start from 1 inch but the current paragraph will start from 1.5 inch.

### Setting First Line Indent

First line Indent Marker provides the facility to start the first line of the paragraph from a place other than that from where rest of the lines of the paragraph start. For example, all the lines of the paragraph may start from the left margin (say 1 inch) but first line for the paragraph may start from 1.5 inch from the left margin. To set the position of the first line of the paragraph, First Line Indent Marker can be dragged to the desired position.

### Setting Left Indent

Left Indent Marker facilitates simultaneous movement of First line Indent Marker and Hanging Indent Marker. When you drag it, both the markers move together in the same direction and by the same distance.

### Setting Tab

Tab button, present in horizontal ruler, facilitates setting of tab positions for the Tab key. To set the tab positions, first click the mouse on Tab button and then click it at-the desired position in the ruler. This will put a mark in the ruler and set the tab position. When you press the tab key, insertion pointer will jump to the set-tab-position. For example, if you set the tab positions at 3, 5 and 7 inches then pressing the Tab key once will take the mouse pointer to a place 3 inches away from left margin. When you press it again it will move 5 inches away and if you press it again, it will move it by 7 inches. To remove a tab position, drag the tab mark and drop it off the ruler.

## TYPING THE TEXT

When you press any key of the keyboard, character gets typed at current cursor position. Thus by pressing different keys of the keyboard, you can type the complete letter, document, or report.

While entering the text, when insertion pointer reaches the right margin, it automatically gets placed in first column of the next line. When current VDU screen becomes full of text, document automatically shifts upwards (this process is called scrolling) and blank line appears at the bottom of the screen, where you could further enter the text.

When current page becomes full, insertion pointer gets automatically placed at the beginning of the next page. Thus, continuing in this way, you can enter the text, page after page.

While entering the text, you can always make use of Del and Backspace keys to erase characters, words or sentences.

To insert the text in between two characters, words, sentences or paragraphs, click the mouse at the place where the text is to be inserted. When you do so, insertion pointer will get placed there. Now whatever you type, will get typed in between and the text on the right hand side will automatically shift.

To overwrite on already entered text, first double click the mouse on OVR button, present in status bar and then type the text. To come back to Insert mode, click the mouse on the same button again.

To start a new paragraph, press Enter key.

## SAVING THE TEXT IN A FILE

To save the text in a file, perform following steps:

1. Select "File" option from Bar menu. When you do so, File submenu will appear on the screen.
2. Now select "Save" option from the File submenu. When you do so, a dialog box, as illustrated in figure 20.4, will appear on the screen.
3. Select the drive and the folder, in which the file is to be saved.
4. Enter the name of the file in "File Name" text box.
5. Click the mouse on "Save" button.

When you do so, current text will get saved on the selected disk, in specified folder and file.

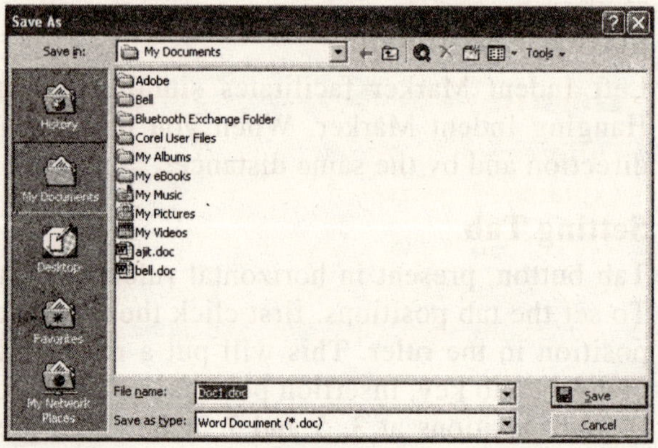


**Figure 20.4**

## CLOSING THE FILE

To close currently opened file, follow the following steps:

1. Select "File" option from the Bar menu. When you do so, File submenu will appear on the screen.
2. Select "Close" option from File submenu.

If no additions or modifications have been made in the document, ever since it was last saved, file will get closed. On the other hand, if they were made, an Alert box, as illustrated in figure 20.5, will appear on the screen.

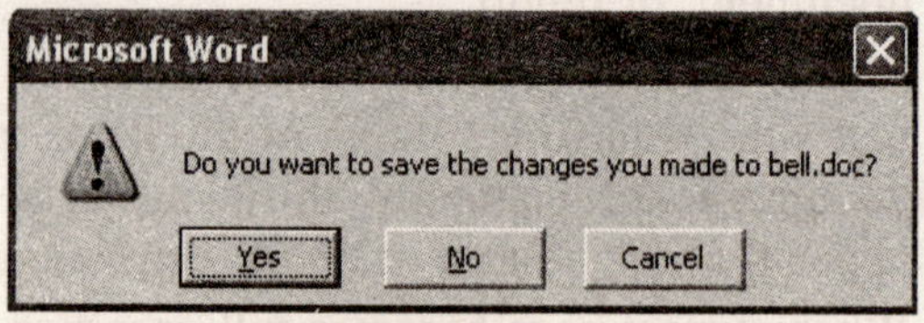


**Figure 20.5**

3. Click the mouse on "YES" button, if you wish to save the additions and modification on the disk. Select "No" button for not saving them. If you wish to cancel the idea of closing the file, click the mouse on "Cancel" button.

Whatever text you save in the file, remains permanently available for use.

# EDITING OPERATIONS

After opening the file, you may wish to make few modifications in it. For example, you may wish to delete few lines or you may wish to copy a part of the text from one place to another or you may wish to move few lines of paragraphs from here to there. For performing all such types of editing operations, you will have to first select the text, on which you wish to perform the operation. MS-Word provides different mechanisms for selecting different portions of the text. They are explained below.

## Selecting The Text

To select any portion of the text, drag the mouse over it. To cancel the selection either click the mouse anywhere in the document or press any arrow key of the keyboard.

## Deleting The Text

Different parts of the text can be deleted differently. Various deletion operations are explained below:

1. To delete a word after the insertion pointer, press CTRL and DEL keys together.
2. To delete a word before the insertion pointer, press CTRL and Backspace keys together.
3. To delete a desired portion of text (either text or graphics), select it, using any of the methods mentioned above and then press DEL key.

## Copying The Text (Copy And Paste)

Often while entering the text, you may wish to copy certain portion of the text (that has already been typed) to another place, so that you don't have to type the matter again. For this, you need to perform following steps:

1. Select the text you wish to copy.
2. Select "Edit" option from the Bar menu. On doing so, Edit submenu, as illustrated in figure 20.6, will appear on the screen.
3. Select "Copy" option from this submenu.
4. Now take the insertion pointer to the place, where you wish to copy the selected text.
5. Select "Edit" option again from the Bar menu and get Edit submenu, as illustrated in figure 20.6, on the screen.
6. Now select "Paste" option from this submenu.

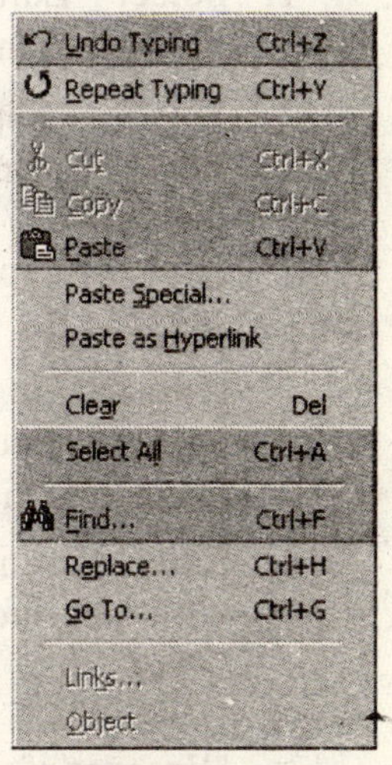


**Figure 20.6**

When you do so, selected text will get copied at current insertion pointer's position.

## Moving The Text (Cut And Paste)

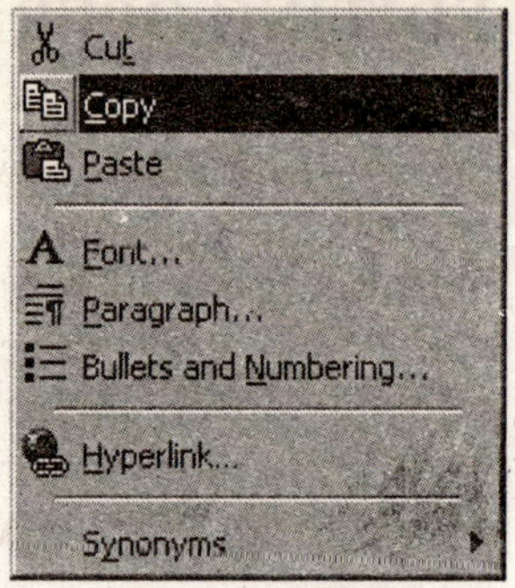


Figure 20.7

Often while entering the text, you may wish to change the location of certain portion of the text. For this, you need to perform Paste operation, as mentioned below.

1. Select the text that you wish to move to other location of the document.
2. Now right click the mouse on selected portion. When you do so, a popup menu, as illustrated in figure 20.7 will appear on the screen.
3. Select "Cut" option from this menu.
4. Now take the insertion pointer to that part of the document, where the text is to be moved.
5. Right click the mouse again and get the popup menu, shown in figure 20.7, on the screen.
6. Select "Paste" option from this menu.

When you do this, selected text will shift its place from its original position to current insertion pointer's position.

## Replacing The Word With Another Word

To find the occurrence of a word and replacing it with another word, MS-Word provides replace facility. Replacement of a word, using this facility can be done as mentioned below:

1. Select "Edit" option from the Bar menu. When you do so, Edit submenu, as illustrated in figure 20.6, will appear on the screen.
2. Select "Replace..." option from Edit submenu. When you do so, a dialog box, as illustrated in figure 20.8, will appear on the screen.

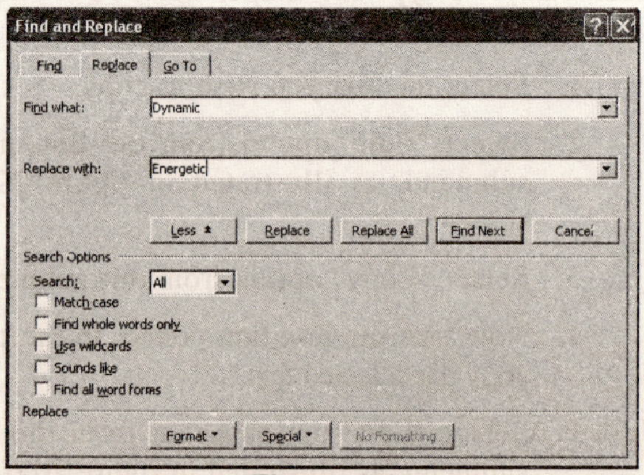


Figure 2.8

3. Enter the word that is to be replaced, in "Find What" text box. For example, you enter the word, "Dynamic" in it.
4. Enter the new word in "Replace with" text box. For example, if you wish to replace the word "Dynamic" with the word "Energetic", enter "Energetic" in "Replace with" text box.
5. If you wish to specify more details related to replacement, click the mouse on "More" button. When you do so, other options will appear on the screen.

6. For replacing the single occurrence of the word, click the mouse on "Replace" button and for changing all the occurrences in one-go, click it on "Replace All" button.

When the Replacement operation is over, "Cancel" button automatically changes to "Close" button. By choosing "Close" button, you can close the search operation and return back to the document.

## FORMATTING THE TEXT

Formatting the document means arranging the contents of the document (such as text, graphics etc.) in such a way that the document looks attractive. For example, changing the font, font size, line spacing, paragraph spacing etc. are the activities, which fall into the category of formatting. How such types of operations are done in MS-Word, is explained below.

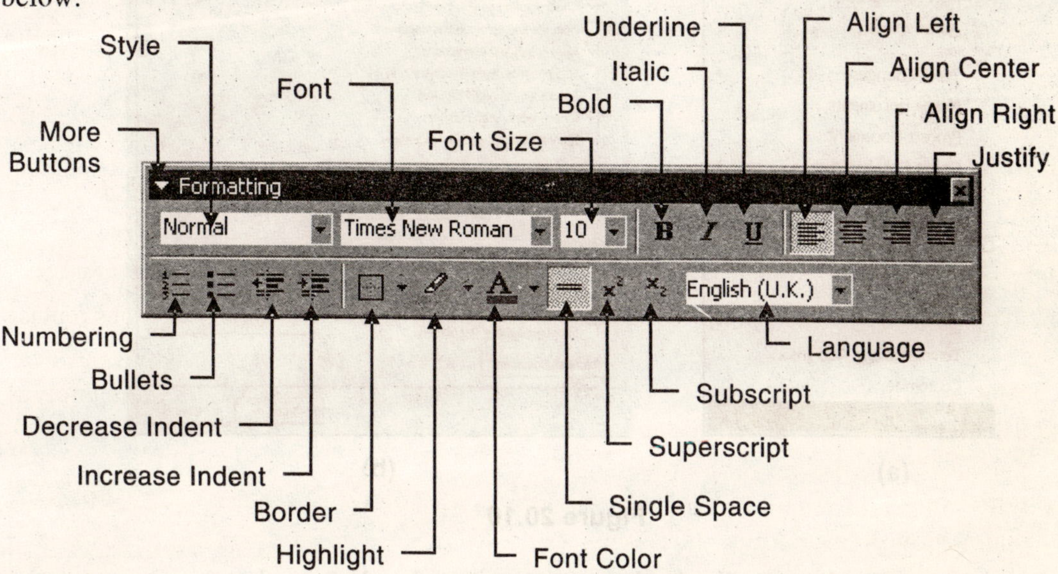


**Figure 20.9**

To perform any formatting function, select the text, which is to be formatted and click the mouse on the corresponding button. For example, if you wish to make the first two lines of the text bold, select them and then click the mouse on "Bold" button present in Format toolbar. Similarly you can perform other formatting operations.

## ACTIVATING ON-LINE SPELL CHECKER

On-line spell checker is a facility, using which the spelling of a word is checked, as soon as it is typed. If the spelling is found incorrect the word is underlined with red zigzag line. If you wish, it could also suggest correct spellings for this word. Note that On-line spell checker marks incorrect spellings, only when it is made on. Perform following steps to switch it on:

1. Select "Tools" option from the Bar menu. When you do so, Tools submenu, as illustrated in figure 20.10 (a), gets displayed on the screen.
2. Select "Options" option from this submenu. When you do so, a dialog box, as illustrated in figure 20.10(b), will appears on the screen. If somehow the options shown in the dialog box are different than those shown in the figure, click the mouse on "Spelling and Grammar" tab.
3. Check "Check Spellings as you type" check box.
4. Now select "OK" button.

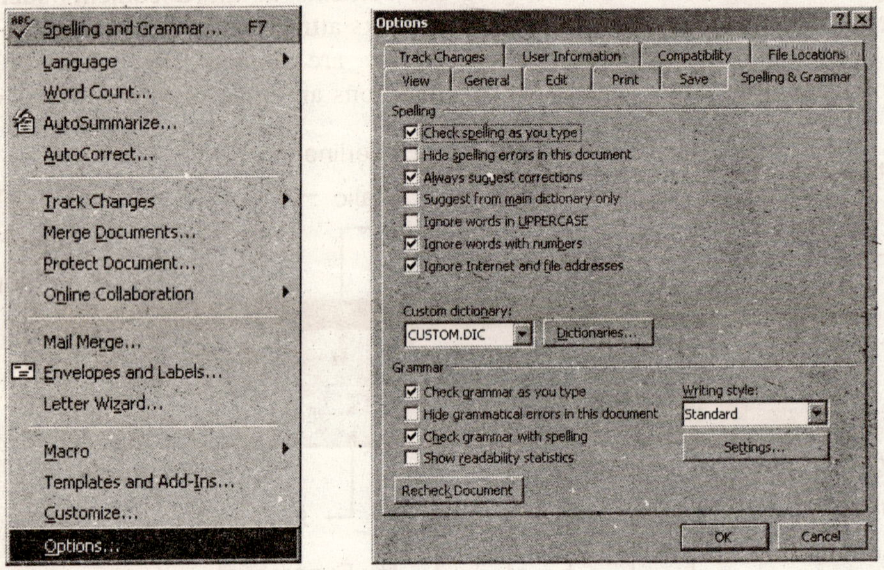


(a) (b)

**Figure 20.10**

When you do so, On-line spell checker gets activated. Now, whenever you type a misspelled word, red colored zigzag line automatically underlines it.

## CORRECTING WRONG SPELLINGS

To correct the spelling, right click the mouse on wrong spelling. When you do so, a popup menu, containing the list of suggestions, appears on the screen. Correct spelling can be chosen from this list by clicking the mouse on appropriate suggestion.

## GRAMMAR CHECKER

MS-Word provides Grammar Checker facility, which can either be activated or deactivated as per requirements. When activated, green color zigzag line underlines all grammatically wrong sentences that have already been entered. After this, when next sentence is typed, grammar checker checks it immediately, on-line.

Grammar checker not only provides the facility to check grammatically wrong sentences but it also displays alternatives to correct them. How grammar checker is activated, how it is made use of is explained below.

Perform following steps to activate the grammar checker:

1. Select "Tools" option from the Bar menu. When you do so, Tools submenu, as illustrated in figure 20.10 (a) appears on the screen.
2. Select "Option" option from this submenu. When you do so, a dialog box, as illustrated in figure 20.10(b) will appear on the screen. If somehow, different options appear in the dialog box, click the mouse on "Spelling & Grammar" tab.
3. Now check "Check grammar as you type" check box.
4. Make sure that "Hide grammatical errors as you type" check box is unchecked.
5. At last click the mouse on "OK" button.

When you do so, Grammar checker will get activated and all grammatically wrong sentences will get, underlined by green color zigzag line.

## CORRECTING INCORRECT SENTENCES

As mentioned earlier, Grammar Checker utility of MS-Word, suggests, few suggestions for correcting grammatically wrong sentences. Perform following steps to make use of this facility:

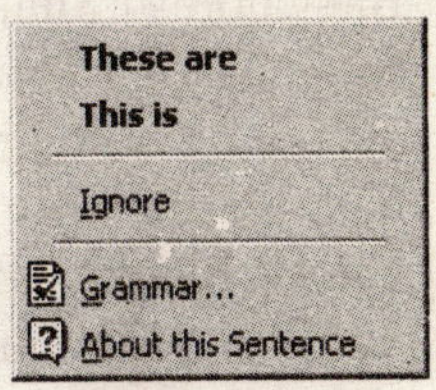


**Figure 20.11**

1. Right click the mouse on grammatically wrong sentence. When you do so, a list of suggestions, as shown in figure 20.11, in the form of popup menu appears on the screen.
2. If the wish to accept a suggestion from the list, click the mouse on that suggestion. When you do so, suggested sentence will replace the incorrect sentence.
3. If you don't want to accept any suggestion from the list and wish to continue with original sentence, click the mouse on Ignore button.

In both the cases (mentioned above), you will return back to the document and green zigzag like will get removed.

Using this procedure, all grammatically wrong sentences can be corrected.

## Practical Assignments

Perform, following activities in lab sessions:

1. Set left and right margins for the entire document.
2. Type the following piece of text in 11 points, Times New Roman, without caring about its formatting:

> The Crossover
>
> A new business opportunity is coming up at the junction of hardware and software embedded technology
>
> An Indian boy selling newspapers at Stockholm railway station was unusual enough for Sunil to investigate. The boy turned out to be a student at the Royal Institute of Technology, Sweden. He talked excitedly about an idea for a pace marker that lasted much longer than the ones available in the market. "Why don't you come and develop this product in my laboratory?" asked Sunil.
>
> The lab is a center of excellence for embedded systems at Tata Consultancy services (TCS) in Bangalore. Sunil had set it up only a year ago, after quitting Sasken Communication Systems. "I had approached TCS with this idea of an embedded systems lab and TCS had agreed," say Sunil. Despite being the largest software services company in India, TCS had not worked on embedded system other than in telecommunications. It was slow off the blocks, but the company was trying to make up for lost time.
>
> Sunil's lab's mandate was not to provide services. It would to research, develop technologies, and build product; not just software but complete devices. Whenever a product came out, it would be sold under the TCS brand name. The first device could be patient health monitoring system. The next one could be a box that would increase spectrum utilisation in CDMA (code division multiple access) networks. The third device could be the boy's pacemaker.
>
> The embedded system lab is also building competencies in three areas: wireless LAN (local area network), multimedia and medical electronics. TCS is about to take over a wireless LAN firm in Demark. This firm has a key technology that will strengthen the TCS portfolio. TCS is also sponsoring research at the University of Demark. Indian software companies have no history of sponsoring research at universities. Neither do they design and sell devices. So is this a short fit of absent-mindedness or a long-term strategic shift?

3. Now perform the following steps to edit and format the text:
   - (i) Find the word utilization in the document and replace it with utilization.
   - (ii) Put the heading, "The Crossover" in 24 points bold, Arial Black.
   - (iii) Make the text, starting with the words "**A new business opportunity......**" till "**embedded technology**" as subheading. Put it in 14 points bold, italic and underlined.

(iv) Find the first appearance of the word "Sunil" and change its color to red and make it italic bold.

(v) Import the clip art from clip art gallery.

(vi) Align the paragraphs.

(vii) Format the document as shown below.

(viii) Save the document in a file.

# The Crossover

***A new business opportunity is coming up at the junction of hardware and software embedded technology***

An Indian boy selling newspapers at Stockholm railway station was unusual enough for Sunil to investigate. The boy turned out to be a student at the Royal Institute of Technology, Sweden. He talked excitedly about an idea for a pace marker that lasted much longer than the ones available in the market. "Why don't you come and develop this product in my laboratory?" asked Sunil.

The lab is a center of excellence for embedded systems at Tata Consultancy services (TCS) in Bangalore. ***Sunil*** had set it up only a year ago, after quitting Sasken Communication Systems. "I had approached TCS with this idea of an embedded systems lab and TCS had agreed," say Sunil. Despite being the largest software services company in India, TCS had not worked on embedded system other than in telecommunications. It was slow off the blocks, but the company was trying to make up for lost time.

Sunil's lab's mandate was not to provide services. It would to research, develop technologies, and build product; not just software but complete devices. Whenever a product came out, it would be sold under the TCS brand name. The first device could be patient health monitoring system. The next one could be a box that would increase spectrum utilisation in CDMA (code division multiple access) networks. The third device could be the boy's pacemaker.

The embedded system lab is also building competencies in three areas: wireless LAN (local area network), multimedia and medical electronics. TCS is about to take over a wireless LAN firm in Demark. This firm has a key technology that will strengthen the TCS portfolio. TCS is also sponsoring research at the University of Demark. Indian software companies have no history of sponsoring research at universities. Neither do they design and sell devices. So is this a short fit of absent-mindedness or a long-term strategic shift?

# PRACTICAL

## CHAPTER 21

# *MS-Excel & Lab Assignments*

Data collection, its organization and presentation in easy-to understand forms like tables, graphs, charts etc. for the purpose of analysis are day-to-day activities of any working setup. For example, a sales organization, at the end of financial year may collect sales figures from all its regional offices to analyze company's sales performance and decide the targets for the next year. Similarly a manufacturing organization may collect all input costs involved in manufacturing an item to estimate fund requirements and to calculate the profitability.

When data volumes are small, such activities are either performed manually or using equipments like calculator etc. But when data volume becomes large and presentation requirements are complex, such tools become a limitation. Thus other sophisticated tools are required to handle the situation and to maintain the accuracy. MS-Excel is one such software-tool, which facilitates maintaining large data volumes on computer, organizing them in the form of tables, charts, reports etc. and performing "What if" analysis. Working details of MS-Excel are described below.

## MS-EXCEL - A WORK BOOK

MS-Excel is workbook software, which has been designed and developed by Microsoft Corporation of USA.

Workbook software comprises of multiple worksheets with data editing facilities and inbuilt calculation abilities. Data editing facilities of workbook ensure proper organization of data and calculation abilities facilitate its automatic updation.

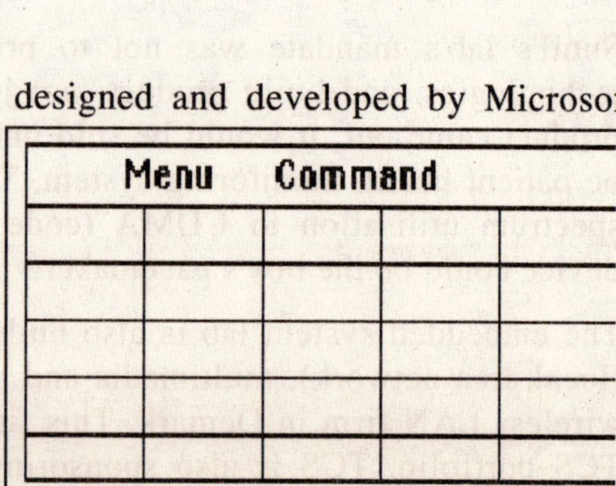


**Figure 21.1**

A simple worksheet is illustrated in figure 21.1. It looks like a page of child's mathematics copy that has squares drawn in it. Squares are used for writing labels, data or formulae in them. When data pertaining to similar entities like organization, item, month, year etc. is put in same column or row, it automatically gets organized in the form of table. Such a table is illustrated in figure 21.2.

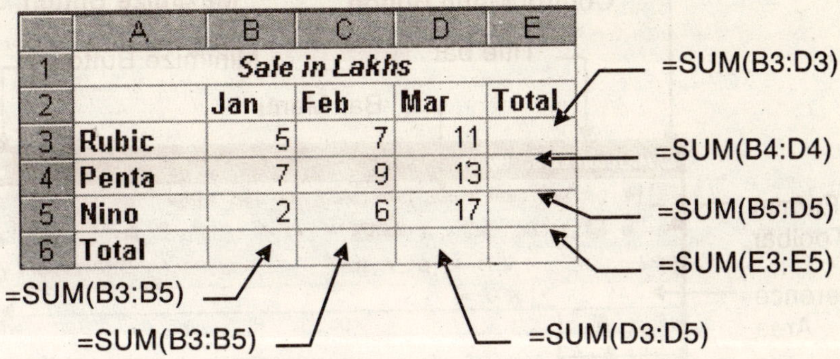


**Figure 21.2**

When formulae are defined in worksheet, automatic calculations are done and figures are generated.

For example, if in a worksheet, formulae are defined as shown in figure 21.2 then figures will be generated automatically, as illustrated in figure 21.3. If somehow participating figures change, all corresponding changes are automatically done. For example, if sales figure of Nino company, for the month of Jan, changes from 2 to 5 then total figure for the month of Jan will change from 14 to 17. Total sales figure for Nino company will change from 25 to 28 and Total sales figure will change from 77 to 80.

| | A | B | C | D | E |
|---|---|---|---|---|---|
| 1 | | *Sale in Lakhs* | | | |
| 2 | | Jan | Feb | Mar | Total |
| 3 | Rubic | 5 | 7 | 11 | **23** |
| 4 | Penta | 7 | 9 | 13 | **29** |
| 5 | Nino | 2 | 6 | 17 | **25** |
| 6 | Total | **14** | **22** | **41** | **77** |

**Figure 21.3**

Note that figure 21.2 or figure 21.3 illustrates a single worksheet. A workbook comprises of many such worksheets. An Excel workbook comprises of 255 such worksheets.

## OPENING A WORKSHEET AND ENTERING DATA

Perform following steps to invoke MS-Excel on your computer:

1. Click the mouse on "Start" button. When you do so, Start menu, as illustrated in figure 20.1 (a) will appear on the screen.
2. Select "All Programs" option from this menu. When you do so, a submenu, as illustrated in figure 20.1 (b) will appear on the screen.
3. Select "Microsoft Excel" option from this submenu.

When you do so, MS-Excel window, as illustrated in figure 21.4, will appear on the screen.

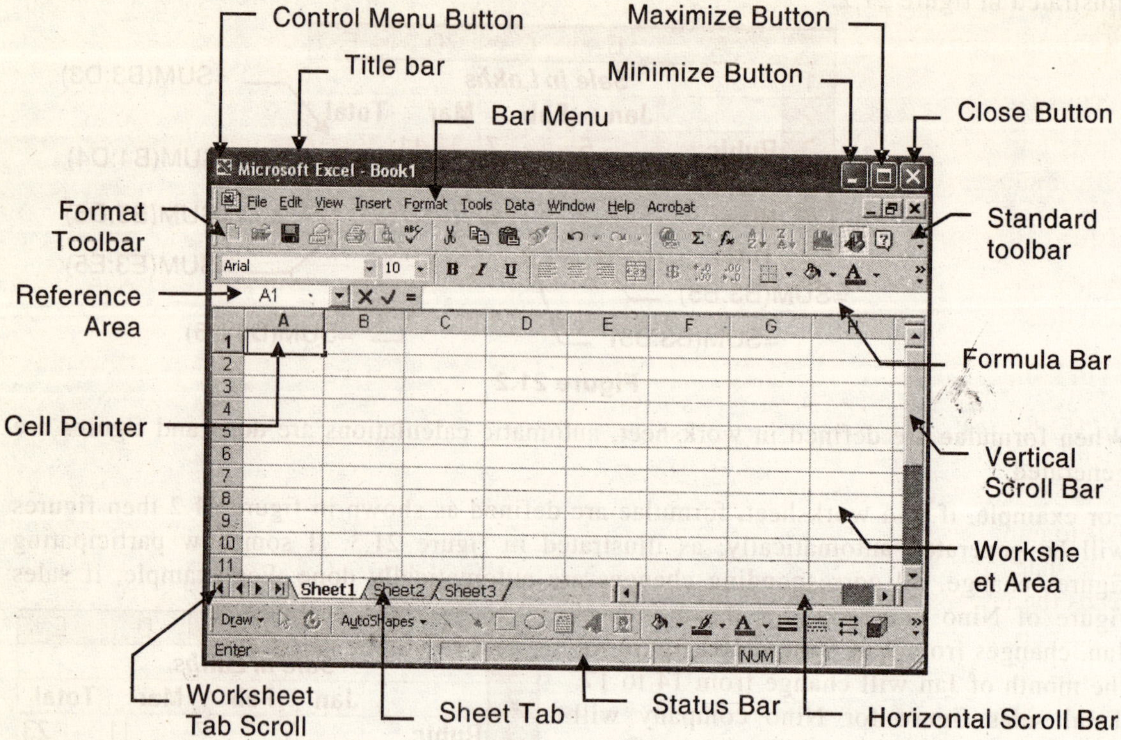


**Figure 21.4**

Few important components of MS-Excel window are described below:

### Formula Bar

Formula bar is a place where you enter the contents of the cell. Whenever you type something in it and press Enter key (or click the mouse on tick symbol), it gets written in the current cell.

### Reference Area

This is the place where the address of current cell gets displayed.

### Drop Down List

Drop down list is basically a list of all those items of the worksheet, which have been assigned a name. When mouse is clicked on the arrow button, names get displayed.

### Formula Entry Area

This is basically the text entry area. Whatever you type here is later transferred to current cell. All modifications in the cell entries are done in this area.

### Tick Button

It works like "Enter" key of the keyboard. When mouse is clicked on this button, contents of formula entry area get transferred to the current cell.

### Cross Button

Cross button works like "Cancel" button. When mouse is clicked on this button, contents of formula entry area get erased and that area becomes blank.

### Equal to Sign Button

When mouse is clicked on this button, function wizard appears on the screen and starts functioning.

### Worksheet Area

Worksheet area comprises of rows and columns. Partial area of the worksheet remains visible on the screen. Other portion of it can be brought on the screen by scrolling the worksheet, using vertical and horizontal scroll bars.

All together there are 65536 rows and 256 columns in a worksheet. Refer figure 21.4 and note that rows are labeled as 1,2,3,4,5 etc. and columns are labeled as A, B, C, D, E etc.. Thus the last row will be labeled as 65536 and the last column will be labeled as IV

Intersection of a row and column forms a rectangular space, called Cell. Each cell is identified by its address and the address of the cell is defined as column number and the name of the row that forms the cell. For example, the cell formed by fifth column (i.e. E) and third row (i.e. 3) will be defined as E3. Similarly the cell formed by fourth column (i.e. D) and second row (i.e.2) will be defined as D2.

Worksheet area is used for data entry. As per standard rules of MS-Excel, contents of a cell are considered as single data item. In other words you can say that a cell of the worksheet can hold single data item only.

### Worksheet Tab

As mentioned earlier, a workbook comprises of 256 worksheets. Each worksheet has a corresponding tab in MS-Excel window. Refer figure 21.4 and note that these tabs are placed just below the worksheet area. Thus tabs are named as Sheet1, Sheet2, Sheet3 ... etc.

When you invoke MS-Excel, first worksheet appears in the window and Sheet1 tab remains highlighted.

### Worksheet Tab Scroll Buttons

Refer figure 21.4 and note that not all the 255 worksheet tabs (as there are 256 worksheets in a workbook) remain visible on the screen. Due to space limitation, only few of them appear on the screen. To scroll the list of these tabs, worksheet tab scroll buttons remain present in Excel window. By making use of these buttons, you can get other worksheets tabs on the screen.

## ENTERING DATA IN WORKSHEET

Data entry in worksheet is quite different from the text entry that is done in MS-Word. In MS-word, you continuously keep on typing the text but in MS-Excel worksheet you need to type each data item in a separate cell. For example, if you wish to enter three data items, say Ravi, 200, 555 then you will have to type Ravi in one cell then move to other cell and type 200 there. Similarly for typing 555, you will have to move to some other desired cell and then type the numbers.

## CHANGING THE CONTENTS OF CELL

Contents of a cell can either be edited in Formula Bar or within the cell itself. Both the procedures are given below:

1. To modify the contents in Formula Bar, click the mouse in the cell, whose contents have to be modified. When you do so, cell contents will appear in the Formula Bar. Now using DEL or Backspace keys etc. you can modify the contents. At last, press Enter key.
2. To modify the contents, within the cell, double click the mouse in that cell, whose contents have to be modified. When you do so, insertion pointer will appear in cell. Now using different keys of the keyboard, modify the contents, within the cell. At last, press Enter key.

## DATA TYPES

The data that can be entered in MS-Excel worksheet is broadly classified into following two categories:

1. Constants
2. Formula

### Constants

The values, which do not change on their own are called constants, For example, if you enter "Ravi" in cell A1, it will not change unless you deliberately modify it (to say, Ravina). Similarly other values like 200, 255 etc. will also be treated as constants because they will not change automatically.

#### Type of Constants

Constants can further be classified into following categories:

1. Numbers (e.g. +176, -200, $735, 10,375 etc.)
2. Text (e.g. Ravi, Delhi, India, etc.)
3. Date (e.g. 22/06/1975, 16/May. 1975, etc.)
4. Time (e.g. 7 AM, 8 PM etc.)

## Formula

Formula is basically a user defined function, which could be formed as per our own requirements by making use of constants, mathematical operators, standard MS-Excel functions (there are numerous standard, predefined functions available in MS-Excel, for use), and cell addresses. For example, (A1*C1)+200 is a valid formula. Similarly SUM(A1:A10) is another valid formula.

To differentiate formulae with text, all formulae should start with = sign in MS-Excel. For example, above mentioned formulae should be entered as =(A1*C1)+200, =SUM(A1:A10)

### Type of Formulae

MS-Excel formulae can be broadly classified into following categories:

1. Numeric Formulae (e.g. =A1+B1+C1, =((A1+B1)* 20) etc.)
2. Logical Formulae (e.g. =A1>B1, B1>=C1 etc.)
3. Text Formulae (e.g. =A1 & B1 Note & operator is text joining operator)

## ENTERING AND EDITING FORMULA

You are now familiar with MS-Excel worksheet and the types of data that can be entered in it. To get familiar with the process of preparing a worksheet, let's take a simple example. Say, we have to prepare a worksheet, in which rate and quantity are entered manually and the amount is calculated automatically, using the formula (rate * quantity)

Perform following steps for preparing this work sheet:

1. Type "Rate" in cell A1. The word, Rate is a text constant and will be treated as label or heading. Its purpose is to communicate that the value written below is rate.
2. Type the word "Quantity" in cell B1. This label will also serve the purpose of heading.
3. Type the word "Amount" in cell C1. This text will also serve the purpose of heading.
4. Since quantity and rate values have to be entered in cell A2, B2 respectively and the amount is to be calculated by multiplying the values in these cells and the result is to be displayed (automatically) in cell C2 hence enter the formula, =(A2*B2) in cell C2.
5. Now, when you enter the values in cell A2 and B2, result will get automatically displayed in cell C2. For example, if you enter 50 in cell A2 and 3 in cell B2 then the result, 150 (i.e. 50*30) will automatically appear in cell C2.

Now, if you change the values either in A2 or B2 or in both the cell, value in cell C2 will automatically update itself, according to the formula. For example, if you change the value in cell B2, from 3 to 5, the value in cell C2 will change from 150 to 250. This worksheet is illustrated in 21.5.

| | A | B | C |
|---|---|---|---|
| 1 | Rate | Quantity | Amount |
| 2 | 50 | 3 | 150 |
| 3 | | | |

**Figure 21.5**

## REFERENCING A GROUP OF CELLS

Refer above mentioned example and recall that in cell C2, we made use of the formula, =(A2*B2), Note that this formula makes use of the cell addresses, which refers single cell. For example, A2 refers a single cell. Similarly B2 refers other cell. But there would be times when group of cells will participate in the formula. For example, column A may contain 12 different sales values i.e. A1 may contain sales value for the month of Jan, A2 may contain sales value for the month of Feb and so on. Thus to add all the 12 values, you will have to write a formula =(A1+A2+A3 -----+A12). This is lengthy and uncomfortable method of writing the formula. To make the task simple, MS-Excel provides many reference operators. Using these operators, you can refer group of cells (more then one cell) and make the formula short. An explanation of these operators is given below.

### Range Operator

As the name indicates, Range operator, refers a group of cells, which fall within the given cell address.

Range operator is denoted by colon sign (:). It requires two operands and is used as Op1:Op2. Here Op1 is the first operand, which refers the address of the cell from where the range starts. Similarly Opt2 is the second operand, which refers the address of the cell, where the range ends. For example, (A1:A12) will signify all the cells starting from cell A1 to cell A12 (i.e.12 cells). Similarly (A5:G5) will signify all the cell from A5 to cell G5 (i.e. 7 cells).

### Union Operator

As the name indicates, Union operator refers to all those cells, which are referred in the expression.

Union operator is denoted by a comma (,). It requires two operands and is used as op1, op2. Here op1 refers a cell address and op2 refers to another cell address. Thus (A7, B10) will refer to two cells i.e. A7 and B10. Similarly (A7, A9, A11) will refer three cells A7, A9 and A11. In the same way (A5:A10), (B15:B20) will refer 12 cells i.e. all the cells starting from A5 to A10 (6 cells) and the cells starting from B15 to B20 (another 6 cells).

### Intersection Operator

Intersection operator refers to all those cells, which are common between the two given ranges.

Intersection operator is denoted by a blank space. It requires two ranges as operands and is used as (R1 R2). Here R1 and R2 are two ranges. For example, (A1:A10 A7:A15) will result into 4 cells i.e. A7, A8, A9 and A10.

## CELL REFERENCING

Consider the use of cell addresses in a formula. For example, in the worksheet shown in figure 21.5, the formula = (A2*B2) was written. Here by writing A2 or B2, You are referring to the values present in these cells. When you refer the contents of the cell, it is called cell referencing. Note that just writing the cell addresses and referring their values is not the only way of cell referencing. MS-Excel provides following three methods of cell referencing:

1. Relative referencing
2. Absolute referencing
3. Mixed referencing

| | A | B | C | D | E | F | G | H |
|---|---|---|---|---|---|---|---|---|
| 1 | Name | Jan | Feb | Mar | | | Total | Tax |
| 2 | LG | 200 | 300 | 500 | | | | |
| 3 | BP | 300 | 700 | 200 | | | | |
| 4 | RK | 400 | 300 | 100 | | | | |
| 5 | | | | | | | | |
| 6 | Tax Rate | 0.1 | | | | | | |

**Figure 21.6**

## Relative Referencing

As the name indicates, in relative referencing, cells are referred in relation to the current cell position. To understand the concept of relative referencing, consider the worksheet shown in figure 21.6. If you put the formula =(B2+C2+D2) in G2, it will show 1000 in G2. but it would, in the background, convey following aspects related to the current cell position (i.e. G2):

1. Take the value of the cell, which is 5 cells away towards left, from current cell.
2. Take the value of the cell, which is 4 cells away towards left, from the current cell.
3. Take the value of the cell, which is 3 cells away towards left, from the current cell.
4. Add all the three values and put the result in the current cell.

Note that in MS-Excel, whenever cell addresses are written in normal way (i.e. A1, C2, D7 etc.), relative referencing automatically takes place.

***Note***

Since in relative referencing all the cell positions are referred in relation to current cell position, hence the same formula, when copied into another cell, (which would then be the current cell) will refer other cells and produce different result.

## Absolute Referencing

As the name indicates, in Absolute Referencing the position of the cell or cells being referred, does not depend upon the current cell position. Such type of referencing is done by prefixing $ sign with row and column names in a cell address.

For example, $B$2, $C$3, $D$4 etc. are examples of absolute referencing. Note that B6 is an example of relative referencing, while $B$6 is an example of absolute referencing.

To understand the concept of relative referencing, consider the worksheet shown in figure 21.6 and assume that values in G2, G3 and G4 are to be multiplied by fixed value of tax rate, present in cell B6 then the formulae as illustrated in the table, will have to be entered in these cells.

| Cell | Formula |
|---|---|
| H2 | =G2*$B$6 |
| H3 | =G3*$B$6 |
| H4 | =G4*$B$6 |

***Note***

Since in absolute referencing, the position of the cell being referred does not depend upon the current cell position hence when a formula containing absolute referencing (only) is copied to other cell, it produces same result.

## Mixed Referencing

As the name indicates, in Mixed Referencing, the cell addresses are partly relative and partly absolute. This is done by either prefixing a $ sign with row name or column name. For example, $B4, B$4 both are examples of mixed referencing (but with a difference). In $B4, column reference is absolute but row reference is relative, while in B$4, it is just opposite i.e. column reference is relative and row reference is absolute.

***Note***

Since in mixed referencing, the position of cell being referred, partially depends upon the position of the current cell and partially doesn't depend upon it hence a formula which contains mixed references, when copied to other cell may or may not produce the same result, depending upon where the formula is copied.

## PREPARING A WORKSHEET

To get familiar with different types of referencing methods, let's prepare a worksheet, which is illustrated in figure 21.6. Say we have to enter labels and data, as shown therein.

Assume that row totals have to be calculated and put in corresponding cells of column G. Tax values have to be calculated by multiplying total by tax rate (present in B6). Perform following steps for this:

1. Enter all the labels, as illustrated in figure 21.6.
2. Enter all the data items, as illustrated in figure 21.6.
3. Enter formula, =(B2+C2+D2) in G2
4. Enter formula, =(B3+C3+D3) in G3
5. Enter formula, =(B4+C4+D4) in G4
6. Enter formula, =(G2*$B$6) in H2
7. Enter formula, =(G3*$B$6) in H3
8. Enter formula, =(G4*$B46) in H4

| | A | B | C | D | E | F | G | H |
|---|---|---|---|---|---|---|---|---|
| 1 | Name | Jan | Feb | Mar | | | Total | Tax |
| 2 | LG | 200 | 300 | 500 | | | 1000 | 100 |
| 3 | BP | 300 | 700 | 200 | | | 1200 | 120 |
| 4 | RK | 400 | 300 | 100 | | | 800 | 80 |
| 5 | | | | | | | | |
| 6 | Tax Rate | 0.1 | | | | | | |

**Figure 21.7**

After entering these formulae, the worksheet will show the values, as illustrated in figure 21.7.

## SAVING AND CLOSING THE WORKBOOK

The method of saving and closing the workbook is same as that of MS-Word document. Recall that procedures for saving and closing the document were described in the previous chapter. For any clarification on this topic refer corresponding section in chapter 20.

# MAKING A CHART

Perform following steps to make a chart of the desired type:

1. Make the Chart Bar appear on the screen. (by selecting View/toolbars/chart option from the Bar menu and the submenus).
2. Select the data, which is to be represented as chart. For example, you select the data, which is illustrated in figure 21.8 (A1:C4).

| | A | B | C | D |
|---|---|---|---|---|
| 1 | | Sale | Expenditure | |
| 2 | Jan | 1200 | 700 | |
| 3 | Feb | 1400 | 800 | |
| 4 | Mar | 1800 | 1100 | |
| 5 | | | | |

**Figure 21.8**

3. Now click the mouse on "Chart Type" button present in Chart Bar. When you do so, a graphical popup menu, as shown in figure 21.9 will appear on the screen.
4. Select the type of chart that you wish to plot for the data, selected. For example you select column chart.

   When you do so, chart will get included in the worksheet as shown in figure 21.10.

Note that chart bar also contains provisions for including other components like legends, Data tables etc. IF you wish to include these items in the chart, make use of these buttons.

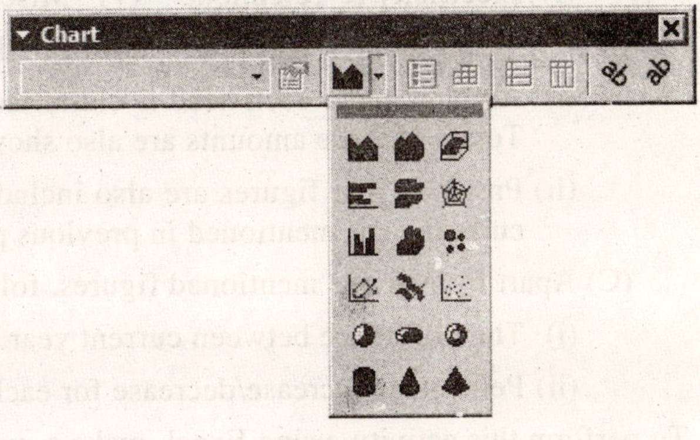


**Figure 21.9**

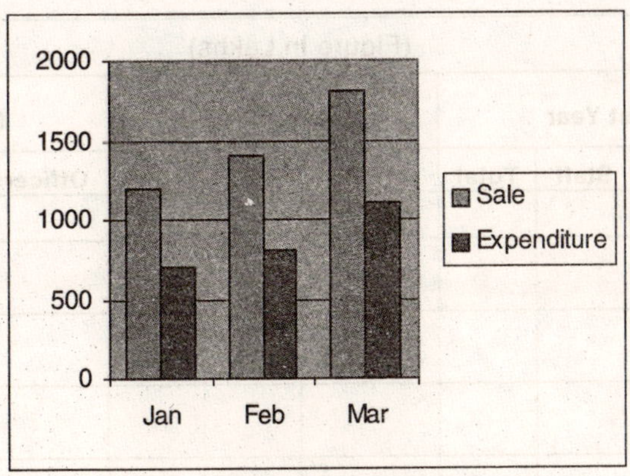


**Figure 21.10**

## Practical Assignments

Perform, following activities in lab sessions:

1. Read the following requirement first:

   Assume that an organization wishes to computerize its Budget Preparation & Analysis activities. It has been decided to computerize the work, using Excel. Say following are the details of the budget activities:

   (A) Organization's budget is prepared under following five heads:

   (i) Salary (ii) Travel & Tour (iii) Rent & Maintenance

   (iv) Electricity & Telephone (v) Miscellaneous

   (B) Following figures are clearly shown in the budget:

   (i) Amount under each head is classified into two categories i.e. Officers and Staff. Totals of these amounts are also shown in the budget.

   (ii) Previous year figures are also included in same format in the report. (as that of current year, mentioned in previous point).

   (C) Apart from above mentioned figures, following analysis is also included in budget:

   (i) The difference between current year figures and last year figures, for each head.

   (ii) Percentage increase/decrease for each head.

To perform this activity, using Excel, make a worksheet as shown below.

| ***Budget For The Year 2006-2007*** | | | | | | | | | | |
|---|---|---|---|---|---|---|---|---|---|---|
| (Figure In Lakhs) | | | | | | | | | | |
| **Head** | **Last Year** | | | **Current Year** | | | **Difference** | | | **Total % Diff.** |
| | **Officers** | **Staff** | **Total** | ***Officers*** | **Staff** | **Total** | **Officers** | **Staff** | **Total** | |
| Salary | | | | | | | | | | |
| Travel & Tour | | | | | | | | | | |
| Rent & Maintenance | | | | | | | | | | |
| Electricity & Telephone | | | | | | | | | | |
| Miscellaneous | | | | | | | | | | |
| **Total** | | | | | | | | | | |

2. Perform following activities, for preparing the worksheet.
   (i) Wherever Totals are there, use formula for them.
   (ii) Make use of formula to calculate differences.
   (iii) Make use of formula to calculate all the percentages.
   (iv) Enter relevant data in the worksheet and test the validity of formulae used and worksheet.
3. Make use of worksheet data and make following two charts.
   (i) Make a Pie chart to illustrate the contribution of each head in current year budget.
   (ii) Make a Column chart to illustrate the comparison of current year figures with last year figures, for each head.
4. Edit the charts and put relevant headings in them.

***

# PRACTICAL

# CHAPTER 22

# *MS-PowerPoint & Lab Assignments*

PowerPoint is presentation software. It has been designed and developed by Microsoft Corporation of USA. Using PowerPoint you can make attractive screens, called slides and display them in sequence so as to create an effect of slide show. Operational details of PowerPoint are described below.

## GETTING STARTED WITH POWERPOINT

Perform following steps to start PowerPoint on your computer:

1. Click the mouse on "Start" button, present on desktop. When you do so, Start menu, as illustrated in figure 20.1(a), appear on the screen.
2. Select "All Programs" option from Start menu. When you do so, another submenu, as illustrated in figure 20.1(b) will appear on the screen.
3. Select "Microsoft PowerPoint" option from this submenu. When you do so, a dialog box, as illustrated in figure 22.1, will appear on the screen.
4. To open a new blank presentation, select "Blank Presentation" option from this dialog box.
5. Now click the mouse on "OK" button. When you do so, a dialog box for selecting the slide layout, as shown in figure 22.2, will appear on the screen.
6. Select the desired layout from this dialog box. Say you select "Text & ClipArt" layout.
7. Now click the mouse on "OK" button. When you do so, selected layout, within the PowerPoint window, as illustrated in figure 22.3, will appear on the screen.

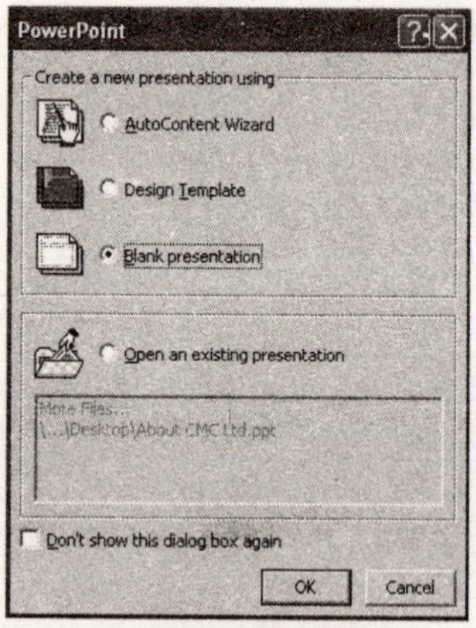


**Figure 22.1**

Now you can add desired contents in the slides. The procedure for including different items in the slides is explained below.

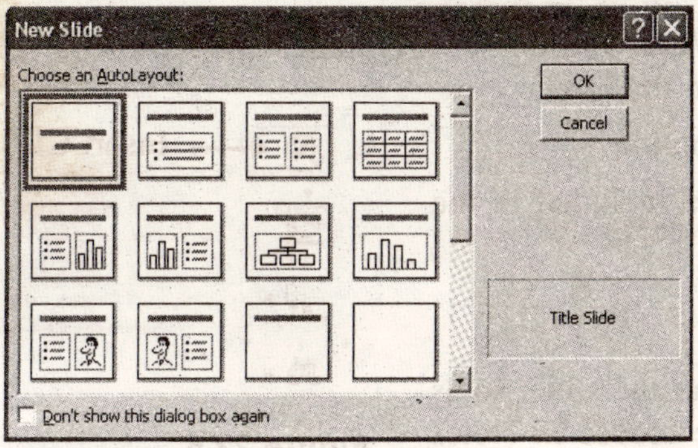


Figure 22.2

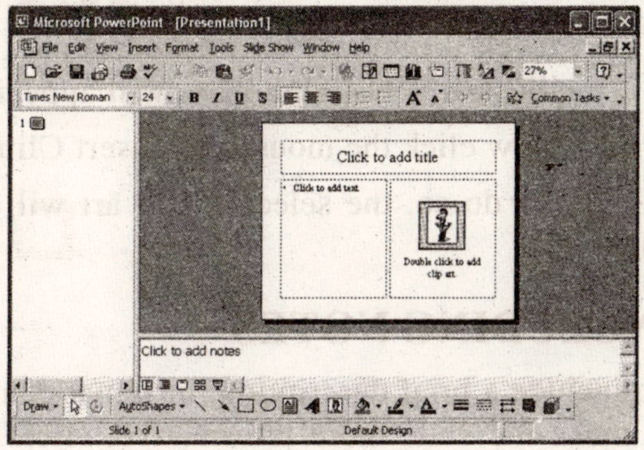


Figure 22.3

## ADDING TITLE IN THE SLIDE

To add title in the slide, click the mouse in Title placeholder and Type the title. If you wish to change the font, font size or font style, drag the mouse on text and select desired specification, using respective object (like drop down lists, buttons etc.), present in Format toolbar.

## ADDING TEXT IN THE SLIDE

Click the mouse within Text placeholder and type the text. To change the font specifications, follow the method mentioned above.

## ADDING CLIPART IN THE SLIDES

PowerPoint has an inbuilt graphics library of its own. Perform following steps to include pictures from that library, into the slide:

1. Double click the mouse within ClipArt placeholder. When you do so, ClipArt dialog box, as shown in figure 22.4, will appear on the screen.
2. Select a category of clipart from this dialog box, by double clicking the mouse on desired category icon. When you do so, all the clip arts available in the selected category will appear on the screen, as shown in figure 22.5.

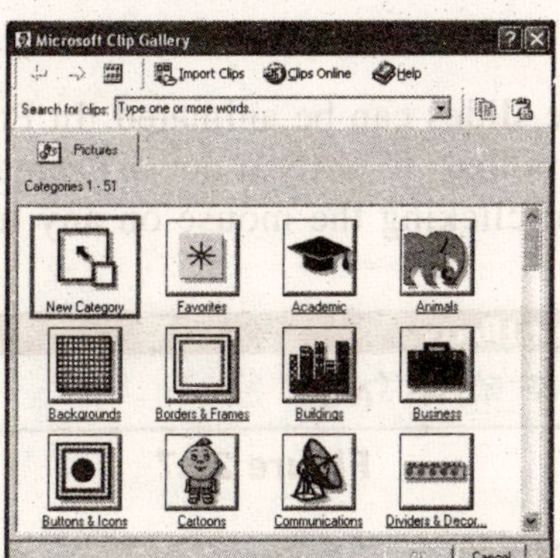


Figure 22.4

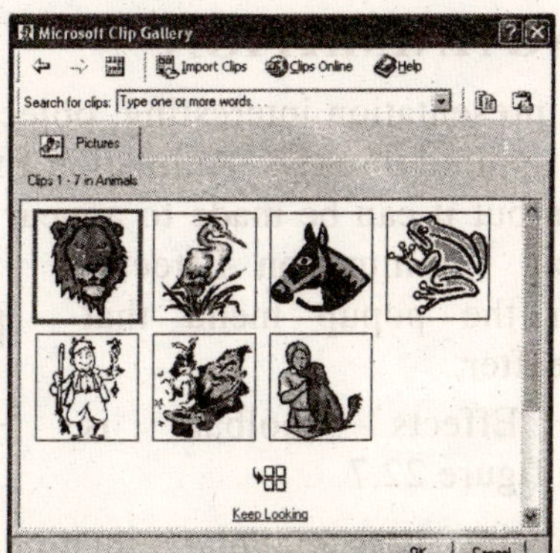


Figure 22.5

3. Click the mouse on desired clipart. When you do so, a graphical popup menu, as illustrated in figure 22.6, will appear on the screen.
4. Now click the mouse on "Insert Clip" option.

When you do so, the selected clip art will get included in the slide.

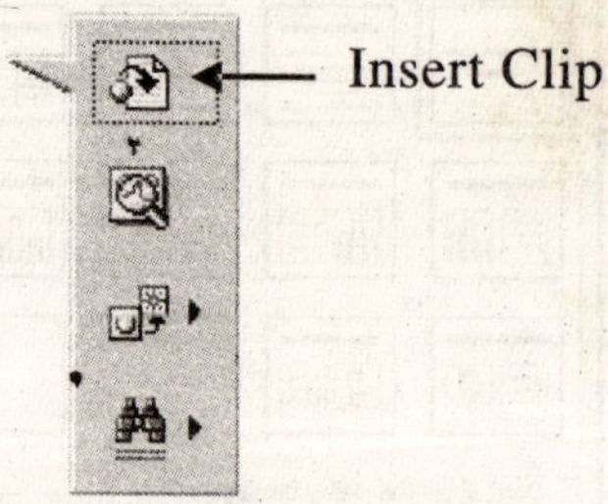

Figure 22.6

## INCLUDING NOTES

To include a brief description, or remarks about the slide, so as to make the slide more informative, click the mouse in Notes placeholder and type the contents.

Note that these contents (notes) are for documentation purpose only. They do not get displayed when the slide show goes on.

## ADDING NEXT SLIDE TO THE PRESENTATION

After completion of the first slide, next slide can be added to the presentation. For this, perform the following steps:

1. Select "Insert" option from the Bar menu. When you do so, Insert submenu will appear on the screen.
2. Select "New Slide" option from Insert submenu. When you do so, a dialog box, as illustrated in figure 22.2 for selecting the slide layout, will appear on the screen.
3. Select the desired layout from this dialog box and click the mouse on "OK" button. When you do so, placeholders as illustrated in figure 22.3, will appear within PowerPoint window.
4. Now add the desired contents, like text, graphics, notes etc. to the slide, following the methods, mentioned above.

By performing above mentioned steps, you can add as many slides to the presentation as you want.

## APPLYING ANIMATION

To make the presentation interesting, objects of the slides can be animated. In PowerPoint, animation is applied through "Animation Effects" toolbar. Usually it doesn't remain present on the screen but it can be made to appear by right clicking the mouse on any toolbar and then selecting "Animation Effects" option from the popup menu that appears thereafter.

Animation Effects toolbar is illustrated in figure 22.7

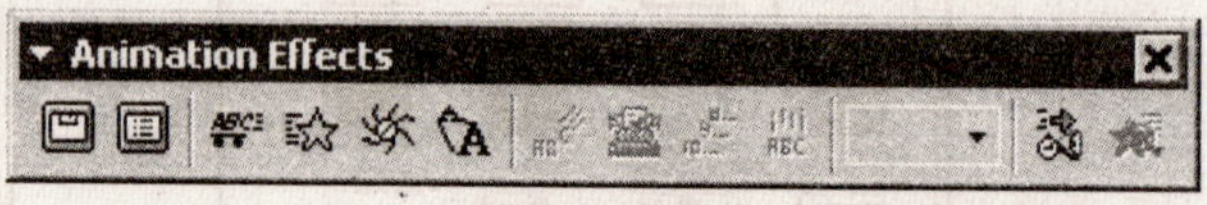

Figure 22.7

Perform following steps to apply animation in any slide:

1. Make the slide appear in Normal view mode.

2. To animate the title of the slide, click the mouse on "Animate Title" button, present in Animation Effects toolbar. On the other hand, if you wish to animate the text of the slide, select "Animate Slide Text" button.
3. Select the desired animation effect by clicking the mouse on corresponding button (such as Drive-In effect, Flying effect etc.) present in Animation Effects toolbar.
4. If more than one object are to be animated on the slide, their animation order can be selected through "Animation Order" drop down list, present in the toolbar.
5. To preview the effect of selected animation effects, click the mouse on "Animation Preview" button, present in toolbar.

When you do so, a small window will appear on the screen and animation effects will get displayed in that window.

Once all the effects have been applied, slide show can be run.

## INTRODUCTION TO SLIDE SHOW

In slide show, all the slides of the presentation get displayed on the screen, one by one. Next slide replaces the previous slide in attractive and eye-catchy manner. Slide shows can be of following types:

1. Manual Slide Show
2. Automatic Slide Show

Basic characteristics of both types of slide shows are described below.

### Manual Slide Show

In manual slide show, one slide replaces the other slide only when space bar is pressed or mouse is clicked. Till the time any of these two events doesn't take place, slide remains visible on the screen.

### Automatic Slide Show

In automatic slide show, each slide has a time duration, called Slide Timing, assigned to it. Slide remains there on the screen, for this duration and as soon as time duration is over, next slide of the presentation replaces it.

### Looped Slide Shows

A slide show, be it automatic or manual, is said to be running in loop, if after the display of last slide, first slide replaces it again and again.

## TRANSITION EFFECT

The way one slide replaces the other slide, during the slide show is called. Transition Effect. For example, the next slide may take entry on the screen from left hand side and gradually replace the previous slide. Similarly, in other transition effect the first siide may tear apart from the middle, as if stage curtains are sliding left and right and the next slide is gradually becoming visible in back.

PowerPoint offers many transition effects, which can be selected and applied to one, many or all slides of the presentation.

## Applying Transition Effect and Slide Timings

Perform following steps to apply slide timing and transition effect to the slides of the presentation:

1. Open the presentation and display the slide, in which the transition effect or slide timing is to be applied, in Normal View mode.
2. Select "Slide Show" option from the Bar menu. When you do so, Slide Show submenu, as illustrated in figure 22.8(a) will appear on the screen.

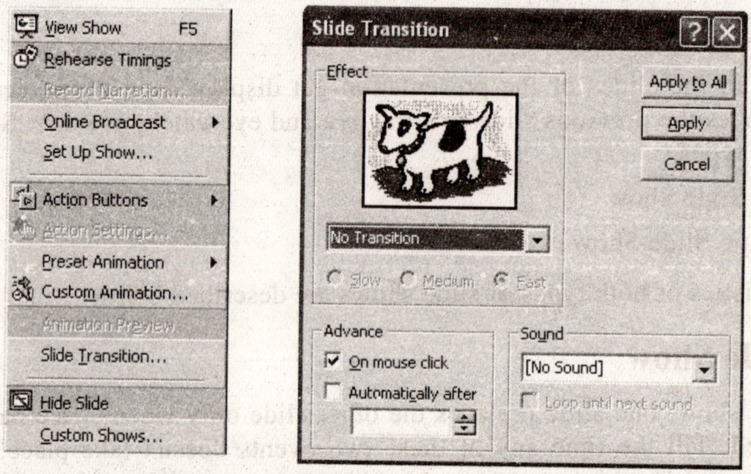


(a) (b)

**Figure 22.8**

3. Select "Slide Transition" option from this submenu. When you do so, a dialog box, as illustrated in figure 22.8 (b) will appear on the screen.
4. Select the desired transition effect from "Effect" dropdown list.
5. Select the speed of the transition (i.e. how fast the replacement of one slide by the other has to be done) by clicking the mouse on Show, Medium or Fast radio button.
6. If the slide show has to advance on the click of the mouse, check "On mouse click" check box.
7. If slide timing is to be assigned to the slide, check "Automatically after" check box and specify the time duration (for which the slide has to stay on the screen), using corresponding spinner.

8. If these settings are to be applied to the current slide only, click the mouse on "Apply" button. On the other hand, if the settings are to be applied to all the slides of the presentation, select "Apply to All" button.

When you do so, selected transition effect and chosen slide timing will get applied to the slides.

## SETTING UP THE SLIDE SHOW

After making the slides and applying transition, certain parameters related to slide show need to be set. Perform following steps to set parameters:

1. Open the presentation, for which the parameters need to be set.
2. Select "Slide Show" option from the Bar menu. When you do so, Slide Show menu, as illustrated in figure 22.8(a), will appear on the screen.
3. Now select "Set up Show" option from this submenu. When you do so, a dialog box, as illustrated in figure 22.9, will appear on the screen.

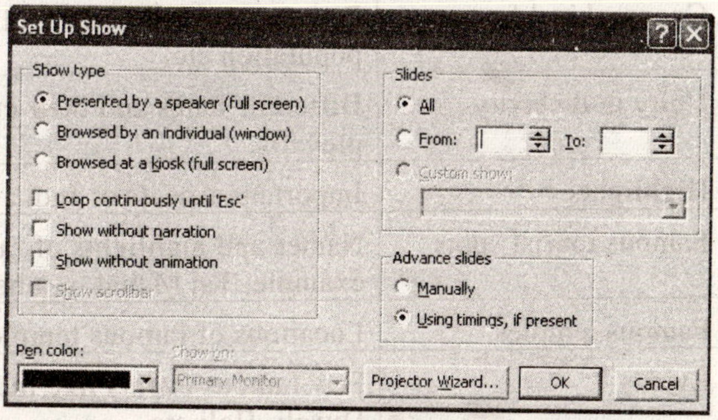


**Figure 22.9**

4. To run the slide show in loop, either select "Browsed at a kiosk (full screen)" radio button or check "Loop Continuously until Esc" check box.
5. If narration and animation, present in the presentation are to be displayed, check the corresponding check boxes.
6. If all the slides of the presentation have to participate in the slide show, select "All" radio button else select "From", button and specify the slide numbers that have to appear in the slide show. For this, make use of the "From" and "To" spinners, present in the dialog box.
7. If the slide show has to run in manual mode (slides advance on the click of mouse or on pressing space bar), select "Manually" radio button. For automatic slide show, select "Using timings, if present" radio button.
8. At last, click the mouse on "OK" button.

When you do so, slide show will get ready to run as per the specifications set by you.

## RUNNING THE SLIDE SHOW

To run the slide show, follow any one of the methods described below:

1. Click the mouse on "Slide Show" view button, present at the left bottom of PowerPoint window.

   or

2. Select "View Show" option from "Slide Show" submenu.

## PRACTICAL Exercises

Assume that you have to talk about India in an international seminar. For this, you have to make PowerPoint presentation comprising of many slides. Titles of these slides and their contents are suggested below:

| Title | Contents |
|---|---|
| India is a great counter | Few sentences in appreciation of India. |
| Geographical facts | Few facts like Himalayas in north, Indian ocean in south, area, population etc. |
| Unity in diversity | Different states and their cultures. Feeling of unity that prevails. |
| Highlights | Important aspects of India and its strong points. |
| Famous tourist spots | Names and highlights of famous tourist spots of India. For example, Taj Mahal, sea beaches of Goa, Palaces of Jaipur etc. |
| Famous temples | Locations of famous temples and stories associated with them. |
| Events | Few famous events like independence day, festivals like Diwali, Holi etc. |
| Personalities | World renowned personalities and their contributions. |
| Indian culture | Greatness of Indian culture and its difference from other cultures. |
| East or west India is best | All those aspects, which make India better than other countries. |
| Developments of India | Progress that India has made in last decade. |
| My dream India | What India should be after 10 years. |

1. Collect above-mentioned details and write them on paper.
2. Create all slides mentioned above.
3. Now run the slide show and test the presentation.
4. Add clip arts in few slides.
5. Run the slide show again.
6. Make Transition Effect toolbar on and apply transition effect on the slides.
7. Run the slide show again.
8. Apply animation in first slide.
9. Run the slide show again.
10. Apply slide timings to the slides and run the presentation in automatic mode.
11. Run the slide show in loop.

****